AF556012

Fundamentals of Livestock and Poultry Management

NIPA® GENX ELECTRONIC RESOURCES & SOLUTIONS P. LTD.
New Delhi-110 034

About the Editors

Dr. Utkarsh Kumar Tripathi is an Assistant Professor with a specialization in Livestock Production Management at the Faculty of Veterinary and Animal Sciences, Institute of Agricultural Sciences, Rajiv Gandhi South Campus, Banaras Hindu University, Barkachha, Mirzapur. He brings a wealth of experience, having spent eight years in teaching, research, and extension activities.

Dr. Tripathi's academic journey began with his graduation from DUVASU, Mathura, followed by his post-graduation and Ph.D. from the prestigious National Dairy Research Institute (NDRI), Karnal. He has also cleared the UGC NET and ICAR-ASRB exams, conducted by UGC, New Delhi, and ICAR-ASRB, New Delhi, respectively.

A creative researcher and writer, Dr. Tripathi has published more than 15 national and international research and review articles. He is also the author of one book and has contributed to numerous book chapters, conference papers, and manual chapters. Additionally, he has prepared a manual for livestock production management and written over 25 popular articles. His contributions reflect his dedication to advancing knowledge in livestock production management and sharing it with both academic and broader audiences.

Dr. Anuradha Kumari is an Assistant Professor specializing in Livestock Production Management. She is a dedicated academic and researcher committed to the advancement of veterinary and animal sciences, with a strong foundation in teaching, research, and extension activities.

Dr. Kumari began her academic journey with a graduation in Veterinary Sciences from the College of Animal Husbandry and Sciences, Kanke, Ranchi. She pursued her post-graduation at the esteemed Bombay Veterinary College under Maharashtra Animal and Fishery Sciences University (MAFSU) Nagpur, and went on to earn her Ph.D. from Lala Lajpat Rai University of Veterinary and Animal Sciences (LUVAS), Hisar, Haryana.

With over eight years of experience, Dr. Kumari has made significant contributions in the field of Livestock Production Management. She has authored more than 15 national and international research and review articles and has presented her work in various national and international conferences. Her scholarly work also includes over 25 popular articles, numerous book chapters, manual chapters, and the authorship of one book. Notably, she has developed a comprehensive manual for the undergraduate course in Livestock Production Management.

Fundamentals of Livestock and Poultry Management

Utkarsh Kumar Tripathi
Assistant Professor
Livestock Production Management
Faculty of Veterinary and Animal Sciences
Institute of Agricultural Sciences, Rajiv Gandhi South Campus
Banaras Hindu University, Barkachha, Mirzapur, Uttar Pradesh

Anuradha Kumari
Assistant Professor
Livestock Production Management
Faculty of Veterinary and Animal Sciences
Institute of Agricultural Sciences, Rajiv Gandhi South Campus
Banaras Hindu University, Barkachha, Mirzapur, Uttar Pradesh

NIPA® GENX ELECTRONIC RESOURCES & SOLUTIONS P. LTD.
New Delhi-110 034

NIPA® GENX ELECTRONIC RESOURCES & SOLUTIONS P. LTD.

101,103, Vikas Surya Plaza, CU Block
L.S.C. Market, Pitam Pura, New Delhi-110 034
Ph : +91-11-43860225, Mob.: +91 9717133558, 9540816132
E-mail: newindiapublishingagency@gmail.com
Website: www.nipaersources.com

Print ISBN: 978-93-58877-30-4
ebook ISBN: 978-93-58875-92-8

Composed and Designed by NIPA®.

Foreword

It gives me great pleasure to write the foreword for the textbook "***Fundamentals of Livestock and Poultry Management***," edited and authored by Dr. Utkarsh Kumar Tripathi and Dr. Anuradha Kumari, Assistant Professors in the Department of Livestock Production Management, Faculty of Veterinary and Animal Sciences, Institute of Agricultural Sciences, Rajiv Gandhi South Campus, Banaras Hindu University, Barkachha, Mirzapur (U.P.).

This book has been designed to provide comprehensive coverage of both theoretical and practical aspects of livestock and poultry management. The content thoughtfully includes vital topics such as the role of livestock in the national economy, principles of housing and feeding, reproductive management, disease control, and breed improvement, along with detailed sections on poultry husbandry, nutrition, and health care.

The authors have made commendable efforts to present the subject matter in a systematic and student-friendly manner, supported by relevant examples, photographs and recent developments in animal science. The book will serve as a valuable resource not only for undergraduate students in veterinary and agricultural sciences but also for postgraduate students, researchers, teachers, and field professionals engaged in livestock and poultry production and management.

I am confident that this book will make a significant contribution to capacity building and skill development in the livestock sector and will inspire readers to pursue excellence in scientific livestock and poultry management.

Dean
Faculty of Veterinary and Animal Sciences
Institute of Agricultural Sciences, BHU

Preface

The book ***Fundamentals of Livestock and Poultry Management*** has been carefully developed to provide a clear and comprehensive guide for students, educators, and professionals engaged in the field of animal husbandry and poultry science. The importance of livestock and poultry in supporting national economy, rural livelihoods, and food security has been growing steadily, and this book aims to equip readers with both theoretical knowledge and practical skills required for effective management and sustainable production.

The content is organized to cover essential topics such as reproduction, housing, nutrition, disease prevention, breed improvement, and species-specific management practices. Special emphasis has been placed on integrating foundational concepts with contemporary advancements and field applications. Each chapter has been crafted to ensure clarity, usefulness, and relevance to current academic curricula and field challenges.

This volume is intended to serve as a valuable academic resource for undergraduate and postgraduate students of veterinary and agricultural sciences, as well as a practical reference for extension workers, researchers, and livestock practitioners. The authors gratefully acknowledge the support of colleagues and institutions that contributed to the development of this book.

It is hoped that this work will contribute positively to education, research, and practice in livestock and poultry management, fostering sustainable growth and welfare in the sector.

Authors

Contents

1

Role of Livestock in the National Economy

Utkarsh Kumar Tripathi[1], Anuradha Kumari[1], Manish Kumar[1] and Shashidhar K.S.[2]

[1]*Faculty of Veterinary and Animal Sciences, I. Ag.Sc., RGSC-Banaras Hindu University (BHU), Barkachha, Mirzapur, Uttar Pradesh-231001*

[2]*Faculty of Agriculture, I.Ag.Sc, RGSC- Banaras Hindu University (BHU) Barkachha Mirzapur, Uttar Pradesh-231001*

The livestock sector plays a vital role in ensuring food security by producing milk, meat, and eggs. It also supports rural development by providing jobs and improving living conditions for many families. With ongoing efforts to boost productivity and sustainability, animal husbandry continues to be a key factor in enhancing rural incomes and overall economic growth in India. Animal husbandry has always been an important part of agriculture in India, and it remains relevant today because many people depend on it for their livelihoods. Dairy farming is the largest agricultural industry in India, contributing about 5% to the national economy and directly employing around 80 million dairy farmers. India is home to a wide variety of livestock, with many breeds adapted to different climates. From 2014-15 to 2020-21, the livestock sector grew at a rate of 7.9% per year, and its contribution to the total Gross Value Added (GVA) from agriculture increased from 24.3% to 30.1%. Besides providing food and income, livestock also creates jobs for rural families and acts as a safety net during crop failures. Moreover, the number of animals a farmer owns can affect their social status in the community.

Contributions of Livestock

- **Food:**
  - Provides milk, meat, and eggs for human consumption.
- **Fibre and Skins:**
  - Contributes to wool, hair, hides, and pelts production.
- **Draft Power:**

- Bullocks are essential for Indian agriculture, especially in rural areas.
- They save fuel costs compared to mechanical equipment like tractors.
- Pack animals (camels, horses, donkeys, ponies, mules) are used for transporting goods, especially in hilly terrains.

- **Dung and Waste Materials:**
  - Animal dung serves as valuable manure and is worth several crores of rupees.
  - Used as fuel (biogas, dung cakes) and in construction as a low-cost cement alternative.
- **Moving assets:**
  - Livestock act as "moving banks," providing capital during emergencies.
  - Serve as assets for landless agricultural labourers and can help secure loans from local lenders.
- **Weed Control:**
  - Used for biological control of weeds and unwanted plants.
- **Cultural Significance:**
  - Livestock provide security and enhance self-esteem for owners, especially prized animals like pedigree bulls or high-yielding cows/buffaloes.
- **Sports and Recreation:**
  - Animals such as cocks, rams, and bulls are used in competitions and sports, despite bans on some animal contests.

India holds a prominent position in global livestock production, reflecting its diverse agricultural practices and significant contributions to food security. The livestock sector in India plays a crucial role in the economy and the livelihoods of millions, employing around 8 percent of the total Indian workforce. Notably, about two-thirds of rural women are actively engaged in livestock rearing, highlighting the sector's importance in empowering women in agricultural communities. Furthermore, nearly 49.8 percent of the population relies on agriculture for their livelihood, with more than 80 percent of livestock owned by marginal, small, and semi-medium farmers. Despite its significance, the area dedicated to all types of livestock farming constitutes only 1.69 percent of the total agricultural land. This demographic dependence on livestock underscores its vital contribution to rural economies and food security in India.

Table 1: Value Output from Livestock Rearing in 2022-23 at Current Prices

Sector	Percentage of Total Output
Milk and Milk Products	66.48%
Meat and Meat Products	23.60%
Eggs	3.69%
Wool and Hair	0.04%
Dung	4.14%
Silkworm Cocoons and Honey	0.76%
Increment in Stock	1.28%

The Economic Survey highlights that the allied sectors of Indian agriculture are increasingly recognized as vital growth centers that significantly enhance farm incomes. From 2014-15 to 2022-23, the livestock sector experienced a remarkable Compound Annual Growth Rate (CAGR) of 7.38% at constant prices. The contribution of livestock to the total Gross Value Added (GVA) in agriculture and allied sectors rose from 24.32% in 2014-15 to 30.38% in 2022-23. In the fiscal year 2022-23, the livestock sector accounted for 4.66% of the total GVA, contributing positively to the per capita availability of milk, eggs, and meat. Additionally, the fisheries sector, which plays a crucial role in the Indian economy, constituted about 6.72% of agricultural GVA and achieved a CAGR of 8.9% from 2014-15 to 2022-23 (at constant prices). This "sunrise sector" supports approximately 30 million people, particularly from marginalized and vulnerable communities (https://pib.gov.in/ ; PRID=2034940).

It also states that the Animal Husbandry Infrastructure Development Fund (AHIDF) is designed to encourage investments from individual entrepreneurs, private companies, Farmer Producer Organizations (FPOs), Section 8 companies, and dairy cooperatives. This fund focuses on critical areas such as dairy processing, meat processing, animal feed production, and breed improvement technology. The government offers a 3% interest subvention to borrowers along with a credit guarantee of up to 25% of the total borrowing amount. As of May 2024, lending banks, NABARD, and NDDB have sanctioned 408 projects valued at INR 13,861 crore, creating 40,000 direct employment opportunities and benefiting over 42 lakh farmers.

Table 2: Countries of the world holding top position in different sectors related to livestock

Country	Achievement
India	Bovines, Cattle, Buffalo, Milk Production, Carabeef, Goat Milk
China	Pigs, Sheep, Duck, Fish, Eggs, Chevon, Mutton, Pork, Poultry, Meat
Ethiopia	Donkeys
Mexico	Mules
Somalia	Camels
USA	Poultry Meat, Beef
Australia	Fibre, Wool

Position of India in Livestock Related Products

1. 1st Place

- **Total Livestock Population**
- **Milk Production:** India is the largest producer of milk globally, accounting for approximately 24% of the world's total milk production.
- **Cattle Population:** With a cattle inventory of about 307.5 million, India leads in this category, representing roughly 33% of the world's cattle.
- **Buffalo Population:** India houses around 57% of the global buffalo population.
- **Goat Population:** India ranks first with a significant goat population, contributing to its diverse meat and dairy products.
- **Carabeef Production**
- **Goat Milk Production**
- **Total Bovine Population**

2. 2nd Place

- **Egg Production:** India is the second-largest producer of eggs globally.
- **Bristle Production:** A by-product of the pig industry, highlighting India's diverse livestock sector.
- **Aquaculture:** Significant contributions to fish farming and production.
- **Goat Meat Production**
- **Duck Production**

3. 3rd Place

- **Sheep Population:** India ranks third in sheep population, contributing to wool and meat production.
- **Fisheries Production:** A vital sector for both domestic consumption and export.

- **Camel Population:**

4. 5th Place:

- **Poultry Meat Production:** India ranks fifth in the production of poultry meat.
- **Meat Production:** Reflecting a robust meat industry across various species.
- **Poultry Production:**

5. 6th Place

- **Clean Wool Production:**

Table 3: Annual Commodity Output and Per Capita Analysis

Commodity	Total Production (per year)	Comparison with Previous Year	Per Capita Availability	ICMR Recommendations
Milk	239.30 MT (million tons)	+ 3.78%	471 grams/day	280 grams/day
Eggs	142.77 billion	+3.18	103 eggs/year	182 eggs/year
Meat	10.25 MT	+4.95% #	7.39 kg/annum	11 kg/year
Wool	33.69 million kg	+0.22%	-	-
Fish	-		-	18 kg/year (WHO)

#Increase in poultry meat production compared to previous year: 0.47%

Source: Basic Animal Husbandry & Fisheries Statistics (2024), Govt. of India

Source: Basic Animal Husbandry Statistics (https://dahd.gov.in/schemes/programmes/animal-husbandry-statistics)

Livestock Census

The Livestock Census in India has a rich history, beginning in 1919, and has been conducted every five years since then. As of the latest census, the 20th Livestock Census was carried out in 2019, revealing a total livestock population of 535.78 million, which marked a 4.6% increase from the previous census in 2012.

The key highlights of livestock census 2019 is as following

- **Total Livestock:** 535.78 million, increased by 4.6%.
- **Total Bovine:** 302.79 million, increased by 1%.
- **Cattle:** 192.49 million, increased by 0.8%.
- **Female Cattle (Cows):** 145.12 million, increased by 18%.
- **Exotic/Crossbred Cattle(male and female):** 50.42 million, increased by 26.9%.

- **Indigenous/Non-descript Cattle (male and female):** 142.11 million, declined by 6%.
- **Indigenous/Non-descript Female Cattle:** Increased by 10%.
- **Buffaloes:** 109.85 million, increased by 1%.
- **Milch Animals (Cows & Buffaloes):** 125.34 million, increased by 6%.
- **Sheep:** 74.26 million, increased by 14.1%.
- **Goats:** 148.88 million, increased by 10.1%.
- **Pigs:** 9.06 million, declined by 12.03%.
- **Other Livestock:** 1.24 million.
- **Total Poultry:** 851.81 million, increased by 16.8%.
- **Backyard Poultry:** 317.07 million, increased by 46%.
- **Commercial Poultry:** 534.74 million, increased by 4.5%.

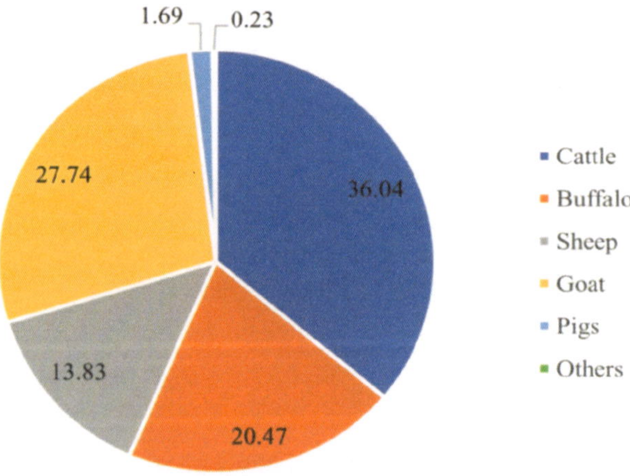

Figure 1: Percentage Distribution of Total Livestock as per 20th Livestock Census (2019)

Source: 20th Livestock census, 2019

Table 4: Percentage of the World Livestock Present in India

Species	Percentage Present in India
Total Livestock	10.23%
Cattle	12.67%
Buffalo	53.84%
Sheep	5.88%
Goat	13.19%
Pig	0.95%
Poultry	2.43%

Share of Agriculture & Allied Sectors in Gross Value Added

In the fiscal year 2022-23, agriculture and its allied sectors made a significant contribution to India's Gross Value Added (GVA), reflecting the sector's vital role in the national economy. Livestock alone accounted for 5.50% of the total GVA, contributing INR 1,355,460 crores to the national GVA of INR 24,659,031 crores. Within the agricultural sector, which represented 18.19% of the total GVA at INR 4,484,268 crores, livestock played a crucial role by contributing 30.23% to agricultural GVA. The breakdown of agricultural contributions reveals that crops generated INR 2,478,925 crores (10.05%), while forestry and logging and fishing and aquaculture contributed INR 324,875 crores (1.32%) and INR 325,007 crores (1.32%), respectively. This data underscores the importance of livestock in enhancing agricultural productivity and overall economic growth in India.

Table 5: Gross Value Added (GVA) and percentage share for various agricultural categories:

Category	**Total GVA (in crores)**	**Percentage Share (%)**
Crops	24,78,925	10.05
Livestock	13,55,460	5.5
Forestry and Logging	3,24,875	1.32
Fishing and Aquaculture	3,25,007	1.32

In the fiscal year 2023-24, India recorded significant figures in the livestock sector, with a total import value of livestock and livestock products amounting to approximately Rs. 56,16,04,237 lakhs. In contrast, the total export value for the same category reached about Rs. 36,18,95,227 lakhs. These figures highlight the substantial economic activity within India's livestock industry, reflecting both the demand for livestock products domestically and the country's role in international markets. The trade dynamics indicate a notable trade imbalance in this sector, with imports exceeding exports by a considerable margin.

Breeds of Livestock in India

Indian Council of Agricultural Research - National Bureau of Animal Genetic Resources (ICAR-NBAGR), located in Karnal, serves as the nodal agency for the registration of indigenous livestock breeds in India.

Table 6: Total number of breeds of different livestock in India

Species of livestock	**Number of Registered Breeds**
Cattle	53
Buffalo	20
Goat	39
Sheep	45

Horses & Ponies	8
Camel	9
Pig	14
Donkey	3
Dog	3
Yak	1
Chicken	20
Duck	3
Geese	1
Synthetic Cattle	1

Source: NBAGR, Karnal (https://nbagr.icar.gov.in/)

Newly registered breeds

This registration was approved by the ICAR Breed Registration Committee during its 11th meeting held on December 5, 2023, emphasizing the importance of conserving and promoting indigenous livestock breeds across the country.

Table 7: Newly registered breeds by NBAGR

Breed Name	Type	Region/State
Aravali Chicken	Poultry	Gujarat
Andamani Duck	Poultry	Andaman and Nicobar Islands
Anjori Goat	Goat	Chhattisgarh
Andamani Goat	Goat	Andaman and Nicobar Islands
Bhimthadi Horse	Horse	Maharashtra
Andamani Pig	Pig	Andaman and Nicobar Islands
Macherla Sheep	Sheep	Andhra Pradesh
Frieswal Cattle	Synthetic Cattle	Uttar Pradesh and Uttarakhand

Source: NBAGR, Karnal (https://nbagr.icar.gov.in/)

Table 8: Leading States in India for Livestock Production (2023-24)

State	Achievements
Uttar Pradesh	- Highest livestock population
	- Highest milk producing state
	- Highest buffalo meat production
	- Highest pig meat production
	- Highest indigenous milch cows
	- Highest buffalo population
	- Highest horses and ponies population

Punjab	- State with highest per capita availability of milk (1245 g)
	- Highest milk yield (exotic/crossbred cow: 13.29 kg/animal/day)
	- Highest milk yield (indigenous/non-descript cow: 8.13 kg/animal/day)
	- Highest milk yield (goats: 1.51 kg/animal/day)
Tamil Nadu	- Highest poultry population
	- Highest cross-bred or exotic cows
Haryana	- State with highest milk yield (buffaloes: 10.83 kg/animal/day)
UT-Chandigarh	- State with highest milk yield (buffaloes: 11.47 kg/animal/day)
	- State with highest milk yield (goats: 1.84 kg/animal/day)
Andhra Pradesh	- Highest egg producing state
	- Highest sheep meat production
	- Highest fish production
Telangana	- State with highest per capita availability of eggs (483 eggs)
	- Highest per capita availability of meat (29.21 kg/annum)
	- Highest sheep population
West Bengal	- Highest meat producing state
	- Highest goat meat production
	- Highest cattle population
Rajasthan	- Highest wool producing state
	- Highest goat milk production
	- Highest camel population
	- Highest donkey population
Kerala	- Highest cattle meat production
Assam	- Highest pig population
Arunachal Pradesh	- Highest mithun population
Jammu & Kashmir	- Highest yak population
Uttarakhand	- Highest number of mules
Maharashtra	- Highest poultry meat production

Table 9: Share of leading states of India in total production of different livestock products (2023-24)

Livestock Product	**Leading State**	**Share (%)**
Total Milk Production	Uttar Pradesh	16.21
	Rajasthan	14.51
Total Egg Production	Andhra Pradesh	17.85
	Tamil Nadu	15.64
Total Meat Production	West Bengal	12.62
	Uttar Pradesh	12.29

Total Wool Production	Rajasthan	45.94
	Jammu & Kashmir	25.24

Milk

The Indian dairy sector plays a crucial role in the country's economy, significantly contributing to both agricultural output and rural livelihoods. As the world's leading producer of milk, India contributes about 25% of the global milk output, with production reaching 230.58 million tonnes in 2022-23. This remarkable growth, driven by a Compound Annual Growth Rate (CAGR) of 6%, highlights the sector's efficiency and productivity.Moreover, the dairy industry not only meets the nutritional needs of India's vast population but also serves as a vital source of income for rural households. The per capita availability of milk stands at 471 grams per day, significantly higher than the global average of 322 grams, showcasing the sector's importance in enhancing food security.

However, unlike the larger herd sizes ofleading milk producing countries in the world, some 95% of milk producers in India hold just 1 to 5 milchanimals per household, which makes this little more than a subsistence-level farming system [Source:Department of Animal Husbandry and Dairy (DAHD), GOI Annual Report 2018/19].Dairy farms sized from 50 to 200 cattle are increasing in some of the major dairy states such as Punjab,Gujarat, Maharashtra, and Telangana/Andhra Pradesh, although they are still few. With the ongoing government support and initiatives aimed at improving productivity, the Indian dairy sector is poised for continued growth, reinforcing its status as a cornerstone of the agricultural economy and a key driver of socio-economic development in rural areas.

Table 10: Species Wise Milk Contribution to Total Milk Production in India

Species	Percentage of Total Milk
Buffalo Indigenous	31.49%
Buffalo Non-Descript	13.83%
Cow Non-Descript	10.11%
Cow Indigenous	11.36%
Cow Cross-Bred	31.11%
Cow Exotic	2.10%
Goat	3.36%

Meat

The meat sector in India is a vital component of the economy, contributing approximately 4.11% to the country's GDP and providing livelihoods for around 20.5 million people, particularly in rural areas. As the fifth largest meat producer globally, India produces about 8.8 million tonnes of meat annually, accounting for roughly 3% of the world's total meat supply. The sector has shown steady growth, with a Compound Annual Growth Rate (CAGR) of 6%, and together with poultry, it contributed around INR 2,51,384 crores, which is 24.08% of the total livestock sector output.

The meat industry not only plays a crucial role in enhancing food security by providing a significant source of protein but also supports rural development through job creation across various segments, including farming, processing, distribution, and retailing. Despite facing challenges such as a largely unorganized sector and quality control issues, ongoing government initiatives and investments aim to improve infrastructure and market access. With India's large livestock population and increasing demand for meat products, the sector is poised for further growth, reinforcing its importance in the national economy and the livelihoods of millions.

India's total meat production for the fiscal year 2023-24 is estimated at 10.25 million tonnes, reflecting a growth of 4.85% over the past decade, compared to 6.69 million tonnes in 2014-15. Additionally, there was an increase of 4.95% in production from the previous year, 2022-23.The majority of this meat production comes from West Bengal, which contributes 12.62% of the total, followed closely by Uttar Pradesh at 12.29%, Maharashtra at 11.28%, Telangana at 10.85%, and Andhra Pradesh at 10.41%. In terms of annual growth rates, Assam recorded the highest growth at 17.93%, followed by Uttarakhand at 15.63% and Chhattisgarh at 11.70%.

Table 11: Species Wise Meat Contribution to Total Meat Production in India

Species	Percentage contribution of total meat
Poultry	48.96%
Buffalo	18.09%
Goat	15.50%
Sheep	11.13%
Pig	3.72%
Cattle	2.60%

Table 12: India's Meat Processing and Slaughter Facility Registration Summary

Facility Type	Number Registered
Integrated Abattoirs cum Meat Processing Plants	73
Meat Processing Plants	11
Stand-alone Slaughter Houses	7
Regular Slaughter Houses	882
Carcass Utilization Centres	8
Small Scale Meat Retail Shops	Approximately 25,000

Table 13: Number of Animals Slaughtered for Meat Production (2023-24)

Species of Animal	Number Slaughtered
Goat	135.17 million
Sheep	80.70 million
Buffalo	14.53 million
Pig	9.62 million
Cattle	2.44 million

- **Major Export Destinations of Buffalo Meat Products (2023-24)**:
  - Vietnam
  - Malaysia
  - Egypt
  - Iraq
  - UAE
- **Global Production:** India accounts for approximately **43%** of the world's buffalo meat production, with **Uttar Pradesh** being the leading state in production.
- **Sheep and Goat Meat Export:** India is the largest exporter of sheep and goat meat globally. Major export destinations for 2023-24 include:
  - United Arab Emirates
  - Qatar
  - Kuwait
  - Maldives
  - Bahrain (in descending order)
- **Poultry Sector:** The organized poultry sector in India constitutes **80%** of the total poultry industry.
- **Non-Vegetarian Population:** The non-vegetarian population in India comprises:

    - **83.4%** of men
    - **70.6%** of women (NFHS-5, 2021)
- **Processing Levels:** Current processing levels are:
    - Poultry: **6%**
    - Meat: **21%**

Government Schemes Related to the Livestock Sector

Animal Husbandry Infrastructure Development Fund (AHIDF): This scheme helps borrowers by providing a 3% interest subsidy and a credit guarantee of up to 25% of the total amount borrowed.

1. **National Livestock Mission (NLM):** This scheme has been updated for the period from 2021-22 to 2025-26. It focuses on developing entrepreneurship and improving breeds in poultry, sheep, goats, and pigs, as well as enhancing feed and fodder production.
2. **Livestock Health and Disease Control (LH&DC) Scheme:** This program aims to support state governments in preventing and controlling animal diseases through vaccination.
3. **National Animal Disease Control Programme (NADCP):** This initiative focuses on controlling diseases like Foot & Mouth Disease and Brucellosis by vaccinating cattle, buffalo, sheep, goats, and pigs.

2

Glossary and Important Concepts in Animal Science

Utkarsh Kumar Tripathi, Anuradha Kumari, Ajeet Singh, and Vipin Maurya

Faculty of Veterinary and Animal Sciences, I. Ag.Sc., RGSC-Banaras Hindu University (BHU), Barkachha, Mirzapur, Uttar Pradesh-231001

Livestock: Livestock refers to domesticated animals raised in agricultural intending for producing commodities like meat, eggs, milk, fur, leather, and wool. It encompasses farm animals used for commercial and domestic purposes.

Livestock Management: This involves combining various elements such as ideas, facilities, processes, materials, and labour to produce market-worthy products or services. It integrates scientific principles of animal breeding, feeding, weeding, and heeding to maximize returns.

Four pillars of Livestock Management

1. **Breeding:** Mating genetically superior animals to produce offspring.
2. **Feeding:** Providing balanced nutrition in terms of quality and quantity.
3. **Weeding:** Culling and eliminating low or unproductive animals.
4. **Heeding:** Daily care and management, including sanitation, hygiene, disease control, health care, and proper housing for animals.

Domestication: It is a multi-generational, mutualistic relationship where one species, such as humans, takes control and care of another species to obtain resources like food, labor, or companionship. This process involves selective breeding, leading to genetic changes that make domesticated organisms more dependent on humans for survival and reproduction. It is a co-evolutionary process that transforms wild animals and plants into domestic forms, often resulting in increased docility, altered morphology, and dependence on human care. Domestication is distinct from taming and involves long-term management of another species for mutual benefit.

Around 8,000 BC, alongside the invention of agriculture, humans began domesticating wild animals. This process, known as domestication, involves altering wild animals and plants to meet human needs through selective breeding. Domesticated species are raised for various uses like food, work, and clothing, and they require continuous human care.

The first attempts at domestication occurred during the Mesolithic Period, with early efforts focusing on dogs, goats, and possibly sheep around 9,000 BC. The Neolithic Period saw significant advancements in both agriculture and domestication, transforming human culture through the Neolithic Revolution. This revolution combined agriculture and animal domestication, leading to a critical shift in human society. Domestication is a biological process where organisms develop traits that increase their utility to humans.

Table 1. Domestication era and the purpose of animals

Species	Where	Why	When (Years Ago)
Dog	Eurasia	Companion	15,000-34,000
Sheep/Goats	Middle East	Food/Clothing	9,000-10,000
Cattle	Middle East	Meat/Labor	9,000-10,000
Horses	Central Asia	Transportation	5,500
Chickens	East Asia/Middle East	Food/Religion	4,000
Ducks	Asia	Meat/Feathers	-
Geese	Europe	Meat/Feathers	-
Turkeys	North America	Food	2,000

Table 2. Taxonomy of livestock

Common Name	Scientific Name	Kingdom	Phylum	Class	Order	Family	Genus	Species	Chromosome Number
Cattle	Bos taurus	Animalia	Chordata	Mammalia	Artiodactyla	Bovidae	Bos	B. taurus	60 (30 pairs)
Buffalo	Bubalus bubalis	Animalia	Chordata	Mammalia	Artiodactyla	Bovidae	Bubalus	B. bubalis	50 (25 pairs)
Sheep	Ovis aries	Animalia	Chordata	Mammalia	Artiodactyla	Bovidae	Ovis	O. aries	54 (27 pairs)
Goat	Capra hircus	Animalia	Chordata	Mammalia	Artiodactyla	Bovidae	Capra	C. hircus	60 (30 pairs)
Pig	Sus scrofa	Animalia	Chordata	Mammalia	Artiodactyla	Suidae	Sus	S. scrofa	38 (19 pairs)
Horse	Equus ferus caballus	Animalia	Chordata	Mammalia	Perissodactyla	Equidae	Equus	E. f. caballus	64 (32 pairs)
Chicken	Gallus gallusdomesticus	Animalia	Chordata	Aves	Galliformes	Phasianidae	Gallus	G. g. domesticus	78 (39 pairs)

Genetic material:A **karyotype** is the complete set of chromosomes in an organism, which can be either **haploid** (n) or **diploid** (2n). In cattle, the diploid number is **2n = 60**, consisting of **58 autosomes** and **2 sex chromosomes** (XX in females, XY in males). The haploid set, found in gametes, is **n = 30** (29 autosomes and 1 sex chromosome). Cattle autosomes are mostly acrocentric, while the X chromosome is typically submetacentric and the Y chromosome is acrocentric. This karyotype is crucial for understanding genetic traits and chromosomal abnormalities in cattle, which can impact fertility and overall health.

Table 3. The scientific name and the diploid (2n) chromosom,e number of domesticated animals

S.N.	Animal	Scientific Name	Chromosome Number (2N)
1	Zebu Cattle	*Bos indicus*	60
2	Exotic Cattle	*Bos taurus*	60
3	Goat	*Capra hircus*	60
4	Sheep	*Ovis aries*	54
5	River Buffalo/Water Buffalo	*Bubalus bubalis*	50
6	Swamp Buffalo	*Bubalus carabanesis*	48
7	Pig	*Sus domesticus*	38
8	Horse	*Equus caballus*	64
9	Donkey	*Equus asinus*	62
10	Camel (Single Humped)	*Camelus dromedarius*	74
11	Camel (Double Humped)	*Camelus bactrianus*	74
12	Cat	*Felis domesticus*	38
13	Dog	*Canis familiaris*	78
14	Chicken	*Gallus domesticus*	78
15	Rabbit	*Oryctolagus cuniculus*	44
16	Yak	*Bos grunniens*	60
17	Mithun	*Bos frontalis*	60

Gestation period: The gestation period is the time interval between successful conception and birth, during which the embryo or fetus develops inside the uterus of a viviparous animal, including mammals. This period is crucial for fetal growth and development, and its length varies significantly among different species.

Table 4. Gestation period of different livestock

S.N.	Animal	Scientific Name	Gestation Period
1	Cattle	*Bos taurus*	283 days (9 months + 9 days)
2	Buffalo	*Bubalus bubalis*	310 days (10 months + 10 days)
3	Sheep	*Ovis aries*	145 days (5 Months + 5 days)
4	Goat	*Capra hircus*	150 days (5 months)
5	Bitch	*Canis familiaris*	62 days (2 months + 2 days)
6	Camel (Dromedary)	*Camelus dromedarius*	384-410 days (12-14 months)
7	Camel (Bactrian)	*Camelus bactrianus*	360-420 days (12-14 months)
8	Pig	*Sus domesticus*	114 days (2 months + 2 days)
9	Cat	*Felis domesticus*	65 days (2 months + 5 days)
10	Rat	*Rattus norvegicus*	21-23 days
11	Rabbit	*Oryctolagus cuniculus*	31 days
12	Horse	*Equus caballus*	320-340 days (11 months + 11 days)
13	Donkey	*Equus asinus*	360-380 days (12 months + 12 days)
14	Monkey (Rhesus Macaque)	*Macaca mulatta*	165 days (5.5 months)
15	Elephant (African)	*Loxodonta africana*	640-660 days (21-22 months)
16	Elephant (Asian)	*Elephas maximus*	640-660 days (21-22 months)

Table 5. Common terms used in Different Species of Livestock for different aspect of animal husbandry

Common Term Used in Different Species of Livestock	Livestock Chromosome No.	Meat Name	Young ones	Immature (Up to maturity)	Matured	Castrated	Act of parturition (Birth)	Place of habitat	Voice Produced	Gestation Period (Days)	Male	Female	Male	Female
Horse (Equus caballus)	64	Basashi Or Chevali-neOr Cheval	Foal	Colt (<3Y) Filly(<3Y)	Stallion	Mare	Gelding Or Geld	Spaying mare	Foaling	Stable	Neighing Or whinny	305-400	Stallion	Mare
Ass (Equus asinus)	62	Foal	Foal	Colt Filly	Jack	Jenny Or Jennet	-	-	Foaling	-	365	Jack	Jenny	-
Cattle (Cow) (Bos taurus Bos indicus)	60	Beef (Calf meat Veal)	Calf	Young bull (1.5-2 Y) Heifer (1.5 - 2Y)	Bull Or Half of the herd	Cow	Bullock Or Steer*	Spayed	Calving	Byre Or Barn	Bellowing	273-294 (283)	Bull	Cow
Buffalo (Bubalus bubalis)	Ind.- 50 & Exo. - 48	Buffen Or Carabeef	Buffalo calf	Buffalo Young bull Buffalo heifer	Buffalo bull Or He buffalo	Buffalo cow Or She buffalo	Buffalo Bullock	Spay ed	Calving	Byre Or Barn	Bellowing	300-310 (305)	Buffalo bull	Buffalo cow

Sheep (Ovis aries)	54	Mutton	Lamb	Ram lamb (6-7 M) Gimmer (5-12 M)	Ram or Tup	Ewe	Wether Or Wedder-OrSeggy*	Spayed	Lambing	Pen Or Bara	Bleating	140-160 (150)	Ram	Ewe
Goat (Capra hircus)	60	Chavon	Kid	Buckling (Up to12 M) Goat-lingOr-Doeling (Up to 12 M)	Buck	Doe Or Nanny	Wether Or Wedder	Spayed	Kidding	Pen Or Bara	Bleating	147-155 (151)	Buck	Doe
Pig (Sus scrofa, Sus indicus, Sus domesticus)	38	Pork (Salted meat Bacon)	Piglet (Shoat) *Crad or Crit or Runt	Runner (7-9 M) Gilt (7-9 M)	Boar	Sow	Hog or Barrow Or Stag*	-	Farrowing	Sty	Grunting	109-125 (114 or 3M3W 3D)	Boar	Sow
Dog (Canis lupus familiaris)	78	**	Puppy (6-12 M)	Puppy (6-12 M)	Dog	Bitch	Altered dog	-	Whelping	Kennels	Barking	55-70	Dog	Bitch
Cat (Felis catus)	38	Felidae	Kitten	-	Tom	Queen	Neuter	-	Queening	Kettery	Mewing	55-70	Tom	Queen

Fowl (Gallus domesticus)	78	Chicken *Squab (Meat of pigeon)	Chick	Cockerel (<1Y) Pullet (<1Y)	Cock Or Rooster	Hen	Capon	-	Laying Or Hatching	Coop (Pen)	Crowing	21 (Incubation Period)	Cock	Hen
Rat (Rattus norvegicus)	42	Rat meat	Pup	Young rat	Male rat	Female rat	-	-	Parturition	Burrow	Squeaking	21-24	Male rat	Female rat
Rabbit (Oryctolagus cuniculus)	44	Rabbit meat	Kit	Young rabbit	Buck	Doe	-	-	Kindling	Warren	Thumping	28-32	Buck	Doe
Camel (Camelus dromedarius)	70	Camel meat	Calf	Young camel	Bull camel	Cow camel	Gelded camel	-	Calving	Desert	Grunting	390-410	Bull camel	Cow camel

** dog meat known as "fragrant meat," "fragrant," and specific names in different cultures like "Nureongi" in Korea or "Gaegogi" in South Korea

Table 6 Water requirement of different livestock

Animal	Drinking Water Requirement per Day	Avg. Daily Dung Production per Animal
Sheep and goat	15-20 litres	1.75 Kg (1-2.5)
Poultry (100 birds)	25 litres (250 ml per bird)	3.0 Kg (2.5-3.5)
Pig	22-25 litres	4.0 Kg (3-5)
Horse	36 litres	21.5 Kg (18-25)
Dairy Cattle	30-35 litres	24 Kg (18-30)
Buffalo	30-35 litres	32.50 Kg (25-40)

Table 7. Water intake of different livestock per Kg of dry matter

S.N.	Animal	Water Intake per Kg Dry Matter Intake
1	Adult Cattle	3-5 liters
2	Milking Cattle	3-5 liters + additional 4-5 liters per Kg milk produced
3	Suckling Calf	6-7 liters
4	Sheep/Goat	4 liters
5	Swine	3 liters
6	Poultry	2 liters

Table 8. Different institutions of India dedicated to livestock and poultry research and development

Institution	Focus	Location
ICAR-Central Institute for Research on Cattle (CIRC)	Cattle Research	Meerut, Uttar Pradesh
ICAR-Central Institute for Research on Buffaloes (CIRB)	Buffalo Research	Hisar, Haryana
ICAR-Central Sheep & Wool Research Institute (CSWRI)	Sheep & Wool Research	Avikanagar, Rajasthan
ICAR-Central Institute for Research on Goats (CIRG)	Goat Research	Makhdoom, Mathura, Uttar Pradesh
ICAR-National Research Centre on Camel (NRCC)	Camel Research	Bikaner, Rajasthan
ICAR-National Research Centre on Equine (NRCE)	Horse Research	Hisar, Haryana
ICAR-Directorate of Poultry Research (DPR)	Poultry Research	Hyderabad, Telangana
Central Poultry Development Organizations (CPDOs)	Poultry Development	Chandigarh, Bhubaneswar, Mumbai, Bengaluru
ICAR-National Dairy Research Institute (NDRI)	Dairy Research	Karnal, Haryana

ICAR-Indian Veterinary Research Institute (IVRI)	Veterinary Research	Izatnagar, Bareilly, Uttar Pradesh
ICAR-Central Avian Research Institute (CARI)	Avian research	Izatnagar, Bareilly, Uttar Pradesh
National Institute of Nutrition (NIN)	Nutrition Research	Hyderabad, Telangana
National Institute of Agricultural Management (MANAGE)	Agricultural Management	Hyderabad, Telangana
National Academy of Agricultural Research Management (NAARM)	Agricultural Research Management	Hyderabad, Telangana
Project Directorate on Poultry (PDP)	Poultry Development	Hyderabad, Telangana
National Research Centre on Mithun (NRCM)	Mithun Research	Nagaland
National Research Centre on Yak (NRCY)	Yak Research	Dirang, Arunachal Pradesh
ICAR-National Research Centre on Pig (NRC-Pig)	Pig Research	Guwahati, Assam

Table 9. Name of the meat of different animals

Livestock/Animal	**Meat Name**
Cattle	Beef
Buffalo	Carabeef
Sheep	Lamb/Mutton
Goats	Chevon
Pigs	Pork
Poultry	Chicken/white meat
Deer	Venison
Guinea Pigs	Cavy meat
Capybaras	Capybara meat
Calves	Veal

Poultry Production

Classification and characteristics of Chicken Breeds: Breeds of chicken are often classified into different classes of chicken like American, Asiatic, English, and Mediterranean based on their origins and characteristics.

Table 10. Representative chicken breeds of different class of poultry chicken

Breed	Type of Comb	Earlobe Color	Skin Color	Shank Color	Shanks Feathered	Egg Color
American						
Plymouth Rock	Single	Red	Yellow	Yellow	No	Brown
Wyandotte	Rose	Red	Yellow	Yellow	No	Brown
Rhode Island Red	Single and Rose	Red	Yellow	Yellow	No	Brown
Asiatic						
Brahma	Pea	Red	Yellow	Yellow	Yes	Brown
Cochin	Single	Red	Yellow	Yellow	Yes	Brown
Langshan	Single	Red	White	Bluish-black	Yes	Brown
English						
Australorp	Single	Red	White	Dark slate	No	Brown
Cornish	Pea	Red	Yellow	Yellow	No	Brown
Dorking	Single	Red	White	White	No	White
Orpington	Single	Red	White	White	No	Brown
Mediterranean						
Leghorn	Single and Rose	White	Yellow	Yellow	No	White
Minorca	Single and Rose	White	White	Dark slate	No	White
Ancona	Single and Rose	White	Yellow	Yellow	No	White

Egg Purpose Breeds

Breeds raised for egg production are typically small in size and light in weight, making them efficient for laying large numbers of eggs. These birds, such as Leghorn, Minorca, and Ancona, are known for producing 250 to 300 eggs per year, with each egg weighing between 50 to 60 grams. Their small size means they require less feed to maintain, but they have a high feed conversion ratio (FCR) or feed conversion efficiency (FCE), which ranges from 2.2 to 2.5 kg of feed per kg of eggs produced. These breeds start laying eggs at around 130 to 150 days of age, making them ideal for commercial egg production due to their early maturity and high yield.

Table/Broiler/Meat Purpose Breeds

Breeds raised for meat production are characterized by their large size and heavy weight. Adult cocks can weigh up to 4.5 kg, while hens typically reach 3.5 kg. These birds, including Cornish, Brahma, Cochin, Aseel, and Chittagong, have blocky bodies with loose plumage and are bred for their ability to fatten quickly. Their feed conversion ratio (FCR) is efficient, requiring only about

1.6 to 1.7 kg of feed per kg of weight gain. This makes them ideal for broiler production, where the goal is to achieve rapid growth and tender meat. These breeds lay fewer than 100 eggs per year, as their primary purpose is meat production.

Dual Purpose Breeds

Dual-purpose breeds are versatile, serving both as egg layers and meat producers. These birds, such as Rhode Island Red, Plymouth Rock, Wyandotte, Australorp, Orpington, and Hamburg, are medium in size, with adult cocks weighing about 3.8 kg and hens around 3 kg. They produce a moderate number of eggs, typically between 150 to 250 per year. Their size and productivity make them suitable for small-scale farming or backyard flocks, where both eggs and meat are desired. These breeds are often preferred for their hardiness and ability to thrive in a variety of conditions.

Miscellaneous Purpose Breeds

Some breeds are kept for their aesthetic appeal or specific purposes beyond egg or meat production. These include Bantam and Hamburg breeds, which are often kept for their fancy plumage and small size. The Andalusian breed is also noted for its beautiful coloration. These birds are not primarily raised for utility but are instead valued for their ornamental qualities or used in game or show contexts. They are often kept by hobbyists who appreciate their unique characteristics and beauty. While they may not contribute significantly to egg or meat production, they add diversity and interest to poultry collections.

Table 11. Characteristics of different classes of poultry chicken

Class	Breeds	Class Characteristics
American	Rhode Island Red, New Hampshire, Plymouth Rock, Wyandotte	Clean shank, yellow skin, red ear lobes, dual purpose, medium size, single comb, brown shelled eggs
English	Australorp, Cornish, Orpington, Sussex, Dorking	Clean shank, white skin, dual purpose, red ear lobes, medium size, single comb, brown shelled eggs
Mediterranean	Leghorn, Minorca, Ancona, Andalusian	Clean shank, yellow or white skin, white ear lobes, egg purpose, small size, light feathering, white shelled eggs
Asiatic	Brahma, Cochin, Langshan, Aseel	Feathered shank, yellow skin, red ear lobes, brown shelled eggs, meat purpose, massive size, loose plumage
Indian breeds	Chittagong, Ghagus, Uttara	Feathered shank, black plumage, single comb, unique to the Kumaon region of Uttarakhand

All-in All-out: A flock of birds enters and exits a facility as a single group. New birds are not introduced and flocks are not mixed. This strategy minimizes the risk of new birds introducing disease agents into a flock.

Bantam: A bantam is a small breed of chicken, often kept as pets or for small-scale egg production due to its compact size.

Biosecurity: Procedures such as sanitation and isolation used to prevent disease exposure to birds or premises.

Boiler/Cull Birds: These are adult hens that are at the end of their laying cycle and are no longer productive, often culled and used for meat. Boiler hens are typically older birds that are removed from the flock to maintain efficiency.

Broiler breeder: Breeders that produce hatching eggs for commercial broiler production. Broody: A hen which has stopped laying eggs temporarily.

Broiler: This term describes chickens raised specifically for meat production, typically marketed at 6 to 8 weeks of age with a body weight of 1.6 to 1.8 kg. Broilers are bred for rapid growth and are a major component of the poultry industry.

Brood: A group of chicks of same age raised in one batch is called as a brood. Brooder: A device for providing artificial heat to the chicks.

Capon: This refers to a castrated male chicken, which is often used for meat production due to its tender and flavorful flesh. Caponization can improve the quality and palatability of the meat.

Chick: A young male or female fowl below 8 weeks of age.

Chicken: the domestic fowl, Gallus domesticus, used for meat, eggs, feathers and showing. There are several hundred recognized breeds of various sizes, only a few of which are important for meat and eggs.

Clutch: The number of eggs laid by a bird on consecutive days. A clutch of 3-4 eggs is preferred.

Cock: This signifies an adult male chicken used for breeding purposes, selected for desirable traits such as fertility and genetic quality. Cocks play a crucial role in maintaining the genetic diversity of poultry flocks.

Cockerel: This refers to a young, maturing male chicken between 18 to 22 weeks of age, similar to pullets but destined for different roles. Cockerels may be used for breeding or raised for meat.

Culling: Removal of unwanted bird from the flock is known as culling e.g. old non-laying birds, sick birds and masculine hens are removed.

Day-old chick: Hatched out chick is called as day-old-chick up to 24 hours.

Designer eggs: eggs produced to contain higher levels of certain constituents, such as omega-3 fatty acids.

Drake: a male duck.

Duck: small to moderate sized fowl (waterfowl) with web feet of the Anatidae family that are used for meat, eggs and feathers.

Duckling: young duck.

Flock: a group of birds that feed or move together. In domestication they will normally be of the same type and age, housed together, and managed the same.

Fowl: domestic birds used for food and other economic purposes, including chickens, turkeys, ducks, geese, guinea, quails, pheasants, pigeons, etc.

Grower: A young chick of 9thweek of 20thweek of age of either sex. Hen: A matured female chicken generally above 20 weeks of age.

Hen: This term describes a female chicken that is about 22 weeks of age and has started laying eggs, marking her transition to reproductive maturity. Hens are the backbone of egg production in poultry farming.

Hen-day-production: This is arrived by dividing total eggs laid in the season by the average number of birds in the house.

Hen-housed-average: This is arrived at by dividing the total number of eggs laid in the season by the number of birds originally placed in the house. No deductions are made for any losses from the flocks.

Layer: An egg laying female chicken up to one year after starting the laying of eggs. Litter: materials used on the floor of poultry houses. Common litter materials are wood shavings, sawdust, rice hulls, peanut hulls, chopped straw and shredded paper.

Laying: This refers to the act of a female bird depositing an egg, a critical process in poultry production. Efficient laying practices are essential for maximizing egg yield and quality.

Moulting: The process of shading old feathers and growth of new feather in their place moulting normally occurs once in a year

Pause: It is the period between two clutches in which eggs are not laid by hen. Post-mortem inspection: inspection performed on the carcass of the bird after slaughter.

Plumage: Plumage is the layer of feathers covering a chicken, providing insulation and visual signals.

Poultry: domestic fowl such as chickens, turkeys, guineas, geese and ducks that are raised for meat, eggs, feathers, or other products.

Pullet: A young female chicken from 9 to 20 weeks of age.

Pullets: These are young, maturing female chickens between 18 to 22 weeks of age. Pullets are typically raised to become laying hens and are monitored for health and growth.

Spur: A spur is a sharp protrusion on a chicken's leg used for fighting and defense.

Shank: The shank is the part of a chicken's leg between the claw and the first joint.

Squab: Squab is the meat of young pigeons, often considered a delicacy.

Cattle buffalo production

A bobby calf is a young calf raised for slaughter and human consumption, often removed from its mother early. These calves are fed a diet to optimize growth and meat quality.

A2 milk: A2 milk is a type of cow's milk that contains only the A2 beta-casein protein, as opposed to regular milk, which contains a mix of A1 and A2 beta-casein proteins. This distinction is believed to make A2 milk easier to digest for some individuals, potentially reducing digestive discomfort and other health issues associated with A1 protein, such as bloating and inflammation.

Abattoir: An abattoir is a facility where farm animals are slaughtered and processed into meat products, ensuring humane treatment and hygiene standards. It plays a critical role in the meat supply chain.

After birth: The fetal membranes that attach the fetus to the membranes of the pregnant female and which are normally expelled from the female within 3 to 6 hr after parturition

Animal Husbandry: Animal husbandry is the branch of agriculture focused on the care and breeding of domestic animals like cattle, goats, and sheep. It involves practices such as feeding, breeding, and disease management to ensure animal well-being and productivity.

Animal Science: Animal science is the scientific study of farm animals, covering their nutrition, genetics, behavior, and health to improve productivity and welfare. This field informs practices in breeding and management.

Artificial Insemination (AI): Artificial insemination involves introducing semen into a female animal without natural mating, used to improve breeding efficiency and genetic traits. This method allows for better control over genetic selection.

Artificial insemination: The injection of mechanically procured semen into the reproductive tract of the female without natural mating and with the aid of mechanical instruments and expert human intervention.

Autopsy: An autopsy is a post-mortem examination performed on a human body to determine the cause, manner, and mode of death. It is typically conducted by a pathologist and can be required for legal or medical reasons

Average Daily Feed Intake

Average daily feed intake is the amount of feed consumed by an animal each day, crucial for managing nutrition and ensuring optimal growth and health. Proper feed intake is essential for productivity.

Average Daily Gain

Average daily gain measures the weight gained by an animal during its growth stages, used to evaluate feeding and management practices. A higher gain indicates better growth rates.

Backcross: A backcross involves mating a hybrid offspring with one of its parents or a genetically similar individual to recover desirable traits. For example, if you cross a Brahman (heat-tolerant) with an Angus (high-quality beef) to get an F1 hybrid, you might backcross this hybrid with a Brahman to enhance heat tolerance while retaining some Angus traits.

Bipedal: Bipedal refers to walking on two legs, not commonly discussed in animal husbandry but relevant in animal locomotion.

Bloat: Bloat is a condition where gases accumulate in the rumen of ruminants, potentially life-threatening if untreated. It often occurs due to feeding practices.

Bovine Somatotropin (BST): Bovine somatotropin is a hormone that naturally stimulates growth and milk production in cows. It is sometimes used in dairy farming to enhance milk yield.

Bred Heifer: A bred heifer is a pregnant female bovine expecting her first calf, managed to optimize health and growth. These animals are essential for dairy and beef production.

Breeding Stock: Breeding stock refers to mature male and female animals retained for producing offspring, selected based on genetic quality. They are crucial for improving future generations.

Bull: A bull is an uncastrated male used for breeding, known for strength and aggression. Bulls play a crucial role in maintaining genetic diversity.

Bullock: A bullock is a castrated male animal, typically used for draft purposes due to reduced aggression and increased growth. Castration is usually performed around one year of age.

Calf: A calf is the young of cattle, typically under six months old and dependent on its mother for milk and care.

Calf: A young cow or bull from birth to about six months of age, requiring special care and nutrition for optimal growth.

Calving Interval: The time between two successive calvings, ideally around 365 days for dairy cattle, to maintain annual calving and optimize milk production.

Calving: Calving is the process of giving birth to a calf, marking the start of a new lactation cycle in dairy animals.

Colostrum: The first milk produced by a cow after calving, rich in antibodies and essential for calf health and immune system development.

Cud: Cud is partly digested food brought back up from the rumen for further chewing by ruminants like cattle and sheep.

Culling: Culling involves removing nonproductive or undesirable animals from a herd to maintain overall health and productivity.

Dam: The dam is the female parent of an animal, important for determining offspring traits.

Dry Cow: A cow not producing milk, typically during the dry period before the next calving, when she is not being milked.

Dry Cow: A dry cow is not currently producing milk, requiring special care to ensure health and future productivity.

Dry Period: The dry period is the nonlactating time between milk production cycles in dairy animals, typically lasting about 60 days.

Dry Period: The period between the end of lactation and the next calving, recommended to be at least 2 to 2.5 months, allowing the cow to rest and prepare for the next lactation.

Drying Off: Drying off is the process of stopping milk production in preparation for the next lactation cycle.

Forage: High-fiber feed such as hay, silage, and grasses used in dairy cattle diets to provide essential nutrients and fiber.

Freemartin: A freemartin is a sterile female bovine born as a twin to a male calf. This condition occurs due to the exchange of blood and hormones between the twins during fetal development, leading to masculinization of the female's reproductive tract. Freemartins are genetically female (XX) but exhibit some male characteristics and have underdeveloped ovaries, making them infertil

Freemartin: A freemartin is a sterile female calf born as a twin to a male calf, resulting from hormonal exchange during fetal development.

Freshening: Freshening occurs when a cow begins producing milk after giving birth, marking the start of a new lactation cycle.

Gestation Period: The gestation period is the time from fertilization to birth, varying among species but typically around 280 days for cattle.

Hay: Dried feed like alfalfa, clover, and grasses used for dairy cattle, providing fiber and nutrients.

Heifer: A female dairy animal that has not yet given birth to a calf, typically raised for future milk production.

Heifer: A heifer is a female bovine that has not yet given birth, typically under two years old.

Heiferettes: Heiferettes are female bovines that have not calved beyond two years of age.

Husbandry: Husbandry involves the care, cultivation, and breeding of crops and animals, crucial for agricultural productivity.

Lactation Period: The duration of milk production after calving, typically around 305 days, during which the cow is milked regularly.

Lactation Period: The lactation period is the time a cow produces milk after giving birth, typically lasting around 300 days.

Litter: A litter is a group of young animals born at the same time, such as piglets or puppies.

Livestock: Livestock includes animals like cows, sheep, and goats raised for profit, providing products like milk, meat, and wool.

Necropsy: A necropsy is the post-mortem examination of a non-human body, such as animals, to determine the cause of death. It is often performed by a veterinary pathologist.

Nymphomaniac: A nymphomaniac cow appears to be constantly in heat, though this term is not commonly used in modern animal husbandry.

Pack: A pack refers to a group of dogs, often used for tasks like hunting or herding.

Parturition: Parturition is the process of giving birth, also known as freshening in dairy animals.

Pasteurization: Pasteurization is a heat treatment process used to kill pathogens and extend the shelf life of foods and beverages. It involves heating liquids to a temperature below 100°C (212°F) for a specified period, followed by rapid cooling. This process is named after Louis Pasteur, who developed it to prevent spoilage in wine and beer.Common methods for milk pasteurization include High-Temperature Short-Time (HTST) pasteurization, where milk is heated to 71.7°C (161°F) for 15 seconds, and Ultra-High Temperature (UHT) pasteurization, where milk is heated to 135-140°C (275-284°F) for 1-2 seconds.

Poddy Calf: A poddy calf is a calf that has lost its mother, requiring special care.

Service Period: The time between calving and successful conception, ideally between 60 to 90 days, during which a cow is bred to become pregnant again.

Silage: Fermented, high-moisture forage made from crops like corn or sorghum, used as a nutritious feed for dairy cattle.

Sire: The sire is the male parent of an animal, selected for genetic quality to improve offspring traits.

Teaser Bull: A teaser bull is used to detect females in heat without breeding them, often vasectomized.

Test Cross: A test cross is used to determine the genotype of an individual by mating it with a homozygous recessive individual. For instance, if you have a Holstein (dairy breed) with an unknown genotype for a specific trait like milk production, you might test cross it with a homozygous recessive individual (e.g., a breed with low milk production) to see if it's homozygous dominant or heterozygous for high milk production.

Trumpeting: Trumpeting is the sound made by elephants, used for communication.

Vealer Calf: A vealer calf is a young calf raised for veal production, typically fed a specialized diet.

Venison: Venison is the meat of deer or other game animals, prized for its flavor and nutritional value.

Weaning: Weaning is the process of transitioning young animals from milk to solid foods.

Weanling: A weanling is a foal that has been weaned but is less than one year old.

Wet Average: Not a standard term; however, it might refer to the average milk production during the lactation period when the cow is actively producing milk.

Yearling: A yearling is a horse approaching or just turning one year old, transitioning from juvenile to adult status.

Sheep Goat production

Accelerated Lambing: A management system where ewes are bred more frequently than once a year, often every six to eight months, to increase lamb production.

Bleating: Bleating is the sound made by animals like sheep or goats, not typically associated with camels. Camels produce different vocalizations

Broken Mouthed: Broken mouthed animals have lost one or more teeth, affecting their ability to eat and digest food properly. Special dietary care may be necessary.

Confinement: A system where animals are kept in enclosures rather than grazing freely. Confinement systems are used for intensive production.

Crone: Thisrefers to an older, experienced ewe that has been retained in the breeding flock due to her excellent reproductive performance, despite being "broken-mouthed" (having lost some teeth). Crones are valued for their genetic contribution to the flock.

Cull: To remove an animal from the breeding flock due to age, health, or reproductive issues. Culling is essential for maintaining herd quality.

Ewe Lamb: This signifies a female lamb that is less than six months old, also dependent on its mother. Ewe lambs are often retained in the flock for future breeding.

Ewe: An adult female sheep, often used for breeding and milk production. Ewes are the foundation of sheep production systems, providing lambs and wool.

Ewe: This term describes an adult female sheep, which is the primary reproductive unit in sheep farming. Ewes are valued for their ability to produce lambs and milk.

Fleece: The wool from a single sheep, often harvested during shearing. Fleece is a valuable product in sheep farming.

Flerd: A mixed group of sheep and cattle. Flerds are managed together for grazing or other purposes.

Flushing: Increasing nutrition for ewes before and during the breeding season to enhance fertility and conception rates.

Fodder: Crops grown specifically for animal feed, such as hay or silage. Fodder is crucial for maintaining the nutritional needs of sheep and goats.

Gimmer: A gimmer is an immature female sheep, typically between 5 to 12 months old.

Gimmer: A young female sheep that has not yet given birth to a lamb. Gimmers are typically one to two years old and are being prepared for breeding.

Gimmer: This term is used for a female sheep that is between one and two years of age and has undergone one or two shearings. Gimmers are transitioning from adolescence to adulthood and are often considered for breeding.

Grazing: Allowing animals to feed on pasture. Grazing is a common practice in sheep and goat production for maintaining health and reducing feed costs.

Gummers: Gummers are animals that have lost all their teeth, requiring special dietary care.

Lamb: A young sheep from birth to about one year of age. Lambs are the offspring of ewes and are raised for meat or further breeding.

Lamb: This denotes a young sheep of either sex, typically up to one year of age. Lambs are vulnerable and require careful management to ensure their survival and growth.

Lambing: This refers to the act of a ewe giving birth to a lamb, a crucial event in sheep farming. Successful lambing practices are essential for maximizing the number of lambs weaned.

Mule: A crossbreed sheep, often referring to a specific type of cross between different breeds. Mules are valued for their hardiness and productivity. The term mule is used to refer to a cross between a Bluefaced Leicester ram and a purebred hill (or mountain) ewe (usually a Swaledale sheep)

Nutrient Management: The process of managing nutrients like nitrogen, phosphorus, and potassium to optimize crop production and minimize environmental impact.

Predator Control: Measures taken to protect sheep and goats from predators such as dogs, coyotes, or foxes. Predator control is essential for maintaining flock health.

Quarantine: Isolating animals to prevent the spread of disease. Quarantine is a critical aspect of flock health management.

Ram Lamb: This refers to a male lamb that is less than six months old, during which time it is still dependent on its mother. Ram lambs may be selected for breeding or raised for meat.

Ram or Tup: This refers to an adult male sheep used for breeding purposes, selected for desirable traits such as fertility and genetic quality. Rams play a crucial role in maintaining the genetic diversity of sheep flocks.

Ram: An adult male sheep used for breeding purposes. Rams are crucial for the reproduction process in sheep farming.

Seggy: This refers to a ram that has been castrated after it has been used for breeding, typically to prevent further mating. Seggies may be used for meat or as companion animals.

Shearing: Shearing is the removal of wool from sheep or other fiber-producing animals.

Shearing: This is the process of removing wool from sheep, a critical practice for maintaining their health and comfort. Shearing is typically done annually or bi-annually, depending on the breed and climate.

Sheep (Common Gender): This term is often used generically to refer to sheep without specifying gender, similar to how "cattle" can refer to both cows and bulls. Sheep are widely raised for wool, milk, and meat, making them a versatile livestock species.

Springer: A ewe close to lambing. Springers are monitored closely to ensure a safe and healthy delivery.

Stag: A male sheep castrated after about six months of age. Stags are often used for meat production.

Staple: A group of wool fibers that form a cluster or lock. Staples are important in evaluating wool quality.

Tup: Another term for a ram, especially in the UK. Tups are used for mating with ewes to produce lambs.

Tupping: This term describes the act of mating between a ram and an ewe, a key process in sheep breeding. Tupping is carefully managed to ensure successful conception and a healthy lamb crop

Wether or Wedder: This term describes an adult male sheep that has been castrated, often used for meat production or as a companion animal. Castration can improve the quality and tenderness of the meat.

Yeld or Eild: This term describes a barren or non-lactating female sheep, which may not be producing milk or offspring. Yeld ewes may be culled from the breeding flock to maintain efficiency.

Swine production

AI (Artificial Insemination): Artificial insemination in swine involves using semen from a boar to breed a sow without natural mating. This method allows for improved genetic selection and reduced need for multiple boars

Back pressure test: The back pressure test in pigs is a method used to detect estrus, particularly in sows and gilts. This test involves applying manual pressure to the back of the female pig to observe her reaction. If she is in estrus, she will typically stand still with her ears erect and may display other signs such as a swollen vulva and increased mucus discharge.

Backfat Thickness: Backfat thickness is a measure used to assess the fatness of pigs, particularly important for determining when they are ready for slaughter. It affects the quality and yield of pork products.

Baconer: A baconer is a pig raised specifically for bacon production, typically weighing between 80-100 kg or more, depending on market requirements.

Barrow: A barrow is a castrated male pig, typically castrated before reaching sexual maturity. Barrows are raised for meat production and are less aggressive than boars.

Barrow: This signifies a castrated adult male pig that is less than one year of age, typically raised for meat production. Castration helps improve meat quality and reduce aggression in these animals.

Boar: A boar is an intact male pig over six months old, used for breeding purposes. Boars can service multiple sows and are typically kept in limited numbers due to their aggression and breeding efficiency.

Boar: This refers to an adult, uncastrated male pig that is used for breeding purposes, emphasizing its role in reproduction. Boars contribute genetic material and are selected for desirable traits.

Breeding Stock: Breeding stock refers to the male and female pigs retained for breeding purposes. These animals are selected based on genetic quality to improve herd traits.

Closed Gilt: This signifies a young female pig that has become pregnant, marking her progression to becoming a productive sow. Pregnancy confirmation is crucial in managing closed gilts.

Crossbreeding: Crossbreeding involves mating pigs of different breeds to combine desirable traits, such as fertility and growth rate. This practice is used to improve overall herd performance.

Culling: Culling involves removing nonproductive or undesirable animals from the herd. In swine production, culling is used to maintain herd health and efficiency by eliminating sows with poor reproductive performance.

Drove/Stock/Herd: This term refers to a collective group of pigs or swine, similar to how "flock" is used for birds or "herd" for cattle. It's the standard nomenclature for describing a group of these animals living or being managed together.

Farrowing: Farrowing refers to the process of a sow giving birth to a litter of piglets. This is a critical event in swine production, requiring careful management to ensure the health of both the sow and piglets.

Farrowing: This is the term used to describe the act of a sow giving birth to her young, a critical event in swine production. Successful farrowing practices are essential for maximizing the number of piglets weaned.

Gestation: Gestation in pigs lasts approximately 114 days, from breeding to birth. Proper management during this period is essential for successful farrowing and piglet health.

Gilt: A gilt is a young female pig that has not yet given birth. Gilts are often used for breeding after reaching maturity or are raised for meat production.

Gilt: This denotes a young female pig that is being kept for breeding purposes but has not yet farrowed. Gilts are carefully selected for traits that will make them productive sows in the future.

Guard rail: A guard rail for pigs is a protective barrier installed in farrowing pens to prevent piglets from being crushed by their mothers. These rails are typically placed around the pen, about 9 to 12 inches from the wall and 9 to 10 inches above the floor, allowing both sows and piglets to move freely while keeping the piglets safe

Inbreeding: Inbreeding involves mating pigs that are closely related, which can lead to inbreeding depression if not managed carefully. It is used to fix specific traits but requires monitoring for genetic diversity.

Linebreeding: Linebreeding is a breeding strategy where pigs from the same family line are mated to concentrate desirable traits. It requires careful management to avoid inbreeding depression.

Litter: This refers to the total number of piglets born in a single farrowing by a sow, indicating her reproductive output. Litter size is a key factor in the efficiency of swine production.

Open Gilt: This term describes a young female pig that is intended for breeding but has not yet been bred or inseminated. Managing open gilts involves careful attention to their reproductive readiness.

Piglet anemia: Piglet anemia, often caused by iron deficiency, is a significant health issue in young pigs. It manifests through symptoms such as pale skin and mucous membranes, rapid breathing, diarrhea, and poor growth, with diagnosis typically involving blood tests to confirm low hemoglobin levels. Treatment and prevention involve iron supplementation, commonly administered via injection of iron dextran between 3 to 5 days of age, to ensure adequate iron reserves and prevent the condition.

Piglet/Pigling: This denotes a young pig that is less than 8 weeks old, a crucial period of early development. Piglets require specialized care during this time to ensure their health and growth.

Porker: A porker is a grower pig intended for pork production, usually weighing between 60-150 kg. Porkers are raised to optimize meat quality and quantity.

Runt/Crad/Crit/Rit/: This refers to the last piglet born in a litter, which is typically smaller and weaker than its siblings. Runts often require extra care to ensure their survival and growth.

Runt:A runt is the smallest piglet in a litter, often culled early to focus resources on larger, healthier piglets.

Sow: A sow is a mature female pig that has given birth to at least one litter of piglets. Sows are crucial for breeding and are often managed based on their parity (number of litters).

Sow: This term is used to describe an adult female pig that is used for breeding and has already given birth to at least one litter of piglets. Sows are the foundation of swine breeding operations.

Stag or Hog or Boar (Castrated): This refers to a castrated male pig that is over one year of age, indicating it has been allowed to grow for an extended period. These animals are often raised for specialized meat products.

Stag: A stag is a male pig castrated after reaching sexual maturity. Stags are less common than barrows but may be used in specific production systems.

Store Pig: This term refers to pigs that are between the weaning stage and the fattening stage, typically between 8 and 15 weeks old. Store pigs are in a transitional phase of growth and management.

Teat Order: It refers to the specific teats that piglets prefer to suckle from on their mother's udder. This preference is established early in life and remains relatively stable throughout the suckling period.

Weaning: Weaning is the practice of separating piglets from their mother to transition them to solid food. This process encourages independent growth and is typically done between four to eight weeks of age in pigs.

3

Male Reproduction in Farm Animals

Manish Kumar, Utkarsh Kumar Tripathi, Anuradha Kumari and Ajeet Singh

Faculty of Veterinary and Animal Sciences, I. Ag.Sc., RGSC-Banaras Hindu University (BHU), Barkachha, Mirzapur, Uttar Pradesh-231001

Male Reproduction

The reproductive functions of the male involve the formation of sperm and the deposition of the sperm into the female reproductive tract. Sperm are produced in the seminiferous tubules of the testes and are then transported through the rete testes to the epididymides, where they are stored and matured. The production of sperm is a continuous process once it has been initiated. The processof male reproduction is assisted by hormones and the autonomicnervous system.

Testes and associated structures: The two testes produce spermatozoa.The seminiferous tubules are convoluted and occupy the greatest portion of each testicle. The testicle is surrounded by a connective tissue capsule called the tunica albuginea. Sertoli Cell (sustentacular or supporting cell) & Leydig Cell (interstitial cell). The Sertoli cell provides a "nurse" function for developing spermatozoa.

The Sertoli cells have their base at the periphery of the seminiferous tubules and extend toward the center.

Epididymis: Epididymis is a collection & storage tubule for the testis. Epididymis is divided into three regions: Head, Body and Tail of the epididymis. Storage in the epididymis allows the spermatozoa to reach maturity and become motile. Reabsorption of much of the seminiferous tubular fluid occurs in the head of the epididymis.

Ductus deferens

The ductus deferens (vas deferens) extends from the epididymis to the pelvic urethra. It travels with the testicular artery, vein, nerve, and lymphatics within the spermatic cord. Passing through the inguinal canal, it terminates in the ampulla of the ductus deferens (absent in boars).

Scrotum

The scrotum is a cutaneous sac containing the testes. The scrotum contains a subcutaneous layer of smooth muscle fibers, the tunica dartos, which contracts in cold weather and holds the testes closer to the abdominal wall.

Descent of the testes

During embryonic development, the testes remain in the abdomen, connected to the scrotum by the gubernaculum testis. As growth progresses, the gubernaculum pulls the testes through the inguinal canal into the scrotum, forming a double-walled peritoneal tube. The inner layer, the visceral vaginal tunic, encloses the testes, epididymis, ductus deferens, and associated vessels, nerves, and lymphatics, which form the spermatic cord. The outer layer, the parietal vaginal tunic, lines the scrotum. The cremaster muscle, located on the spermatic cord, aids in testicular movement. The narrow vaginal cavity remains between the two tunics, and the spermatic cord passes through the superficial and deep vaginal rings into the abdomen.

Cryptorchid testes are those that fail to descend. This condition seems to be most prevalent in pigs and horses. When the testis is in the inguinal canal but not in the scrotum, the horse is referred to as a HIGH FLANKER. Often the testes are retained entirely within the abdominal cavity.

Accessory sex glands

The accessory sex glands provide secretions that empty into the pelvic urethra near their origin. They vary in size and shape among species and can be absent in some. The accessory sex glands are composed of the ampullae of the ductus deferentes, the vesicular glands (seminal vesicles), the prostate gland, and the bulbourethral glands (Cowper glands). The ampullae (absent in the boar & dog) are enlargements of terminal part of the ductus deferentes. The vesicular glands (absent in the dog) are paired glands that empty into the pelvic urethra along with the ductus deferentes.The prostate gland is present in all domestic animals.

The paired bulbourethral glands (absent in dog) are the most caudal of the accessory glands. At the time of ejaculation, the accessory sex gland secretions (seminal plasma) are mixed with sperm and fluid from the epididymides to form semen. The seminal plasma provides an environment conducive to the survival of sperm within the female reproductive tract. It is rich in electrolytes, fructose, ascorbic acid, and other vitamins.

Penis and prepuce

The penis is the male copulatory organ for urine and semen passage via the penile urethra. It consists of the roots (crura) at the pelvic ischial arch, the body extending forward, and the glans at the free end. Internally, it contains cavernous (erectile) tissue—blood sinusoids separated by connective tissue. Stallions have more erectile tissue, allowing greater enlargement during erection than bulls. The urethra runs along the ventral aspect of the penis.

The ram has a visible urethral process that sprays semen in a circular motion during ejaculation. In dogs, the bulbus glandis at the glans' caudal part enlarges, facilitating prolonged penile retention during coitus, aided by female vaginal muscle contraction, known as the "tie."

The bull, buck, ram, boar and llama have a sigmoid flexure of their penis, resulting in an S shape when not erect. Boar glans penis is Corkscrew shaped.

Prepuce

The prepuce is an invaginated fold of skin that surrounds the free extremity of the penis. The stallion has a double-folded prepuce. The boar has a preputial diverticulum (pouch) on the dorsal wall, which often contains decomposing urine and macerated epithelium. The fluid in the diverticulum also contains a pheromone that induces sows to assume the immobile mating stance.

Unique Features of Male Reproductive Organs

In rams, the penis features a filiform appendage containing the urethra. Pigs have large accessory glands that produce a high semen volume. In stallions, the urethra protrudes a few centimeters from the glans penis. Bulls, rams, and boars have a sigmoid flexure that straightens during erection. Dogs possess an os penis and experience penile locking during coitus. Except for the prostate and ampulla, other accessory glands are absent in dogs. Cats have penile spines.

Androgens

Leydig cells are the primary site of androgen production, including testosterone, dihydrotestosterone, and androstenedione. Androgens are also produced by the adrenal cortex, ovary, and placenta. In horses, seminiferous tubules and the epididymis also produce testosterone. About 97-99% of testosterone is protein-bound, while the free portion enters cells. In target tissues, testosterone converts to biologically active dihydrotestosterone in the cytoplasm.

Functions of Androgens

In involves in the various functions such as development of fetal testicles and descent of testes before birth. Spermatogenesis and prolong the life span of epididymal spermatozoa.growth and development, and secretary activity of accessory glands maintenance of secondary sex characters, such as development of muscle mass, male voice, Hair growth, Thickness of skin,

Increase bone growth, Aggressiveness, Sexual behavior and libido, Protein anabolic activity by nitrogen retention.

SPERMATOGENESIS

Spermatogenesis is defined as process by which the spermatogonia undergo a series of cell divisions followed by a metamorphosis resulting in the production of highly differentiated and potentially motile cells called spermatozoa and can be divided into two phases: **spermatocytogenesis** (mitotic proliferation) and **spermiogenesis** (meiosis, and packaging). Spermatogenesis begins at the wall of the seminiferous tubules with spermatogonia and ends with the release of mature spermatozoa into the lumen. It starts at puberty and continues throughout life unless infertility occurs, leading to tubule atrophy. In seasonal breeders, testes regress during the non-breeding season. As sperm pass through the epididymis, they gain motility upon exposure to secretions from the ampulla, vesicular glands, bulbourethral glands, and prostate. The seminiferous epithelium consists of developing germ cells and Sertoli cells. Spermatogonia divide by mitosis, then undergo meiosis to form haploid spermatids. These spermatids transform into spermatozoa through metamorphic changes. Sertoli cells support and nourish the developing germ cells throughout this process.

Spermatogenesis

The other cell becomes a type A spermatogonium, which moves into the adluminal compartment through the Sertoli cell barrier. It undergoes several mitotic divisions, producing many type B spermatogonia. Type B spermatogonia then complete a final mitotic division to form primary spermatocytes (2n). These undergo meiosis I to form secondary spermatocytes, which then undergo meiosis II to produce spermatids (n). In the bulls, one type A spermatogonium ultimately gives rise to 64 spermatids.

SPERMIOGENESIS

The second phase of spermatogenesis, involves maturation of the spermatids while they are still in the adluminal compartment. Spermiogenesis comprises a series of nuclear and cytoplasmic changes and transformation from a nonmotile cell to a potentially motile cell in which a flagellum (tail) has formed. The mature spermatids produced during the final phase of spermiogenesis are released into the lumen of the seminiferous tubules as spermatozoa. Spermiogenesis starts in the seminiferous tubules and completes in the epididymis. Spermiogenesis occurs within the cytoplasm of Sertoli cells and then released into the tubular lumen. The release of matured spermatids into the lumen of the seminiferous tubules is known as spermiation. The various stages of development of spermatid transformation of are divided into 4 phases.

Golgi phase

The process begins with the formation of proacrosomal granules within the Golgi apparatus, which then coalesce into a single acrosomal granule. This granule adheres to the nuclear envelope, marking the initial formation of the acrosome. Simultaneously, early tail development begins at the pole opposite the acrosomal attachment. The proximal centriole moves toward the nucleus, where it likely serves as the anchoring point for tail attachment to the head.

Cap phase

The acrosomal granule spreads over the surface of the spermatid nucleus, eventually covering approximately two-thirds of the anterior portion with a thin, double-layered membranous sac closely adhered to the nuclear envelope. Meanwhile, the axonemal components of the developing tail, derived from the distal centriole, extend and elongate well beyond the edge of the cytoplasm.

Acrosomal phase

During the acrosomal phase of spermiogenesis, the spermatid undergoes key structural transformations in the nucleus, acrosome, and tail. The spermatid rotates so that the acrosome is oriented toward the basement membrane and the tail toward the lumen. The nucleus condenses and reshapes from spherical to elongated and flattened, while the acrosome elongates in tandem, maintaining close contact with the nucleus. These changes are orchestrated by the Sertoli cells.

The cytoplasm shifts to the posterior side of the nucleus, surrounding the developing tail's base, where microtubules form a temporary sheath. The annulus, a ring-like structure, forms near the proximal centriole and migrates distally along the tail. Mitochondria cluster around the axoneme, forming the mitochondrial sheath that characterizes the midpiece of the tail.

Maturation phase

This stage marks the final transformation of elongated spermatids into mature spermatozoa ready for release into the lumen of the seminiferous tubules. The sperm head, composed of the nucleus and acrosome, assumes its species-specific elongated shape, and chromatin condensation is completed. A fibrous sheath, supported by nine coarse fibers, forms around the axoneme, extending from the neck to the start of the end piece.

The annulus migrates distally along the tail to its final position, where it separates the middle piece from the principal piece. Mitochondria become tightly packed into a continuous sheath surrounding the axoneme from the neck to the annulus. The remaining cytoplasm forms a residual body, completing the maturation process, and the sperm is ready for release (spermiation).

Epididymal Transport

Newly formed spermatozoa are initially immotile and are transported to the epididymis via fluid secretions from the seminiferous tubules and rete testis, assisted by contractile elements within the testicular tissue. As sperm move through the epididymis, they progressively acquire the ability to fertilize, undergoing key changes such as the development of forward motility, chromatin condensation, and modifications to the plasma membrane. The primary storage site for sperm in the male reproductive tract is the distal (tail) region of the epididymis, which holds about 70% of sperm located outside the rete testis. In sexually inactive rams, a significant portion—up to 85%—of daily sperm production is lost through phagocytosis in the duct system or excretion in urine. Spermatogenesis takes approximately 64 days in domestic males, followed by an additional 8–14 days for sperm to transit through the epididymis. Thus, the total duration from a type A spermatogonium to ejaculated sperm is around 60–70 days in rams and bulls, and about 50–60 days in boars, dogs, and stallions.

Spermatozoa

A spermatozoon has three parts: head, midpiece, and tail. The head contains the nucleus with genetic material and is capped by the acrosome, an enzyme-filled vesicle formed from the Golgi complex, which helps penetrate the ovum. A long, whip like tail that grows out of one of the centrioles provides motility for the spermatozoon. Acrosomal enzymes activate upon egg contact.

Spermatogenic Cycle

In a cross-section of seminiferous epithelium, developing sperm cells progress through distinct, synchronized stages alongside neighboring cells. These repeating patterns of cell associations form the spermatogenic cycle or cycle of the seminiferous epithelium. Each new spermatogonium initiates division at species-specific time intervals, maintaining the cycle. The spermatogenic cycle for:Pig – 8 days; Sheep – 10 days; Cattle – 14 days; Human – 16 days. The spermatogenesis is approximately 4 times the duration of the spermatogenic cycle.

Spermatogenic wave

The stages of the spermatogenic cycle vary both over time and along the length of the seminiferous tubule. Sections of the tubule at one stage are typically next to sections at earlier or later stages. This orderly progression of stages along the tubule is known as the wave of spermatogenesis.

Spermiation

The release of formed germ cells into the lumen of the seminiferous tubules is known as spermiation. The elongated spermatids, which are oriented perpendicularly to the tubular wall are gradually extruded into the lumen of the tubule.

Sperm Transport in Male Duct System

Testicular spermatozoa are transported from the testis through a high convoluted duct known as the Epididymis. The Epididymis mainly serves three processes.

Transport

The passage of spermatozoa is aided by the localized contractions of the duct smooth wall at a frequency of 3/min. The spermatozoa transport through the epididymis requires about - 7 days in the bull and 16 days in the ram. The transit time may be reduced by 10 -20% by an increased frequency of ejaculation.

Maturation

During their transit through the epididymis, spermatozoa undergo maturation, gaining the potential to fertilize ova. Key functional changes include the presence of forward motility protein, acrosome maturation, reduced metabolic activity, distal migration and loss of the protoplasmic droplet, acquisition of motility, and storage. The head of the epididymis absorbs much of the fluid from the seminiferous tubules, concentrating sperm in the tail, which holds about 70% of sperm in the duct system. In contrast, the vas deferens contains only around 2%. Spermatozoa within the epididymisfrom head to tailare considered extragonadal reserves, though only those in the distal tail are available for ejaculation. Unejaculated sperm are either excreted in urine or gradually disintegrate.

Function of sertoli cell

These cells are also called sustentacular cells or nurse cells. The Sertoli cell provides a "nurse" function for developing spermatozoa. Sertoli cells are the somatic cells of the testis that are essential for testis formation and spermatogenesis. The Sertoli cells have their base at the periphery of the seminiferous tubules and extend toward the center. The basal junction (tight junction) with adjacent Sertoli cells forms a blood–testis barrier that allows control of the environment within the tubule and also prevents spermatozoa from entering the interstitium. The Sertoli cells divide the seminiferous tubules into two compartments:

i) Basal compartment, which communicates with interstitial fluid and provides space for germinal epithelial cells; and

ii) Adluminal compartment, which is the space between Sertoli cells that communicates centrally with the lumen of the tubule.

Sertoli cells secrete fluid into the adluminal compartment that supports the development of spermatozoa. They aid germ cell maturation through direct contact and by regulating the microenvironment within the seminiferous tubules. Additionally, Sertoli cells and the epididymal epithelium produce a nutrient-rich fluidcontaining hormones like testosterone and estrogen, along with enzymesthat is ejaculated with sperm and plays a key role in sperm maturation.

Sertoli Cells Secrete the Following

a) Enzyme aromatase, which converts androgens into estrogen.

b) Androgen-binding protein (ABP), which is essential for testosterone activity, especially during spermatogenesis.

c) Inhibin, which inhibits FSH release from anterior pituitary.

d) Mullerian regression factor (MRF) in fetal testes. MRF is also called mullerian inhibiting substance (MIS). MRF is responsible for the regression of mullerian duct during sex differentiation in fetus.

Leydig Cell

Leydig cells, also known as interstitial cells of Leydig, are polyhedral in shape with a prominent nucleus, eosinophilic cytoplasm, and lipid-filled vesicles. Located in the interstitial tissue around the seminiferous tubules, their primary function is to synthesize and secrete testosterone in response to LH stimulation. LH promotes Leydig cell hypertrophy, while prolactin enhances LH receptor binding. Testosterone, essential for spermatogenesis particularly meiosisenters the seminiferous tubules via simple or facilitated diffusion. FSH supports meiosis indirectly by acting on Sertoli cells, and GH aids early spermato gonial divisions and testicular metabolism. Leydig cells also produce C-16 unsaturated androgens, which act as pheromones (excreted in saliva) and contribute to "boar taint" in urine.

Functions of Testosterone

In addition to its spermatogenic activity, testosterone fulfills other functions in the peripheral circulation.

a) The development and maintenance of libido,

b) Secretory activity of the accessory sex glands, and

c) General body features associated with the male.

Libido, or sexual drive, is largely influenced by testosterone and is typically lost after castration, though in some animals, small amounts of testosterone from sources like the adrenal glands may preserve it. Testosterone also regulates the development and function of accessory sex glands and drives the expression of male secondary sexual characteristics such as increased bone mass, muscle growth, thicker skin, and a deeper voice (notably in bulls).

During fetal development, testosterone is essential for testicular descent and determines the formation of male genitalia (penis and scrotum), while its absence leads to female structures (clitoris and vagina). Initially, embryos possess both Wolffian and Mullerian ducts; under male hormonal influence, the Wolffian ducts develop into the male reproductive tract and the Mullerian ducts regress. In females, the reverse occurs.

Testosterone also plays a key metabolic role, promoting protein synthesis and contributing to greater muscularity, thicker skin, and voice changes in males.

1. **Effects before Birth:** Masculinizes the reproductive tract and external genitalia, Promotes descent of the testes into the scrotum of most mammals.
2. **Effects on Sex-Specific Tissues:** Promotes growth and maturation of the reproductive system at puberty, Essential for spermatogenesis, and maintains the reproductive tract throughout adulthood.
3. **Other Reproduction-Related Effects:** Develops the sex drive at puberty (sexual maturity) and controls gonadotropic hormone secretion.
4. **Effects on Secondary Sexual Characteristics:** Induces the male pattern of hair or feather growth, Causes the voice to deepen because of thickening of the vocal cords and Promotes muscle growth responsible for the male body configuration.
5. **Non-reproductive Actions:** Exerts a protein anabolic effect, promotes bone growth at puberty (sexual maturity), Closes the epiphyseal plates after being converted to estrogen by aromatase and induces aggressive behavior.

Cryptorchidism

Cryptorchidism refers to the condition where one or both testes fail to descend into the scrotum. In horses, if the testis is retained in the inguinal canal but not in the scrotum, the animal is called a "high flanker." In some cases, the testes may remain entirely within the abdominal cavity. During fetal development, the testes originate from the gonadal ridge at the rear of the abdominal cavity. Toward the end of gestation, they normally descend through the inguinal canal into the scrotum, guided by testosterone produced by the fetal testes. When descent is incomplete at birth, it may occur naturally before puberty or be stimulated by testosterone administration. If a testis remains undescended into adulthood condition termed cryptorchidismit typically produces lower levels of androgens, and the higher abdominal temperature disrupts spermatogenesis by impairing mitotic division of spermatogonia. As a result, sperm production is usually absent. Cryptorchidism is more commonly observed in humans, horses, and pigs.

Puberty

Puberty marks the onset of sexual capability, when an animal can produce fertile gametes and begins to show adult reproductive behavior. In males, it is defined by the presence of sufficient sperm in the ejaculate to fertilize females, though full sexual maturity takes several more weeks. In females, puberty begins with the onset of ovarian activity and is typically indicated by the first estrus or heat.

However, puberty is distinct from sexual maturity, which refers to the point at which an animal reaches its maximum reproductive potential. The period between puberty and sexual maturity is known as adolescence. Puberty is a gradual process resulting from the maturation of the reproductive system, accompanied by growth of reproductive organs and influenced by body weight more than age.

In females, the formation of Graafian follicles begins at puberty due to increased levels of LH and FSH, with rising gonadotropin secretion. Puberty also brings behavioral changes, such as responsiveness to reproductive stimuli and visible signs of estrus.

Environmental factors, such as light, play a role in reproductive activity. In sheep and goats, increasing daylight triggers testicular regression, while decreasing light affects stallions by reducing, but not eliminating, fertility. Minor seasonal effects are also observed in cattle and swine. These responses are regulated by the hypothalamus and pineal gland.

GnRH release from the hypothalamus increases in frequency as puberty approaches, stimulating LH secretion from the anterior pituitary, which supports testicular or ovarian function. The presence of adult males can accelerate puberty in females through pheromonal effects—exposure to boars in gilts or rams in ewe lambs is commonly used for this purpose. Poor nutrition delays puberty in both sexes.

In bulls, the first ejaculate considered fertile must contain at least 50 million sperm with 10% progressive motility. Dairy heifers typically reach puberty at 35–45% of their mature weight, but breeding is advised at 55–60%. Sheep generally reach puberty at 40–50% of their adult weight, with breeding recommended at 60–65%.

Puberty is initiated by the maturation of the hypothalamus and the increased pulsatile release of GnRH, leading to FSH and LH secretion. These gonadotropins stimulate the gonads to produce hormones, which drive the development of genital organs and secondary sexual characteristics. The first estrus in pubertal animals is often silent. Recommended service weights are around 240 kg for bulls and 280 kg for buffalo bulls, usually achieved by 18 months of age. In most domestic species, puberty strongly correlates with achieving a specific body weight.

Table 1. Puberty and sexual maturity in domestic and laboratory animals:

I	**Domestic animals**	**Puberty**	**Sexual maturity**
	Dog	6-12	8
	Tom	5-7	9
	Boar	5-6	8-9
	Bull	10-12	3-4 yr
	Stallion	18-24	3-4 yr
	Ram	6-10	18-20
	Buck	5-6	8-12
II	**Laboratory animals; (age in weeks except as indicated)**		
	Guinea pig	7-8	8-10
	Hamster	4-6	6-7
	Mouse	4-6	6-8
	Rabbit	4 months	6-9 months
	Rat	6	10-11

FACTORS AFFECTING PUBERTY

Several factors promote the onset of puberty, including interaction with the opposite sex, adequate to high nutritional levels, a favorable climate, and a low-stress environment. Conversely, factors such as the confinement of femaleslikely due to the absence of pheromonal stimulationpoor nutrition, and unfavorable environmental or climatic conditions can delay or inhibit the onset of puberty.

Breed and Genetic Influences: Breed plays a significant role in determining the age and size at which puberty occurs. Generally, smaller breeds tend to reach puberty earlier than larger ones. For instance, small-breed bitches often enter their first estrus several months before large-breed bitches. Among cattle, Jersey heifers typically reach puberty around 8 months of age, while Holsteins average around 11 months and Ayrshires around 13 months. Dairy heifers, in general, attain puberty earlierat about 11–12 months of age and at approximately 35% of their mature body weightcompared to beef heifers, which reach puberty at 14–15 months and 45–55% of their adult weight.

Climatic Effects

Puberty in man occurs earlier in the tropics than in the temperate zones. However, temperate climates, including the interaction of temperature, humidity, diurnal variation and daylight, favor early puberty in all animals.

Environmental Effects: Heat stress will delay puberty in heifers.

Seasonal Effects

In seasonal breeders like sheep, the timing of puberty can be influenced by the breeding season, sometimes overriding age-related expectations. If the hypothalamo-pituitary-ovarian axis is sufficiently mature, puberty can be triggered earlier than usual. However, for seasonal breeders to reach puberty on time, they must attain the appropriate body size during the correct season; otherwise, puberty may be delayed.

Studies have shown that ewe lambs require exposure to a period of long daylight followed by short daylight to initiate puberty. For example, ewe lambs born early in the spring may exhibit their first estrus at around 180 days of age. In contrast, those born later in the spring or early summer may not experience their first estrus until the breeding season in the fall of the following year—reaching puberty at 400 to 500 days of age.

Confinement raised gilts reach puberty at much older ages and heavier weights than those raised outdoors.

Effect of Nutrition

A high plane of nutrition accelerates puberty, while poor nutrition delays it-especially in non-seasonal breeders. Nutrition, body weight gain, and age interact, with well-nourished animals reaching puberty earlier. However, poor nutrition, though unable to prevent puberty entirely, can cause significant delays, potentially doubling the age at onset.

Sex effect: Females generally reach puberty earlier than males across all species.

Disease, Hygiene and cleanliness: Poor health and unhygienic conditions are un-favourable for onset of puberty.

Exposure to opposite sex: Favors the early onset of puberty.

Table 2. Average Age of Puberty in the Male and Female of Various Mammals.

Species	Male	Female
Cat	9 months (8–10)	8 months (4–12)
Cow	11 months (7–18)	11 months (9–24)
Camel	3–5 years	3 years
Dog	9 months (5–12)	12 months (6–24)
Sheep	7 months (6–9)	7 months (6–16)
Swine	7 months (5–8)	6 months (5–7)
Horse	14 months (10–24)	18 months (12–19)
Elephant	13–15 years	11–14 years

Functions of Testosterone

Testosterone plays a crucial role in male reproductive and physiological development. It maintains spermatogenesis by supporting meiosis and promotes the differentiation of the fetal male reproductive tract, as well as testicular descent. It is essential for the development and maintenance of libido and sustains the secretory function of accessory sex glands. Testosterone is also responsible for the development and maintenance of secondary male sexual characteristics, such as heavier bones, increased muscle mass, thicker skin, a deeper voice, and aggressive behavior particularly notable in bulls. As an anabolic hormone, it promotes a positive nitrogen balance, stimulates erythropoiesis, increases basal metabolic rate (BMR), and enhances pheromone secretion in species like cats and boars. Overall, testosterone is key to maintaining libido, accessory gland function, and general male body characteristics.

Libido

Testosterone is crucial for libido, and castration-removal of the testes-effectively eliminates the male sex drive and makes animals more docile. A vasectomy, by contrast, involves removing a section of the vas deferens to block sperm transport from the epididymis without affecting libido.

Secretary activity of accessory organs

Testosterone supports the development and maintenance of the secretory epithelium in accessory sex organs. It reaches these organs through general circulation, unlike the epididymis, which is maintained via direct transfer from the seminiferous tubules.

General body features

Testosterone influences general body characteristics, including increased muscle mass through nitrogen retention (myotropic effect), thickening of vocal cord muscles (lowering voice pitch), promotion of hair growth patterns, and affecting feather pigmentation and growth in poultry. It also enhances body size, bone growth, and basal metabolic rate (BMR).

Behavioral changes

Male dogs typically exhibit a raised-leg urination posture, increased aggressiveness, and territorial marking using pheromones substances produced by the kidneys under the influence of testosterone.

4

Female Reproduction in Farm Animals

Manish Kumar, Anuradha Kumari
Utkarsh Kumar Tripathi, and Kaustubh Kishor Saraf

Faculty of Veterinary and Animal Sciences, I. Ag.Sc., RGSC-Banaras Hindu University (BHU), Barkachha, Mirzapur, Uttar Pradesh-231001

Unlike most other body systems, which are essentially similar in both sexes, the reproductive systems of males and females are remarkably sexually dimorphic, reflecting their distinct roles in reproduction.

Animals reproduce either asexually or sexually. Asexual reproduction includes **mitosis**, seen in prokaryotes and some eukaryotes, where a cell divides to form genetically identical offspring. **Budding or fission** allows animals like sea anemones to produce copies of themselves without embryos. **Parthenogenesis** involves development of an unfertilized egg into an embryo. In **sexual reproduction**, male and female gametes (haploid) unite to form a genetically unique diploid offspring.

Intersexuality

A freemartin heifer is a sterile, intersex female born as a co-twin with a male. Due to placental blood vessel anastomosis, male hormones like MIH from the male twin enter the female's circulation, disrupting normal female development. This leads to masculinization, with testis-like gonads, inhibition of the Müllerian ducts, and retention of Wolffian ducts. Freemartinism is most common in cattle, but also occurs in sheep, pigs, and goats.

Hermaphrodite: The presence of both male and female gonads in one individual animal. Female pseudo hermaphrodites have normal internal genitalia but intermediate external genitalia. Clitoris may be enlarged or modified to penis. Male hermaphrodites are common in animals. True hermaphrodites have both ovarian and testicular tissue with intermediate genitalia and are very rare in higher animals.

Sexual Differentiation

Reproduction is vital for species continuity. In mammals, it involves the fusion of anisogametes from morphologically and physiologically distinct males and females. Sexual differentiation encompasses chromosomal, gonadal, somatic, and psychological aspects. Fertilization forms a zygote, which undergoes mitotic divisions. The embryonic period spans from blastocyst formation to organ system differentiation.

Chromosomal Sex Differentiation: Chromosomal sex is determined by a pair of sex chromosomes. In mammals, females are homogametic (XX) and males are heterogametic (XY), while in poultry, males are homogametic (ZZ) and females are heterogametic (ZW). Sex is established at fertilization, though no physical differences are present then. Female somatic cells show a Barr body (chromatin mass), making them chromatin positive, while males are chromatin negative.

Gonadal Sex Differentiation: Although genetic sex is fixed at fertilization, early embryos can develop either male or female genitalia, as the primitive gonad contains elements for both. Female-determining genes are mainly on the X chromosome, requiring XX for ovary development, while male genes are on the Y chromosome and autosomes, with the Y triggering testis formation. Mammalian embryos tend to develop as female unless influenced by male genes; early castration of male embryos leads to female development. Gonadal development starts with genital ridges, where primordial germ cells migrate from the yolk sac. Initially bipotential, the gonad differentiates based on genetic signals: XY embryos form testes from the medulla, and XX embryos form ovaries from the cortex. Once differentiation is complete, the organism is identified as male or female.

Somatic Sex Differentiation

After gonadal differentiation, the development of accessory genitalia is guided by the gonad. In the undifferentiated stage, both the Müllerian (female) and Wolffian (male) duct systems are present. If testes develop, the Müllerian ducts regress due to testicular secretions, while the Wolffian ducts mature into male reproductive structures. In contrast, if ovaries form, the absence of testicular factors allows the Müllerian ducts to develop into female reproductive organs, and the Wolffian system regresses. Ultimately, one of the duct systems gives rise to the complete male or female reproductive tract, resulting in a clearly distinguishable sex at birth.

Psychic Sex Differentiation

At birth, although an individual has male or female reproductive organs, they lack sex drive or sexual behaviour. The final stage of sexual differentiation-psychic sex-involves the development of sexual behaviour, which is primarily regulated by gonadotropic and gonadal hormones in specific amounts and ratios. External factors like season also influence this. Prepubertal removal of gonads prevents mating behaviour. In mature females, ovariectomy halts mating behavior, but estrogen replacement can restore it. In males, early castration prevents normal mating patterns, while adult castration reduces sex drive and copulatory behaviour.

Reproduction

In vertebrates, the primary reproductive organs, testes in males and ovaries in female sexist in pairs and serve two main functions: gamete production (sperm in males, ova in females) and sex hormone secretion (testosterone in males; estrogen and progesterone in females). Each sex also has a reproductive tract to transport or house gametes, along with accessory glands like the prostate or mammary glands. External genitalia are visible, while internal genitalia are housed within the body. The structure of reproductive organs varies by species, depending on reproductive strategies, fertilization type (internal or external), and yolk volume in eggs.

Anatomy of reproductive tract

Female

Female reproductive system consists of Pair of ovaries, Duct system, A pair of oviducts / Fallopian tubes, uterus, cervix, internal vagina, and External genitalia. External genetalia includes, Outer vagina, vestibular glands, clitoris, labia majora and minora. Primary Sex Organs – the ovaries (Gametogenic). Accessory Sex Organs – the fallopian tubes, uterus, cervix, vagina and vulva (Essential for reproduction but not gametogenic).

Ovary

The paired ovaries function both as gametogenic (producing ova) and endocrine organs. Their shape varies by species and estrous cycle stage. In polytocous animals (e.g., sow, bitch, cat), ovaries appear like a cluster of grapes due to multiple follicles or corpora lutea. Monotocous animals (e.g., cow, ewe, mare) have ovoid ovaries, while the mare's kidney-shaped ovary has an ovulation fossa, the exclusive site of ovulation. Structurally, the ovary has an inner vascular medulla and an outer cortex with connective and epithelial cells. A single germinal epithelium layer covers the ovary, except in mares, where it's restricted to the ovulation fossa.

Oviduct

Uterine tubes or fallopian tubes are paired, convoluted tubes which has the unique function of conveying the eggs and sperms in opposite direction. It connects the ovary to the uterus. Oviduct functionally divided into four parts:1. Fimbriae, 2. Infundibulum, 3. Ampulla, 4. Isthmus

1. **Fimbriae:** Ovarian end of the oviduct with fringed edges moves freely and sweep on the surface of the ovary at ovulation. It forms ovarian bursa with mesosalpinx. Ovarin bursa open in all farm animals, it completely covers the ovary in the pig and in mares it only covers ovulation fossa.
2. **Infundibulum:** It funnels shaped, size varies with age and size of the animal.
3. **Ampulla:** It forms more than half the length of the oviduct and connects infundibulum to isthmus.
4. **Isthmus:** Connected to uterus at the utero-tubal junction.

Musculature of the oviduct: It has outer longitudinal and inner circular smooth muscles. Contraction of longitudinal muscles shorten the oviduct and contraction of circular muscles cause annular constrictions. Oviduct exhibits three types of contractions, Localized peristalsis, Segmental contractions and Worm like writhing contraction.

Functions of the oviduct:Pick up of the ovum: Fimbriae pick up the ovulated egg from the ovarian surface and capable of picking up of ovum from the contralateral side.

Provide Oviducal secretions: The oviduct provides a conducive environment for fertilization, supports sperm capacitation, protects gametes and the early embryo from maternal immune attack, and nourishes the embryo. It secretes cleavage-inducing proteins and aids in gamete interaction through ciliary movement, muscular contractions, and fluid flow. It also denudes follicular cells, transports the zygote to the uterus, prevents tubal implantation, and coordinates opposite-direction transport of sperm and egg.

Uterus

The uterus is a muscular, membranous organ composed of two horns (cornua), a body (corpus), and a cervix. It supports fetal development post-fertilization. Uterine structure varies by species: mares have a large uterine body (bipartite), cows and pigs have a small body with long horns (bicornuate), and rodents have a duplex uterus with two separate cervices. Humans and primates have a simplex uterus with a single body and no horns. Sows have the longest uterine horns, while buffalo have more tortuous and muscular horns with closely adhered intercornual ligaments.

Cervix

The cervix is a thick-walled, sphincter-like structure that separates the uterus from the vagina. Its narrow lumen features annular rings-prominent and interlocking in cows and ewes, less defined in mares, and corkscrew-shaped in sows. The folded mucosa enhances the secretory surface. Cervical mucus has distinct properties-elasticity, viscosity, and stickiness-which vary with the estrous cycle.

Functions of cervix

It involved in the **transport of spermatozoa**.

Storage site for the spermatozoa: Large number of sperm lodged into cervical crypts, and being released slowly later. Makes the sperm available at the site of fertilization for up to 24h after mating.

Sperm selection centre: Several sperm lodged in cervical crypts never get released, particularly those that are less motile. Thus, cervix selects only vigorously motile sperm for transport to the site of fertilization.

Provides Protection to upper reproductive tract

Cervical mucus becomes thick and viscous during the luteal phase and seals. This seal persists till next estrus or parturition.

Helps in parturition: Gross biochemical changes increase the size of the cervix enormously to allow the passage of the calf.

Vagina: The vagina has two parts: the outer vestibular portion and the internal portion. Its wall comprises surface epithelium, a less developed muscular layer, and serosa, with no truemucosa. Vaginal epithelium changes with the estrous cycle-stratified squamous at estrus and low cuboidal during the luteal phase. Vaginal smears can detect cycle phases in rats and bitches.

Functions of vagina

The vagina serves as the female copulatory organ and semen reservoir for the cervix. It plays a role in sperm transport, absorption of seminal plasma, and psychosexual satisfaction. Though it lacks secretory glands, fluids exude from its vasculature, and most mucus is derived from the cervix. Cervical and vaginal fluids help buffer vaginal pH. In cows, estrus-related vaginal secretions have a specific smell recognized by bulls and dogs. Sperm capacitation occurs in the vagina in some species, such as rabbits.

External genetalia

External genetalia consists of outer vagina, vestibular glands, clitoris, labia majora & minora

Clitoris: Embryological homologue of penis. It is burried in the vestibular mucosa in the cow but prominent in the mare. It has large number of nerve endings and is important in psycho-sexual satisfaction.

Labia: Labia is homologues of scrotum. Labia minora poorly developed in animals. Labia major is rich in sebaceous and tubular glands. It is swollen, congested and edematous at estrus.

Vestibular glands: It has Sebaceous and Bartholin glands which resemble bulbo-urethral glands in the male and secrete lubricating mucus to boost copulation.

Follicular Development

Ovarian follicle: An ovarian follicle consists of an oocyte and its surrounding follicular cells at various developmental stages. It is the fundamental unit of female reproductive biology, supporting oocyte growth, maturation, and preparation for fertilization. Females are born with a fixed number of oocytes in primordial follicles, which decline over time. Mammalian ovaries have two follicle pools: a non-growing pool (primordial follicles) and a growing pool (primary, secondary, and tertiary follicles). Folliculogenesis-the progression from primordial to Graafian follicle-starts independently of gonadotropins but becomes increasingly gonadotropin-dependent, with ovulation fully reliant on these hormones.

Primordial follicle: The Primordial (quiescent) follicles are non-growing (inactive) follicles and consist of an oocyte, surrounded by a single layer flattened squamous epithelial cells (pregranulosa cells).

Primary follicle: Primary follicles are active and growing follicles in which an oocyte is surrounded by a single layer of cuboidal epithelial cells. As development proceeds, the proliferation of granulosa cells not only increases the number of cells but also granulosa cell layers around the oocyte.

Secondary follicle: The follicle having two or more than two layers of granulosa cells but not more than six layers are called as secondary follicle. A thecal cell layer also develops around the membrana granulosa.Zona pellucid is absent. The secondary follicles are also called Preantral Follicles because of absence of antrum (a fluid filled cavity).

Tertiary follicle: Tertiary follicles, also known as antral or vesicular follicles, are marked by six or more granulosa cell layers, a zona pellucida, and well-defined theca layers. The theca interna (large, round, foamy cells) and theca externa (smaller cells) are separated from the granulosa layer by a basement membrane. Granulosa cells secrete follicular fluid, which forms small cavities that later merge into a single large cavity called the antrum. The zona pellucida, a glycoprotein layer, lies between the oocyte membrane and granulosa cells.

Graafian follicle / Preovulatory follicle

The Graafian follicle, named after Regnier de Graaf, features a large antrum filled with viscous liquor folliculi. The oocyte is surrounded by cumulus cells, with the innermost layer termed the corona radiata. The oocyte rests on a cluster of cumulus cells called the cumulus oophorus or discus proligerus, projecting into the antrum. The fluid present in the antrum is called liquor folliculi (viscous fluid).

Reproductive Physiology of Animals: In mammals, there are two separate but coordinated cycles, i.e. Ovarian & Uterine cycle are present.

Ovarian cycle: In which an oocyte matures in the ovary and is ovulated to travel to the uterus. Ovaries perform the dual function of producing ova (oogenesis) and secreting the female sex hormones estrogens& progesterone.

Uterine cycle: The uterine cycle prepares the uterine lining for embryo implantation. If fertilization occurs, the cycle halts, and the female system supports the embryo until birth. Post-birth, the female produces milk for nourishment. The estrous cycle, comprising ovarian and uterine cycles, spans from one period of sexual receptivity to the next, interrupted only by pregnancy.

Sexual receptivity, known as estrus or heat, occurs periodically, often aligning with ovulation. While ovarian cycles are generally similar across mammals, uterine cycles differ. In most mammals with an estrous cycle, the uterine lining is reabsorbed if pregnancy doesn't occur. In contrast, primates and humans have a menstrual cycle where the uterine lining is shed as menstruation. These cycles begin at the end of each menstruation and continue consistently throughout the year. The length of both estrous and menstrual cycles is species-specific but can vary within a species.

- The frequency of estrous cycles differs between species:
- **Continuously polyestrous** (cycles occurring uniformly throughout the year) e.g. Cattle, swine, & rodents.
- **Seasonally polyestrous** (estrous cycles are restricted to a particular time of year).
- **Short-day breeders:** exhibit estrous cycles as day length decreases e.g. deer, sheep, & goats.
- **Long-day breeders:** they come into estrous as day length increases, e.g. bears, hamsters, &horses.
- **Seasonally monoestrous females** (most carnivores, including bears, dogs, foxes, and wolves) are characterized as having a single estrous event followed by a long period of anestrus. In these animals the period of estrus is prolonged and lasts for several days.

- **Spontaneous ovulators** that ovulate with a regular frequency and do not require copulation, most animals. Some species of birds, for example, ovulate daily for extended periods of time without any contact with the male.
- **Reflex (induced) ovulation:** ovulation is induced by stimulation of sensory receptors in the vagina and cervix during coitus, either mechanically or by semen components. e.g. cat, mink, ferret, rabbit, llama, alpaca and camel.

Oogenesis

The undifferentiated primordial germ cells in the fetal ovaries, the oogonia, divide mitotically during gestation and/ or neonatally. This ends at the time of birth.Formation of haploid ovum in ovary by meiosis is called as oogenesis. Oogonia undergo mitotic division until the final generation of oogonia enters the prophase of meiosis-1.

At this point it is called primary oocyte. This is under the influence of meiosis initiating factor.

In domestic animals (except Bicth& Queen), oogonia develops into primary oocyte shortly before or after birth. The nucleus of primary oocyte enters the dictyate / resting stage of meiotic prophase I and does not compete until the animal reaches maturity. Resumption of meiosis, beyond dictyate stage depends on preovulatory surge of LH. After puberty, the rising level of FSH causes follicular cells to develop. The primary oocyte grows by RNA synthesis. Contact between oocyte and granulose cells prevent maturation of oocyte beyond dictyate stage through oocyte maturation inhibitor (OMI) produced by granulosa cell. OMI concentration declines as follicle matures. Ovulatory LH surge block the transfer of OMI from cumulus cells to the oocyte allowing meiosis to resume.

Formation of Primary Oocytes & Primary Follicles

During late fetal life, oogonia enter the first meiotic division but remain arrested as primary oocytes, containing a diploid set of chromosomes (2N). Each primary oocyte is enclosed within a primordial follicle. At birth, only a fraction of the original follicles remain, serving as a fixed reservoir since no new oocytes or follicles develop postnatally in most mammals. Throughout reproductive life, only a small percentage of these follicles mature and ovulate, while the rest undergo degeneration.

Before puberty, all developing follicles undergo atresia without ovulating. Even after puberty, many remain anovulatory. Most follicles from the initial pool never ovulate and instead degenerate at various stages. By the end of reproductive life, only a few follicles remain, eventually succumbing to atresia.

The primary oocyte in a primary follicle remains diploid with 2N doubled chromosomes. Some resting follicles develop into secondary (antral) follicles, with the number developing proportional to the pool size. Secondary follicle development involves oocyte growth and proliferation of surrounding cells. The extent of oocyte enlargement varies by species and results from cytoplasmic or yolk accumulation needed for early embryonic development. As differentiation occurs, theca cells form around granulosa cells, making the follicle dependent on gonadotropins and significantly increasing estrogen production.

Just before ovulation, the primary oocyte, which has been in meiotic arrest for years, completes its first meiotic division a few hours before ovulation, except in bitches and some mares, where this occurs post-ovulation. This division produces two haploid daughter cells, but most of the cytoplasm remains with the secondary oocyte, which will become the ovum. The other cell, the first polar body, contains half the chromosomes but minimal cytoplasm and soon degenerates, ensuring the ovum retains essential nutrients.

Formation of a Mature Ovum

The secondary oocyte, not the mature ovum, is ovulated and fertilized. It enters meiosis II, which completes only if a sperm penetrates the zona pellucida, triggering the final division. Unfertilized oocytes never complete this process. During this division, half of the chromosomes and a thin layer of cytoplasm are extruded as the second polar body, while the remaining maternal chromosomes combine with paternal chromosomes from the sperm, completing fertilization. If the first polar body is still present, it also undergoes meiosis II, forming two polar bodies.

Follicular wave

Follicular waves involve the growth of 3–6 follicles under rising FSH levels. The first wave during the mid-luteal phase reaches the antral stage but regresses, followed by a second wave. If CL regression occurs, one follicle becomes dominant, reaching the ovulatory stage while others become atretic. The dominant follicle secretes inhibin and estradiol, blocking further follicular growth. In sheep, cattle, and horses, 2–3 waves occur per cycle, with ovulation in the last wave. In pigs, rats, and humans, ovulatory follicles develop during the follicular phase. Progesterone from the CL inhibits gonadotropin surges until luteolysis, when falling progesterone and rising estradiol trigger ovulation. Monotocous animals ovulate a single dominant follicle, while polytocous species release multiple ova. After ovulation, the follicular antrum fills with blood, forming the corpus hemorrhagicum, which matures into the corpus luteum. The CL eventually regresses into the corpus albicans.

Ovarian Cycle

- The shape, size and functionality of the ovaries vary both with the **species** of animal as well as with the **stage of the estrous cycle**.
- In cattle and sheep it is **almond shaped**,
- Horse the ovary is **bean shaped**,
- Sow and bird the ovary resembles a **cluster of grapes**.
- In **chickens, only the left ovary & oviduct normally mature**. The right ovary remains vestigial, although it develops if the left ovary is removed or becomes diseased.
- After the onset of puberty, the mammalian ovary alternates between two phases: i.e. Follicular phase and Luteal phase.

1. **The follicular phase**, which is dominated by the presence of maturing follicles, which produce a mature egg ready for ovulation and the steroids responsible for maturing the oocyte.
2. **The luteal phase**, which is characterized by the presence of the corpus luteum, the dominant ovarian structure during this phase.

- In estrous mammals, the follicular phase is comparatively small, usually encompassing no more than **20%** of the estrous cycle, with the luteal phase encompassing the remaining **80%**.

Ovulation

Although ovarian cycles are similar in many ways among vertebrates, differences can be found in the type of ovulation. Most animals are **spontaneous ovulators** that ovulate with a regular frequency and do not require copulation. Some species of birds, for example, ovulate daily for extended periods of time without any contact with the male.

Reflex (induced) ovulation has been described for a large number of mammals including the cat, mink, ferret, rabbit, and camel. In this strategy ovulation is induced by stimulation of sensory receptors in the vagina and cervix during coitus, either mechanically or by semen components.

Estrous cycle

The term estrous cycle refers to the rhythmic phenomenon observed in all mammals involving regular but limited periods of sexual receptivity (estrus) that occur at intervals characteristic of a species. One cycle interval is defined as the time from the onset of one period of sexual receptivity to the next.

Stages of estrous cycle

The estrous cycle can be divided into several stages according to behavioral or ovarian changes.

1. **Estrus:** The time of sexual receptivity, sometimes referred to as heat. Ovulation usually, but not always, occurs at the end of estrus.
2. **Metestrus:** The early postovulatory period, during which the CL begins development.
3. **Diestrus:** The period of mature luteal activity, which begins about 4 days after ovulation and ends with regression of the CL.
4. **Proestrus:** The period beginning after CL regression and ending at the onset of estrus.

During proestrus, rapid follicle development leads to ovulation and to the onset of sexual receptivity.

Based on changes occurring in the ovary, estrous cycle consists of

Follicular phase- also known as proliferative or estrogenic phase - oocyte matures, ovulation occurs, estradiol from the follicle is dominant - includes proestrus and estrus phases.

Luteal phase (secretory phase)- corpus luteum (CL) forms, fertilization and embryo development, progesterone from the CL is dominant - includes metestrus and diestrus.

Proestrus

Proestrus is a phase of rapid follicular growth under gonadotropic stimulation, coinciding with the regression of the previous cycle's CL in polyestrous species. Follicle development progresses from small to large follicles under the influence of GnRH, FSH, and LH. Estrogen levels rise while progesterone declines due to luteolysis triggered by PGF2α. In most domestic animals, except mares, folliculogenesis is suppressed during pregnancy due to high progesterone levels.

This phase is marked by increased epithelial tissue growth, mucus secretion, and enhanced vascularity of the endometrium and vaginal mucosa. In cows and mares, a clear, stringy mucus discharge appears late in proestrus. In most domestic species, proestrus lasts 2–3 days and is not clinically evident. However, in bitches, it lasts 7–9 days, characterized by noticeable genital changes, increased sexual excitement, vulvar swelling, and bloody discharge due to increased endometrial vascularity. Mucus secretion from the cervix, anterior vagina, and uterine glands also rises.

Estrus

Estrus is the period of sexual receptivity, mating, and ovulation in most species, marked by the maturation of the Graafian follicle and initial corpus luteum formation. Estrogen levels rise due to theca and granulosa cell synergy, while inhibin and estrogen suppress FSH via negative feedback. Meanwhile, high estrogen levels trigger hypothalamic LHRH release, leading to an ovulatory LH surge.

During estrus, the oviducts are tonic with active cilia and contractions near the Graafian follicle. The uterus is turgid and edematous with increased blood supply. Vaginal, vulval, and cervical mucosa are pink and congested, with increased mucus secretion, forming characteristic discharge in cows. Ovulation typically occurs at the end of estrus. External signs include mounting, bellowing, frequent urination, vaginal discharge, and reduced feed intake and milk yield.

Signs of Estrus in Different Animals

Cow

Smaller breeds of cows usually reach puberty at an earlier age than larger breeds (Jersey, 8 months; Holstein, 11 months). Behavioral changes associated with estrus include restlessness, mounting activity, standing to be mounted, being more alert to other animals, & decreased appetite. At the same time, decreased milk production, mucus discharge from the vulva, & redness and relaxation of the vulva are noted. Most domestic animals ovulate toward the end of estrus, but the cow ovulates 12–14 hours after estrus.

Mare

The onset of puberty in the mare occurs during the breeding season after birth. A wide range of age for puberty is seen in the mare, from 12 to 18 months. Ovulation occurs about 24 hours before the end of estrus. Signs of estrus in the mare are elevation of the tail, standing with the hindlegs apart, squatting and urinating, and rhythmically erecting the clitoris.

Ewe

A prominent sign of estrus is fluttering of the tail or wagging of tail (move it laterally). Also, females separated from males by a barrier often assume a close proximity to the barrier. The vulva is slightly swollen and congested, and there is often a slight discharge of clear mucus. Estrus ewe stand when mounted by ram (most easily noticed sign).

Doe

Estrus does are restless. Peculiar continuous bleating sound (vocalization). Wagging of tail from side to side and up and down (most reliable sign). The vulva is slightly swollen and congested, and there is often a slight discharge of clear mucus. Homosexual behavior (occasionally). Reduced appetite and milk yield. Male seeking behavior present. Estrus does stand when mounted by bucks. Sometime rub her neck and body against the male. Buck shows interest and will follow does 3 to 5 days before standing estrus occurs, suggesting proestrual activity.

Sow

Swelling of the vulva, restlessness, and decreased appetite. Application of pressure on the sow's back during estrus elicits the rigidity reflex that occurs during natural mating with a boar. Ovulation occurs from both ovaries, and 14–16 oocytes can be released.

Bitch: The bitch might be sexually attractive during proestrus but is not sexually receptive until after the LH surge. Vaginal cytologic changes seem to be more pronounced in bitches than in other domestic species and have been correlated with each estrous cycle stage. Vaginal smears are useful for assessing the stage of estrus and for predicting the most suitable time for breeding.

The principal cytologic changes are

i) Thickening and cornification of the vaginal epithelium;
ii) Loss of leukocytes because of the thickened epithelium; and
iii) Appearance of erythrocytes from the developing vascular system of the endometrium.

Queen

Signs of estrus in queens include an increase in affection, which can be shown to almost any object – humans, table legs, or other pieces of furniture. They also crawl with their thorax against the floor, roll about, and vocalize for prolonged periods. Several coital contacts might be made, with intromission and ejaculation occupying only 10–15s each time. A refractory period or lack of sexual receptivity occurs for 10–15 min after each intromission. During the first hour of contact, four or five intromissions and ejaculations might occur.

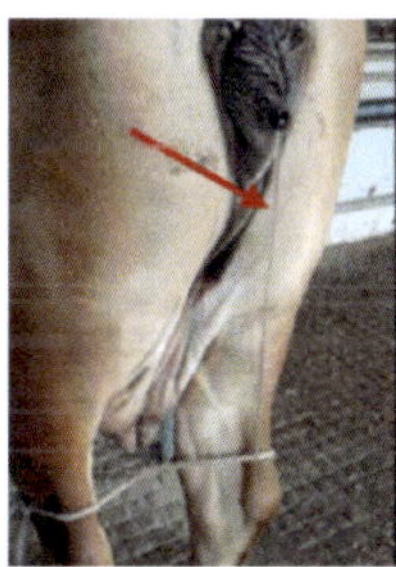
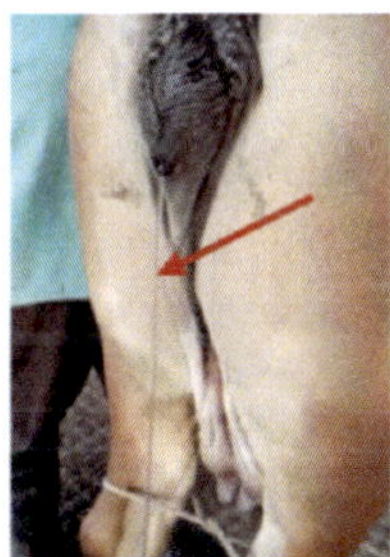

Fig. 2. Showing the discharge from the vulva of cow

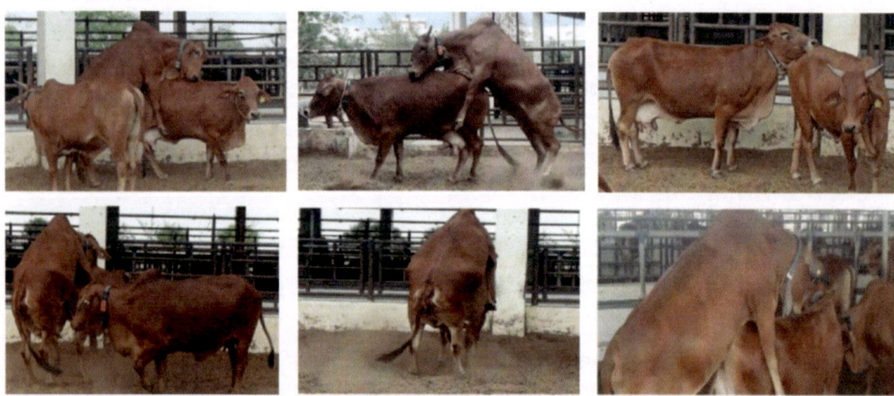

Fig. 3. Showing the mounting behaviour of cows

Metestrus

Metestrus is the transitional phase between ovulation and full corpus luteum (CL) development, marking the shift from estrogen to progesterone dominance. Granulosa and thecal cells reorganize into the CL under LH influence, producing progesterone to inhibit FSH and prevent further follicular development. In cows, metestral bleeding occurs due to estrogen withdrawal. Mucus secretion decreases, while endometrial glands grow rapidly. Metestrus lasts about 3-4 days in cows, sheep, sows, and mares, aligning with the time for ova to reach the uterus. In dogs and cats, pseudopregnancy extends for 50-60 and 30-40 days, respectively, before CL regression leads to anestrus.

Diestrus

Diestrus is the longest phase of the estrous cycle in domestic animals like cows, sheep, goats, sows, and mares. It is a mid-luteal period of sexual quiescence between successive estrous cycles in polyestrous animals. During this phase, the corpus luteum (CL) is fully developed and functional, with progesterone dominating to prepare the uterus for implantation and fetal growth. LH maintains the CL in domestic animals, while prolactin is luteotropic in rodents. The endometrium thickens, glands hypertrophy, the cervix constricts, and vaginal mucus becomes scanty and sticky. Late in diestrus, the CL regresses, endometrial glands atrophy, and in non-pregnant animals, prostaglandin F2α (PGF2α) induces luteolysis by disrupting LH action and reducing blood supply. PGF2α serves as the natural luteolytic agent in cows, mares, sows, ewes, and does.

Table 1. Reproductive cycle and related phenomena in domestic animals.

Species	Onset of puberty (average)	Recommended age for first service	Oestrous cycle length	Follicular diameter (mm)	Oestrus duration	Time of ovulation	Optimum time for service at oestrus	Gestation period (days)	Optimum time for breeding after parturition
Mare	10-25 month (18m)	2-3 year	19-23 day (21d)	25-35	4-8 d (5.5)	1-2 before end of estrum	2-4 d before end of estrum or 2nd-3rd of estrum	325	About 25-35 d or 2nd estrum; about 9 d or 1st estrum in normal foaling
Cattle	4-25 m (6-18 m)	14-22 m	18-24 d (21 d)	10-20	12-24 hr (18 hr) estrum	10-12 hr	after end of estrum Mid to late	278	60 days
Buffalo	18-24 m	24-364	18-24 d (21 d)	10-15	12-24 hr (18 hr) estrum	13-15 hr after end of estrum	Mid to late	310	60 days
Sheep	4-12 m (first fall)	12-18 m	16-17 d (16.5 d)	15-19	24-36 hr (1-2 hr)	12-24 hr before the of estrum	18-24 hr after the onset	145 end of estrum	Usually, the following the fall
Goat	4-12 m (first fall)	12-18 m	17-21 d (20 d)	5-8	30-40 hr (36-48 hr) estrum	About the last day of of estrum	24-36 hr after onset	150	Usually, the following the fall
Pig	5-8 m	8-9 m	19-20 d (21 d)	7-10	48-72 hr (2-3 d) of estrum	30-40 hr after onset of estrum	12-30 hr after onset	114	First estrum 4-9 d after weaning

Dog	6-12 m (7-10 m)	12-18 m	1-4 cycle per year (2 cycle per year)	6-8	4-13 d (9 d)	1-2 d after onset of true estrum or 10-14 d after onset of proestrus bleeding	2-3 d after onset of true estrum	62	Usually the first estrum or 3-4 weeks after weaning

Table 2. Duration of different phases of estrous cycle:

Animal species	Proestrus	Estrs	Metestrus	Diestrs	Length of estrous cycle
Cattle and buffalo	3 days	12-24 hr	3-5 days	12-13 days	20-22 days
Mare	3 days	4-8 days	3-5 days	6-10 days	20-23 days
Sheep	2 days	24-36 hr	3-5 days	7-10 days	16-17 days
Goat	2 days	30-40 hr	3-5 days	7-10 days	17-21 days
Sow	3 days	48-72 hr	3-4 days	9-13 days	19-20 days
Bitch	9 days	9 days	**Pregnancy / Pseudopregnancy**		

Anestrus

It is a stage of sexual quiescence characterized by the lack of estrus behaviour. Anestrus is a normal stage of reproductive function in the prepubertal and in aged animals of all species. Pregnancy is the most common cause of anestrus in polyestrus species. In non-pregnant animals in monoestric species such as bitch.In seasonally polyestrus species during non-breeding season. During the lactation in females.In all domestic species, anestrus may occur as a pathological condition caused by a variety of factors, including nutritional deficiencies, environmental influences. Anestrum is characterized by the quiescent, function less ovaries & reproductive tract.Uterus is small and flaccid. The vaginal mucosa is pale, scanty and sticky and cervix tightly closed.

Transport of Oocytes

After ovulation, the fimbriated end of the infundibulum captures the ovum, which is then transported through the uterine tube into the uterus by ciliary action and muscular contractions. These contractions depend on estrogen-to-progesterone ratios (with high estrogen enhancing activity), prostaglandin levels, and sympathetic nervous stimulation.

The ovum, surrounded by the zona pellucida and cumulus oophorus cells, reaches the ampullary-isthmus junction within 8–30 minutes. The corona radiata is well-defined in bitches but absent in species like cows, ewes, and sows. Cilia in the rabbit oviduct beat rapidly (1,500/min), aiding transport. In most domestic species, cumulus and corona cells shed within hours post-ovulation, though they may persist for days in dogs and cats. These cells retract their projections due to fibrinolytic enzymes in the oviductal fluid and eventually die.

The zona pellucida, a semipermeable glycoprotein layer, surrounds the vitelline membrane (similar to a somatic cell's plasma membrane) and dissolves under proteolytic enzymes like trypsin and chymotrypsin. After fertilization, the perivitelline space forms between the zona and vitelline membrane, where polar bodies are extruded.

The fertile life of ovulated ova in domestic animals is relatively short, 12 – 24 hours except in Bitch (4 – 8 days)

Cow = 20 - 24 hours

Horse = 6 – 8 hours

Sheep = 16 – 24 hours

Swine = 8 – 10 hours

In cows, ewes, and sows, the ovum is ovulated as a secondary oocytehaving completed the first meiotic division and arrested at metaphase of the second, which only completes after fertilization. In contrast, horses, dogs, and foxes ovulate a primary oocyte, which matures in the uterine tube post-ovulation.

Capacitation

Freshly ejaculated mammalian spermatozoa cannot fertilize an ovum until they undergo capacitation, a process of physicochemical changes within the female reproductive tract. Capacitation starts in the uterus and is completed in the oviduct, enabling sperm to fertilize oocytes. Unlike mammalian sperm, non-mammalian sperm are ready for fertilization immediately after ejaculation.

First described by Austin and Chang in 1951, capacitation requires sperm to reside in the female tract for a specific period. It involves modifications to the acrosome, allowing the release of enzymes like hyaluronidase and proteases, which aid in ovum penetration. The process also removes stabilizing factors such as cholesterol, glycosaminoglycans, and seminal plasma proteins collectively known as decapacitation factors.

Estrogen promotes capacitation, whereas progesterone inhibits it. This selective process ensures only healthy sperm reach the oviduct's ampulla for fertilization. In vivo, interspecies sperm capacitation is possible, indicating that decapacitation factors are not species-specific.

Sperm Attachment

The attachment of sperm head to the zona pellucida is regulated by receptor sites on the zona surface. The mature oocyte synthesis a glycoprotein ZP3, which functions as the sperm receptor to which the sperm with intact acrosome can bind. The presence of glycosyl transferases proteinases and glycosidases on the plasma membrane covering help the sperm head to bind with ZP3.

Fertilization

Fertilization is the fusion of male and female gametes to form a zygote, occurring at the ampullary-isthmic junction of the oviduct. The process begins with the sperm penetrating the zona pellucida using motility and enzymes like hyaluronidase and acrosin. Once contact is made with the oocyte, sperm motility ceases.

In most domestic species, the second meiotic division is triggered by sperm entry, whereas the first occurs shortly before ovulation. The zona reaction follows penetration, preventing additional sperm from entering. Polyspermy, or penetration by multiple sperm, disrupts normal zygote development.

After fertilization, pronuclei develop from the sperm and oocyte nuclei, then fuse to form a diploid zygote. Fertilization is complete when the fused pronuclei disappear, replaced by chromosomes in prophase of the first mitotic division. The zygote remains in the uterine tube for 3–4 days before moving to the uterus.

To support survival, estrogen dominance at estrus shifts to progesterone dominance with corpus luteum formation. Progesterone stabilizes the uterus and promotes glandular endometrium development, which secretes uterine milk to nourish the embryo before implantation. Cell division forms a 16–32 cell morula, which develops into a blastocyst by 6–8 days. The embryonic period begins with blastocyst attachment, marked by rapid growth and organ development. The fetal period, starting around day 45 in cows, continues until birth.

Table 3. Time of ovulation and Al /service:

Animal species	Time of ovulation	Time of Al/service
Cattle & Buffalo	10-15 hr after end of estrus	10-12 hr after onset of estrus (Mid to late estrus)
Mare	1-2 days before end of estrus	2-4 days after onset of estrus
Sheep	12-14 hr before end of estrus (Towards the end of estrus)	12-24 hr after onset of estrus
Goat	12-14 hr before end of estrus (Towards the end of estrus)	24-36 hr after onset of estrus
Sow	30-40 hr after onset of estrus (Towards the end of estrus)	24-36 hr after onset of estrus
Bitch	1-2 day after onset of true estrus	2-3 d after onset of true estrus

Placenta

As the embryo increases in size, the extra embryonic or fetal membrane of placenta develops to meet the increasing need for more nutrients to nourish the developing embryo. It also acts as an organ of Respiration and Excretion.

Classification of Placenta

Table 4: Based on The Number Layers That Separate Maternal &Fetal Blood.

Layers	Epitheliocho-rial	Syndesmo-chorial	Hemocho-rial	Endothelio-chorial	Hemoendo-thelial
Maternal Layers					
Endotheilum	+	-	-	+	-
Connective tissue	+	+	-	-	-
Epithelium	+	+	-	-	-

Fetal Layers					
Epithelium	+	+	+	+	-
Connective tissue	+	+	+	+	-
Endothelium	+	+	+	+	+
Examples	Sow, Mare, Cow & donkey	Ewe & Doe	Monkey & Women	Bitch & Queen	Guinea pig, rabbit & rat

1. Based on the loss of maternal tissues occurring at birth of fetus

i) **Deciduate:** In this type, the deciduate composed of portions of maternal epithelium, submucosa, deciduate cells and the fetal placenta are shed at parturition leaving a portion of the endothelium denuded. In bitches and queens, there is moderate loss of maternal tissues at birth of the fetus. In monkey and women, there may be extensive loss of maternal tissue during parturition.

ii) **Non- deciduate type:** There is no loss of maternal tissue at the birth of the fetus (sow, mare, ewe, doe & cow). In this type, placenta and fetel membranes are expelled at the time of parturition leaving the endometrium intact except in ruminants in which only the surface of the caruncles are devoid of epithelium. The caruncle sloughs about 6-10 days following parturition.

Table 5. Based on chorionic villi pattern and their shape of attachment to the uterus.

Types of Placenta	**Distribution of villi**	**Animals**
Diffuse	Chorionic villi cover most of the fetal placenta, projects into the crypts, scattered over the entire endometrium of the uterus.	Sow, Mare
Discoid or Spherical	The placenta gets attached to the uterus in a disk- shaped area only.	Guinea pigs, Rabbit & Rat
Zonary	The chorion of the fetal placenta is attached in a grill like band with the uterus of the dam.	Bitch & Queen
Cotyledonary	• Here the attachment of the portion of the chorion of the fetus (Cotyledons) with the uterine epithelium of the dam in mushroom like area is called as the caruncles. • Endometrial caruncle and fetal cotyledon together constitutes Placentomes. • There are 75-120 placentomes in the pregnant cow and 80- 90 placentomes in the pregnant ewe	Ewe, Doe, Cow & Buffalo
Bidiscoid	In the form of two disc	Monkey

Functions of placenta

The fetal and maternal blood do not mix, but the placenta facilitates nutrient transfer, gaseous exchange, and waste removal. It transports carbohydrates, proteins, fatty acids, water, and minerals, though vitamins A, D, and E are in lower concentrations. The placenta also acts as a temporary endocrine organ, producing progesterone, placental lactogen, gonadotropins (eCG and hCG), and relaxin near term. It supports pregnancy maintenance and plays a role in parturition.

Parturition

Parturition (labor, delivery, or birth) is the process by which the uterus expels the fetus and placenta from the female. Parturition, sometimes called labor, is the physiologic process by which the pregnant uterus delivers the fetus & fetal membranes from the mother.

Parturition is divided into three stages

i) Dilation of the cervical canal to accommodate passage of the fetus from the uterus through the vagina and to the outside.

ii) Contractions of the uterine myometrium that are sufficiently strong to expel the fetus.

iii) Expulsion of the placenta.

Signs of Approaching Parturition

Several events take place near the end of pregnancy in preparation for parturition. As parturition approaches, the cervix begins to soften (or "ripen") as a result of the dissociation of its tough connective tissue (collagen) fibers. Most signs of approaching parturition relate to changes in the pelvic ligaments, enlargement &edema of vulva & mammary activity.

1. **Physical signs:** During pregnancy, the abdomen enlarges, reaching its peak before parturition. The mammary glands swell, and "waxing" occurs, especially in mares, 6–48 hours before foaling, followed by milk secretion. Signs of impending birth include vulvar swelling, mucus discharge, abdominal muscle relaxation, and increased respiratory rates in sows. Relaxin and estrogen loosen ligaments for birth, while PGF2α aids cervical relaxation. A drop in rectal temperature signals approaching labor, particularly in bitches, where a 2–3°C decrease occurs 6–8 hours before parturition.
2. **Behavioral signs:** Restlessness, Frequent lying down and getting up, Frequent urination, the bitch & sow often attempt to build elaborate nests. Nest building is a feature of impending parturition in polytocous. Cattle & ewe seek isolation just prior to the onset of parturition.

Mechanism of parturition

Successful parturition depends on Myometrial contraction and Capacity of cervix to dilate.

1. **Myometrial contraction:** Progesterone blocks myometrial contraction during gestation.As parturition approaches, E2 rises causes contraction and release of PGF2α.
2. **Dilation of cervix:** Cervix firm & rigid due to high collagen content. It retains fetus in the uterus. Under hormonal influence changes in physical characteristics of collagen at parturition.

Initiation ofparturitionis triggered by fetus & completed by complex interaction of Endocrine, Neural & Mechanical factors.

Hormone Changes

Just before parturition, estrogen production increases, with estrone from the fetoplacental unit rising as fetal maturity advances (around 3–4 weeks prepartum in cows). Fetal cortisol triggers this estrogen surge, which promotes uterine muscle contractile protein production. Estrogen also signals PGF2α release (24–36 hours prepartum in cows), leading to CL regression and a drop in progesterone. This hormonal shift transitions the uterus from quiescence to contractility, preparing for labor.

Changes in maternal hormonal levels do not seem to play a major role in parturition in the mare. At parturition the mare has relatively high levels of progestogens and low levels of estrogens. However, PGF2α level increases during foaling. The progesterone concentration does not decrease in the mare after PGF2α secretion because no CL is present after about 150 days of pregnancy. PGF2α is also believed to increase the contractility of the uterus by permitting greater mobility of sarcoplasmic calcium. These early contraction increases might be important in positioning the fetus for delivery (presentation) through the pelvic canal. The presence of the fetus in the pelvic canal causes oxytocin to be released from the posterior pituitary. In the presence of an estrogen primed uterus, the muscle contractions increase in intensity to assist in expelling the fetus. PGF2α also increases the sensitivity of the uterus to oxytocin, which enhances the rhythmic contractions of the uterine musculature during delivery.

The uterus can only assist in the expulsion of the fetus and must have the coordinated contraction of the abdominal muscles.

The presence of the feet in the pelvic canal and the consequent stimulation of the vagina provides for reflex contraction of the abdominal muscles, similar to the straining that occurs when one attempts to replace a prolapsed uterus.

Ripening of cervix: Cervical ripening, essential for dilation and fetal passage, begins before labor and is hormonally regulated. It is influenced by increased estrogen, relaxin (in pigs), and PGF2α at parturition onset. In CL-dependent species (goats, pigs, cattle), parturition starts with CL regression. In placental-dependent species (mares, sheep), fetal cortisol activates 17α-hydroxylase, converting progesterone to estrogen. Rising estrogen and declining progesterone stimulate PGF2α, which induces strong uterine contractions and cervical dilation during the second stage of labor.

Ferguson's reflex

Oxytocin mediates the neurohumoral reflex (Ferguson's reflex) stimulation of the birth canal by the conceptus during labor. *Fetus determines the day of parturition, whereas the mother decides the hour of parturition.* In sheep, cortisol stimulates the placenta to convert progesterone to estrogen.The elevated levels of estrogen stimulate secretion of PGF2 alpha. In CL dependent species (goat, pig & cattle), cortisol in addition to the synthesis of estrogen causes a release of PGF2 alpha from the endometrium, which in turn causes regression of the corpora lutea. Anxiety, stress or fear prolong the act of parturition in several species through a decrease in myometrial contractility induced by a release of epinephrine. The abdominal and uterine muscle contraction, coupled with relaxed pelvic ligaments, separation of the pelvic symphysis, and dilatation of the cervix, provide for expulsion of the fetus. The combined forces of intra-abdominal and intra uterine pressure mark the beginning of the second stage of labor. The greatest effort is associated with the emergence of the head & chest. The expulsion of the placenta is rapid in the mare but is slower in ruminants due to the cotyledonary type of placentation. In the pig contraction begin at both ends of the uterine horns and subsequently are propagated toward the cervix or in the opposite direction. The largest mass of placenta, however, is usually expelled 3 to 4 hours after the delivery of the last piglet.

Stages of Parturition

Parturition is a continuous process but for convenience, it is divided in to 3 stages.

1. **First stage:** Contraction of uterine muscle and dilation of cervix. Uterine peristalsis starts at apex of the horn. Uterine contraction forces the fetal membrane and their fluid into the dilated cervix. Rotation of fetus in (mare, bitch) taking place. By end of this stage cervix is completely dilated. This stage is apparently larger in primiparous than pluriparous animals. Once the portion of fetus enters pelvis, reflex stimuli results in straining. Species Duration; Cow and Ewe 2-5 hrs; Horse 1-4 hrs; Sow and Bitch 2 – 12 hrs.

2. **Second Stage:** Entrance of fetus into the dilated birth canal. Rupture of one or both water bags. Abdominal contraction takes place. Almost in all species, when straining commence it will lie down. As the fetal parts touches cervix – Fergusan's reflex initiates. Combination of uterine and abdominal contraction leads to expulsion of fetus. Cow 3-4 hrs; Ewe, goat 0.5 – 2 hrs; Mare 5 – 40min; Dog 1 hr. In multiparous animals, the length of second stage is variable depending on the number of fetusus.
3. **Third stage:** Expulsion of foetal membranes and involution of uterus.

5

Housing Principles and Space Requirements for Different Species of Livestock and Poultry

Vipin Maury, Utkarsh Kumar Tripathi, Anuradha Kumari and Ajeet Singh

Faculty of Veterinary and Animal Sciences, Institute of Agricultural Sciences, Banaras Hindu University, Barrkachha Campus, Mirzapur, Uttar Pradesh

The provision of appropriate housing is a corner stone of effective livestock and poultry management. Housing systems directly influence the health, productivity, and overall welfare of animals. A well-designed animal shelter not only protects livestock and poultry from environmental extremes but also facilitates efficient management practices, waste disposal, and disease control. Inadequate housing can result in heat stress, leading to reduced feed intake, lower milk yield, and compromised immune function in dairy cattle. Poorly ventilated poultry housing causes respiratory issues and lower growth rates. Investing in scientifically designed shelters tailored to species-specific needs ensures sustainability and profitability. Beyond economic returns, proper housing underscores the ethical responsibility of farmers to provide humane living conditions, aligning with global standards for animal welfare. As we delve deeper into this chapter, the critical aspects of housing principles and space requirements for different species will be elaborated.

Importance of Animal Shelter

Animal shelters are indispensable for maintaining the health and productivity of livestock and poultry. They offer a controlled environment that mitigates the adverse effects of climate, disease, and management inefficiencies.

Protection from harsh weather: Animals exposed to extreme weather conditions experience significant stress, which adversely affects their physiological functions. Heat stress in cattle reduces milk production drastically. Similarly, cold stress in young livestock can increase mortality rates due to hypothermia. Shelters provide insulation against such extremes, maintaining optimal ambient conditions for the animals.

Health and hygiene: Scientific evidence underscores the role of housing in disease prevention. Properly designed shelters reduce the incidence of mastitis in dairy cows, respiratory diseases in poultry, and parasitic infections in sheep and goats. Well-drained floors and adequate ventilation in poultry houses reduced ammonia levels, minimizing respiratory disorders.

Enhanced productivity

Animals housed in comfortable environments show improved feed conversion efficiency (FCE), growth rates, and reproductive performance. The broilers kept in temperature-controlled environments achieve up to 10% higher growth rates compared to those in poorly ventilated setups. In dairy farming, comfortable housing can increase lying time, which is directly linked to milk production.

Animal welfare

Providing shelter reflects a commitment to animal welfare. Humane housing ensures that animals can exhibit natural behaviors, such as resting, grooming, or perching.

Economic benefits

Proper housing minimizes animal losses, reduces veterinary expenses, and enhances productivity. For instance, investing in insulated roofing materials can reduce heat stress in poultry, leading to better feed efficiency and higher egg production, ultimately improving profitability.The importance of animal shelters transcends basic protection. It encompasses a comprehensive approach to animal management, emphasizing health, welfare, productivity, and sustainability.

Housing systems

Efficient animal housing plays a pivotal role in optimizing livestock productivity, ensuring animal welfare, and maintaining hygienic conditions for both animals and farm workers. Animal housing systems are designed based on various factors such as geographical location, climatic conditions, land availability, economic constraints, and species-specific behavioral needs. Broadly, animal housing systems in dairy farming are categorized into Conventional (Enclosed) Barns and Loose Housing Systems. Each system has distinct structural and functional features tailored to meet the needs of different production environments.

Conventional (enclosed) barns

Also referred to as stanchion barns or tie-stall barns, conventional barns are enclosed structures where dairy cattle or buffaloes are confined individually.

Each animal is tied at the neck using neck chains, ropes, or stanchions, and remains in its designated place throughout the day. Feeding, watering, and milking are done in the same area. This system is particularly suitable for temperate and cooler regions such as the Himalayan zones, where protection from cold weather is a priority.

The enclosed design of barns minimizes the exposure of animals to external environmental stressors such as cold winds, rain, and snow. The walls are often constructed with insulation materials, and ventilation is managed through windows, ridge vents, or mechanical systems to maintain optimal air quality. However, in tropical regions, barns are often modified with open or semi-open sidewalls to enhance air movement and prevent heat stress. Scientific studies have shown that tie-stall barns enable better control over individual feeding, milking, and health monitoring. For instance, milk yield and feed efficiency may improve due to individualized management. However, limited mobility can lead to welfare concerns such as lameness and hock lesions. Additionally, labor requirements and maintenance costs are higher compared to loose systems, and animal behavior is significantly restricted.

Loose housing system

Loose housing is a more extensive and welfare-friendly system where animals are allowed to move freely in an open paddock or yard during the day and night, except during milking. A shaded area with a common feed manger is provided for resting and feeding, while a roofed shed offers protection from solar radiation, rain, and cold. Milking is usually carried out in a separate milking parlor, and concentrates are provided individually to animals before milking. Common watering tanks or troughs ensure easy access to clean drinking water.

This system is particularly suited for tropical and subtropical climates, especially in semi-arid to arid regions and moderate rainfall zones. However, it may not be ideal in regions with excessive rainfall or extremely cold weather, where animals require more shelter and protection. Loose housing systems promote better thermoregulation due to voluntary movement of animals between the open area and shade/shelter (Fig: 01). Animals can express their natural behaviors such as lying, ruminating, grooming, and socializing, which contributes significantly to their overall welfare and reduces stress-related problems. A study by West (2003) confirmed that animals in well-ventilated, open housing systems exhibit lower cortisol levels, indicating reduced stress and better adaptability to heat.

Additionally, loose housing offers economic and labor advantages. Construction costs are relatively lower due to minimal use of building materials, and the system is easier to expand, modify, or upgrade. Farmworkers also benefit from simplified cleaning, feeding, and manure management procedures. Moreover, in large-scale operations, grouping animals based on physiological status (e.g., dry cows, lactating cows) becomes easier in a loose housing layout.

From an animal health perspective, animals in loose housing systems show reduced incidence of mastitis, better hoof health, and higher fertility rates, especially in tropical countries. For instance, buffaloes housed in open paddocks in the northwestern Indian plains have demonstrated improved heat tolerance and reproductive performance during summer, helping mitigate issues of summer anestrus and subfertility. Their voluntary access to shelter or open space, based on weather conditions, provides them with thermoregulatory choices that enhance their comfort and productivity.

The design of dairy housing structures and the materials used in their construction significantly influence the thermal comfort of animals. Well-planned dairy shelters help mitigate heat stress, thereby improving feed intake, milk production, and reproductive efficiency. Given India's diverse climatic, geographical, and economic conditions, it is impractical to design a universally ideal dairy housing system that suits all regions.

Fig. 1. Low cost loose housing for sheep and goats

Site Selection for animal housing

Selecting an ideal site for animal housing is a critical step that determines the long-term success and sustainability of livestock and poultry operations. The livestock management can be operated safely, and economically and with less disease occurrence and less parasitic load if proper site selection is done when planning an entirely new farm. For the construction of farm buildings selection of site is most important for the safety operation of a farm, better economics of a farm, ease of farm operations, availability of resources, ease of management, less disease occurrence in the animals, and less parasitic load in farm animals. Certain points must be considered before site selection for the livestock housing:

Topography	**Drainage**
Soil type	Accessibility
Marketing facility	Transport facilities
Water availability	Labour availability
Electricity	Climate
Ventilation	Thermo-neutral zone
Miscellaneous points	

Topography: A dairy building should be at a higher elevation than the surrounding ground to offer a good slope for rainfall and drainage for the wastes of the dairy to avoid stagnation within. A levelled area requires less site preparation and thus lesser cost of building. Low lands and depressions and proximity to places of bad odour should be avoided. It also prevents the parasitic infestation due to water logging and high moisture.

Drainage: The soil should be porous and the gentle slope so that drainage is efficient and the farm premises remain dry. Proper drainage of rain and subsoil water should be provided to keep healthy environment and to protect the building from dampness.

Soil type: Fertile soil should be selected for cultivation. Foundation soil as far as possible should not be too dehydrated or desiccated. Firm, well-draining soil is essential to support the foundation of the housing. Clay-loam soils, for example, provide stability but require drainage systems to prevent water accumulation. Sandy loam soil should be preferred. Soil must be suitable for strong foundation. Marcy, clay, sandy, rock soils are not suitable. Loamy and gravely soils are best suited for building construction.

Availability of land: There should be vast area to construct all building and should give way to future expansion of farm. Atleast 0.5 to 1 hectare land is required for accommodation of 100 to 200 dairy animals. For 2 cows 1-acre land is essential for fodder production.

Exposure to the sun and protection from wind: A dairy building should be located to a maximum exposure to the sun in the north and minimum exposure to the sun in the south and protection from prevailing strong wind currents whether hot or cold. Buildings should be placed so that direct sunlight can reach the platforms, gutters and mangers in the cattle shed. As far as possible, the long axis of the dairy barns should be set in the East-West direction in the tropical regions (reduces direct exposure to harsh sunlight and maximizes natural ventilation) and north-south direction in the temperate regions to have the maximum benefit of the sun. If the farm building in open or exposed area, the wind breaks in the farm of tall quick growing trees should be grown near the building. This will reduce the wind velocity and solar radiation.

Accessibility: Situation of a cattle shed by the side of the main road preferably at a distance of about 100 meters should be aimed at. Easy accessibility to the buildings is always desirable. The housing should be close to essential resources such as feed storage, water sources, and transportation routes. However, it should maintain a safe distance from residential areas to prevent nuisance and disease transmission.

Marketing: The farm should be away from the city but at the same time it should be nearer to city thereby the products produced from the farm could be marketed easily. Dairy buildings should only be in those areas from where the owner can sell his products profitably and regularly. The animal products are usually of perishable nature and so it needs quick and remunerative disposal. The marketing facility is also related with the availability of good quality and cheap inputs like feed, fodder and labour.

Durability and attractiveness: It is always attractive when the buildings open up to a scenic view along with this, durability of the structure is obviously an important criterion in building a dairy.

Availability of water: Plenty of water is needed for farm operations like washing, fodder cultivation, processing of milk and byproducts and for drinking. Abundant supply of fresh, clean and soft water should be available at a cheap rate.

Transport facility: The farm buildings should be provided with good road and also have the accessibility to reach the market. This will reduce the transport cost and avoid spoilage of products.

Surroundings: Areas infested with wild animals should be avoided. Narrow gates, high manger curbs, and loose hinges, protruding nails, smooth finished floor in the areas where the cows move and other such hazards should be eliminated.

Facilities, labour and feed: Cattle yards should be so constructed and situated in relation to feed storages, hay stacks, silo and manure pits as to effect the most efficient utilization of labour. Sufficient space per cow and well-arranged feeding mangers and resting areas contribute not only to greater milk yield of cows and make the work of the operator easier but also minimizes feed expenses.

Electricity: It is needed for operating various machines used in the farm and is the light source to the animals. Electricity is the most important sanitary method of lighting a dairy. So, it is desirable to have an adequate supply of electricity.

Labour: Honest, economic and regular supply of labour should be available. The labour cost comes next to the feed cost in livestock keeping.

Climate: Suitable climate improves the production and reproduction of animals, reduces the managemental cost and the disease occurrences. The climatic factors influencing the production are

High and low ambient temperature	Atmospheric humidity
Solar radiation	Reflected radiation
Rains/ Hail storm	Winds
Dust	Light
Gases	Sound

Miscellaneous

Other facilities like availability of network, nearby school for children of farm workers, post office, shopping area/ market and entertainment facilities. Protection from noise and other disturbance. The farm sitthee should be away from noise producing factory/chemical industry, sewage disposing area.

Principles of Housing

The principles of animal housing are rooted in scientific understanding of animal behavior, physiology, and environmental interactions. Proper housing design not only safeguards animal health but also boosts productivity and reduces operational costs. The following principles are central to effective livestock and poultry housing:

Ventilation: Adequate ventilation is essential to maintain air quality and regulate temperature and humidity within the housing. Poor ventilation can lead to the accumulation of ammonia and other harmful gases, causing respiratory issues and reduced growth rates. Well-ventilated barns improve feed efficiency by 10-15% and decrease the prevalence of respiratory diseases by 25-30%.

Lighting: Natural and artificial lighting play a crucial role in supporting animal physiology and behavior. For example, poultry exposed to 16 hours of light and 8 hours of darkness exhibit better egg production and improved immune responses. Optimal lighting schedules increase laying rates by 5-10% in layers and enhance growth in broilers.

Space Allocation: Providing adequate space per animal is vital to prevent overcrowding, stress, and aggression. Overcrowding has been linked to reduced feed intake, poor growth, and increased susceptibility to diseases. Increasing space allowance by 20% can result in a 15% improvement in weight gain among broilers.

Drainage: Effective drainage systems are necessary to keep housing dry and sanitary. Floors should be sloped (1:60 gradient) to facilitate water runoff and prevent pooling. Proper drainage reduces the incidence of hoof diseases, such as laminitis, by 40%.

Durability: Using durable and cost-effective materials ensures the longevity of the housing structure. For instance, galvanized iron sheets and concrete floors withstand harsh weather conditions and heavy use. Long-term studies have shown that durable materials reduce maintenance costs by 30-40% over a 10-year period.

Flexibility: Housing designs should be adaptable to accommodate changes in herd size, animal types, or farming practices. Modular designs, for example, allow for easy expansion or reconfiguration, making them ideal for dynamic farming operations.

Design considerations

A well-planned housing layout is vital for ensuring efficiency, animal comfort, and economic feasibility. The design should accommodate the physiological and behavioral needs of the animals while minimizing labor and maintenance costs. In India, shelter of animals need initial capital to the extent the dairy farmers can afford. Therefore, the house design should be cost effective and economical. The following factors may be considered while designing a livestock farm:

1. **House design:** Designing a shed which can accommodation a workable unit, is suitable for effective management. It also allows to study the need of smaller group with regard to floor space, feeding space required for different kinds of animals.
2. **Foundation:** Building`s foundation should be on firm ground, suitable to structural requirements and its depth depends on the soil conditions. It is usually 2 to 4 times wider than the structural walls
3. **Housing structure:** Shape and design of building should meet the needs of all classes of livestock. Uniformity in the appearance should be maintained. We have to decide the number of animals to be housed in the building and number of buildings to be constructed.

4. **Flexibility:** Animal building has to be designed to meet the requirement of changing enterprises. This will increase the utility of buildings. Spacious building without pillars can be easily being adopted for different enterprises with little modifications in the building.
5. **Roof:** It is designed to suit the local climatic conditions. Gable with roof ventilator is necessary for hot condition (Fig. 02). The roof slope should be around 1:4 to ensure proper rainwater runoff and prevent accumulation. Monitor roof is suitable for building with smaller width (Fig. 03). Materials such as galvanized iron sheets, asbestos, or polycarbonate reduce temperature fluctuations and provide durability. Insulated roofing materials tend to lower heat stress in animals.The pitch of the roof should be about 22 to 30 degrees. The pitch should be 35°, 25-30° and 12-18° in case of thatch, tiled and sheet roof, respectively. Slope is generally kept in heavy rainfall area. In any case the pitch of the roof should not exceed more than 35°. The height of the eaves should be about 1.8 to 2.0 meters, but in heavy rainfall area height shall be as lower as possible but not lower than 1.6 meters. A height of 8 feet at the sides and 15 feet at the ridge will be sufficient to give the necessary air space to the cows. An adult cow requires at least about 800 cubic feet of air space under tropical conditions. To make ventilation more effective continuous ridge ventilation is considered most desirable.

Types of Roofs

Different types of roofs (Fig. 2-5) (anyone can be selected on the basis of need)

Fig. 2. Gable Roof

Fig. 3. Monitor roof

Fig. 4. Shed type roof

Fig. 5. Flat type roof with open area

Roofing materials

- ***Thatch:*** Thatch is an excellent and cheap material to reduce the heat stress as it has lower thermal conductivity, but less durability and fire hazard makes it less acceptable at an organized farm.
- ***Clay tiles:*** They are mostly used in rural areas in animal shed roofs. Clay tiles roofs are relatively heavy thus demanding sufficiently strong frame. Thermal conductivity of clay tiles is higher than that of thatch material and requires frequent maintenance.
- ***Galvanized Iron (GI) sheet:*** It is a good roofing material for preventing rain water entry inside the shed. Moreover, it is easy to install, light weight and durable; but higher thermal conductivity outweighs its merits.
- ***Asbestos sheets:*** Asbestos sheets roofs are generally used at organized farms as they are comparatively cheaper than RCC, durable than thatch, and have intermediate value of thermal conductivity. The major demerit of asbestos sheet is its radiation emission property, due to which on getting heated up during peak hours of summer, these sheets start emitting radiations which not only increase the surface temperature of animal but also alter microclimate of the shed.

- ***Reinforced Cement Concrete (RCC):*** RCC roofs are popular in human housing due to their pest (termite) resistance, natural calamity (cyclones) resistance, availability and cost effectiveness of concrete ingredients. RCC roofs are durable and provide favourable microclimate inside the animal shed (Fig. 05)
- ***Polythene sheets:*** They are generally used in temporary sheds because they can be fixed easily, modifiable as per need and economical and commonly used to make thatch waterproof. Although thermal conductivity is at par with the thatch but its extreme thinness makes it ineffective in protection from solar radiation. Its main drawbacks are shorter life and fire prone nature.
- ***Polycarbonate sheets:*** Polycarbonate is a strong thermoplastic material that is lightweight and can withstand extremely low and high temperatures.

6. **Standard width of buildings:** Single row cow shed should have length of 3. 80 to 4.25 metres and Double row cow shed should have 7. 90 to 8.70 metres length.
7. **Standard height of the building:** The standard height of the building may differ according to the roofing material and agro climatic condition.
8. **Length of building:** The standard length of building may be of any. It may vary depends upon the number of animals housed. Length can be determined based on the total stock to be housed within the building, e.g., 10-20 animals in single row system and 20-50 animals in double row system and above 50 animals a separate shed should be provided.
9. **Walls:** Wall height should range between 1.5 - 2.0 m for livestock housing, allowing for adequate ventilation.Materials such as brick, concrete, or bamboo are commonly used, with variations based on climatic conditions.Windows should be strategically placed to facilitate natural ventilation, reducing ammonia/ methane buildup.
10. **Floor:** Flooring should be made of non-slippery, durable, and easy-to-clean materials like concrete. A slope of 1:60 is recommended for effective drainage, preventing water stagnation and maintaining hygiene. Bedding materials such as straw or sawdust contribute to thermal insulation and animal comfort.

Slatted Flooring for Sheep and Goats

- In regions with high rainfall, slatted flooring is often preferred for sheep and goat housing due to its superior drainage and cleanliness.

- For slatted floors, each wooden slat should have a width between 7.5 cm and 10.0 cm, with a thickness ranging from 2.5 cm to 4.0 cm to ensure durability under animal load.
- The sides of the slats must be well-rounded to prevent injury to the animals, and a gap of 1.0 cm to 1.5 cm should be maintained between slats (Fig. 6). This spacing allows for effective disposal of dung and urine, keeping the floor dry and hygienic.
- The entire slatted platform should be raised to a minimum height of 1 metre above ground level to enhance ventilation and waste management.
- To allow easy access for the animals, a gentle ramp or sturdy steps made of wooden planks should be provided leading up to the slatted floor.

Fig. 6. Slatted floor system

Table 1. Merits of Slatted floor system

1	**Hygiene & Disease Control**	Slatted floors reduce direct contact between goats and feces/urine, lowering parasite loads and incidence of diseases like **coccidiosis** and **foot rot** (Devendra & Burns, 1983). Faecal egg counts (FECs) reduced by **35–40%** compared to earthen floors.
2	**Dry & Clean Flooring**	Urine and manure fall through slats, keeping the surface dry. Moisture content of flooring is typically <10%, minimizing bacterial growth (Tamang et al., 2021).
3	**Better Hoof Health**	Elevated dry flooring reduces hoof softening and injuries. Studies show 60–70% lower foot problems compared to muddy floors (Devendra & Burns, 1983).
4	**Improved Feed Efficiency**	Clean environment improves feed intake and growth rates. Studies show 8–12% increase in average daily gain (ADG) in slatted floor-raised goats (Singh et al., 2020).

5	**Easy Waste Management**	Droppings can be collected below and used for vermi-composting or biogas. Ideal for integrated farming systems.
6	**Labor Saving**	Reduced need for cleaning and bedding. Time spent on cleaning is reduced by 40–60% (Tamang et al., 2021).
7	**Ideal for Rainy Regions**	Raised floors prevent water stagnation, particularly beneficial in humid or flood-prone areas.

11. **Water and Feed Facilities:** Placement of automatic waterers and feed troughs should be optimized to minimize feed wastage and encourage uniform growth.

Other important structures

- ***Pillars***: Pillars may be either of hard wooden post, cast iron pipes. Columns of bricks of RCC or cement. Each of them shall be placed at intervals of 2.50 to 2.75 m.
- ***Fencing***: The fencing material should be cheap and locally available. The effective height of the fence for calf and adult may be 1 meter and 1.2 to 1.5 meters, respectively. The fence may be made by brick wall or iron railing or iron wire. Iron railing wire fencing 33.7 mm iron pipe or 5 mm iron wire may be provided horizontally and placed at 30, 60 and 100 cm for calves and 40, 80 and 120 cm for adult cows from ground level with the support of posts made of 6x4 cm angles iron pillars 5 cm (dia), ground iron pillar/ 10x10 cm timber pillar or brick pillar (40x30 cm) placed 2 meters apart.
- ***Gate***: The gate in dairy farm varies in sizes. The width of the gate leading from sheds to sheds to be about 1.0 to 1.2 meters. The gate which leads from paddock to road is to be 2.5 meter. The main gate of the farm premises should be bigger in width i.e. 5.5 to 6 meters for each entrance and exist of tractors, trollies and other heavy vehicles.
- ***Manger***: Cement concrete continuous manger with removable partitions is the best from the point of view of durability and cleanliness. A height of 1'-4" for a high front manger and 6" to 9" for a low front manger is considered sufficient. Low front mangers are more comfortable for cattle but high front mangers prevent feed wastage. The height at the back of the manger should be kept at 2'-6" to 3". An overall width of 2' to 2.5' is sufficient for a good manger.
- ***Alleys***: The central walk should have a width of 5'-6' exclusive of gutters when cows face out, and 4'-5' when they face in. The feed alley, in case of a face out system should be 4' wide, and the central walk should show a slope of 1" from the centre towards the two gutters running parallel to each other, thus forming a crown at the centre.

- ***Manure Gutter:*** The manure gutter should be wide enough to hold all dung without getting blocked, and be easy to clean/ Suitable dimensions are 2" width with a cross-fall of 1" away from standing. The gutter should have a gradient of 1" for every 10' length. This will permit a free flow of liquid excreta.

Animal Sheds

***Calving Boxes*:** Allowing cows to calve in the milking cowshed is highly undesirable and objectionable. It leads to unsanitary milk production and spread of disease like contagious abortion in the herd. Special accommodation in the form of loose-boxes enclosed from all sides with a door should be furnished to all parturient cows. The calving pen should have an area of about 100 to 150 sq. ft., it should be provided with sufficient ventilation through windows and ridge ventilation with ample soft bedding (Fig. 7)

Fig. 7. Calving shed

***Isolation Boxes*:** Animals suffering from infectious disease must be segregated soon from the rest of the herd. Loose boxes of about 150 sq. feet. are very suitable for this purpose. They should be situated at some distance from the other barns. Every isolation box should be self-contained and should have a separate connection to the drainage disposal system.

***Sheds for Young Stocks*:** Calves should never be accommodated with adults in the cow shed. The calf house must have provision for daylight ventilation and proper drainage. Damp and ill-drained floors cause respiratory trouble in calves to which they are susceptible. For an efficient management and housing, the young stock should be divided into three groups, viz., young calves aged tip to one year bull calves, female calves.

- Each group should be sheltered in a separate calf house or calf shed.
- As far as possible the shed for the young calves should be quite close to the cow shed.
- Each calf shed should have an open paddock or exercise yard.

A suitable interior lay-out of a calf shed will be to arrange the standing space along each side of a 4 feet wide central passage having a shallow gutter along its length on both sides. Provision of water troughs inside each calf shed and exercise yard should never be neglected.

Dry animal shed: In large farms, milch and dry cows are housed separately. The floor in the covered area should preferably be made of cement concrete. Under Indian conditions, in smaller farms, milch and dry animals can be housed together. Normally, one third of the animals in a farm will be in dry or in dry cum pregnant stage.

***Bull or Bullock Shed*:** Safety and ease in handling a comfortable shed protection from weather and a provision for exercise are the key points while planning accommodation for bulls or bullocks. A bull should never be kept in confinement particularly on hard floors. Such a confinement without adequate exercise leads to overgrowth of the hoofs creating difficulty in mounting and loss in the breeding power of the bull.

- A loose box with rough cement concrete floor about 15’ by 10’ in dimensions having an adequate arrangement of light and ventilation and an entrance 4’ in width and 7’ in height will make a comfortable housing for a bull. If possible, the arrangement should be such that water and feed can be served without actually entering the bull house.
- The bull should have a free access to an exercise yard provided with a strong fence or a boundary wall of about 2’ in height, i.e., too high for the bull to jump over.
- The exercise yard should also communicate with a service crate via a swing gate which saves the use of an attendant to bring the bull to the service crate.

***Feeding Space*:** If the feeding space is inadequate, and every animal is not provided equal chance feeding at the same time, the dominant animals do not permit the subordinates to feed unless their need is satisfied. The subordinates have to follow the path of avoidance. The feeding space should be at least equal to the barrel diameter of dairy animals.

***Feeding trough*:** The height, width and depth of feeding trough are most important and its design should also be proper. If the dimensions are not as per the size of animals sometimes, they attempt to stand in the trough spoiling the feed and sometimes they defecate and urinate in the feeding trough. The heights of inner edge should range between shoulder point and elbow height. The depth should be equal to the length between throat to muffle. The width of trough should be just enough so that the feed is within the reach of animal. The bottom of the through should be rounded and the maximum depth should be to the animal side.

Handling place and device: A place and device by which the animals are restrained for vaccination, miner treatment etc. should follow the approved standard. If the animals are getting experience of excessive pain, either they develop the aggressiveness or avoidance. The handling crush should be round, smooth narrow to accommodate only one animal. It should not have the sharp projections to avoid the injury and distress.

Quarantine shed: It should be located at the entrance of the farm. The newly purchased animals entering into the farm should be kept in quarantine shed for a minimum period of 30 to 40 days to watch out for any disease occurrence.

Space requirements for livestock and poultry

The provision of optimum and comfortable space and microclimate for the fullest expression of the milk production potential of a cow is the primary objective of dairy housing. Providing adequate space for livestock and poultry is essential for their welfare, productivity, and health. Space allocation must consider animal behavior, movement, and social interactions. Overcrowding leads to stress, aggressive behavior, and reduced feed intake, thereby negatively impacting growth rates and reproduction. (Table 2-10). To provide clean and comfortable shelter at a cheaper cost may be the other important aspect in dairy animal housing. There are well defined standards for the space requirements for various categories of dairy animals (Table 2 and 7) and it is essential to follow those standards while constructing the dairy cattle houses for efficient and economic management of the stock.

Table 2. Floor space requirement for cattle and buffaloes (BIS standards)

Type of animal	Floor space (sq. m.)		Maximum no. of animals per pen	Height of the sheds at eaves (cm)
	Covered area	Open area		
Cow	3.5	7.0	50	175 in medium and heavy rainfall areas and 220 in semi-arid and arid areas
Buffalo	4.0	8.0	50	
Down-calver	12.0	12.0	1	
Bull	12.0	12.0	1	
Heifers	2.5	5.0	30	
Young calves	1.0	2.0	5 – 10	
Older calves	2.0	4.0	25	

Table 3. Recommended floor space requirements for Sheep and Goat

Age groups	Covered space (m^2)	Open (m^2)
Up to 3 months	0.2-0.25	0.4-0.5
3 months to 6 months	0.5-0.75	1.0-1.5
6 months to 12 months	0.75-1.0	1.5-2.0
Adult animal	1.5	3
Male, Pregnant or lactating ewe/ doe	1.5-2.0	3.0- 4.0

Table 4. Floor space requirement per animal in Sheep and Goats (BIS standard)

Types of animals	Minimum floor space per animal (m^2)
Ram or buck in groups	1.8
Ram or buck - individual	3.2
Lambs or kids - in group	0.4
Weaner in groups	0.8
yearling or goatlings	0.9
Ewe or doe in groups	1
Ewe with lamb	1.5

Table 5. Dimensions of different sheds in a sheep and goat farm

Name of the shed	Lx W x H (m)	No. of animals housed	Remarks
Ewe/ doe shed	15 x 4 x 3	60	-
Ram/ Buck shed	4 x 2.5 x 3	8	Make partition length wise
Lamb/ kid shed	7.5 x 4 x 3	75	Make partition width wise
Lambing/ kidding shed	1.5 x 1.2 x 3	1	Provide manger and waterer
Isolation / sick animal shed	3 x 2 x 3	1	Provide proper ventilation and bedding materials
Shearing shed	6 x 2.5 x 3	1	Make arrangement for storage of wool
Shepherd house	6 x 4 x 3	-	It should be located nearer to flock
Milch doe shed	1.2 x 0.8 x 3	1	-

Table 6. Space requirement of pigs

Type of animal	Floor space requirement (m^2 per animal)		Maximum number of animals per pen
	Covered area	Open paddock	
Boar	6.0-7.0	8.8-12.0	Individual pens
Farrowing pen	7.0-9.0	8.8-12.0	Individual pens
Fattener (3-5 months old)	0.9-1.2	0.9-1.2	30

Fattener (above five months)	1.3-1.8	1.3.1.8	30
Dry sow/gilt	1.8-2.7	1.4-1.8	3-10

(*Source:* National Bank for Agriculture and Rural Development)

- Farrowing sows may be housed individually in a farrowing pen of 2.5 x 4.0 = 10.0 m^2 having guard rails, creep area, feed and water troughs.
- Housing of growing and finishing pigs: A covered concrete yard for feeding and resting having feed and water trough arranged in the front side and an open yard in the rear will suffice for fatteners. The total space requirement may be 2 m^2 per grower/fattener pig.
- Housing of Boars: Boar pen should have covered area of 6.25-7.5 m^2 and open area of 8.8-12 m^2 for exercise. The walls should have a minimum height of 1.5 m.
- Housing of Female: Open yard type with partial roofing as in the case of boar may be provided. A total of 10-15 females can be grouped in a pen. An area of 2 m^2 per animal may be provided.

Table 7. Space requirements for feeding (cm) per dairy animal (BIS standards)

Type of animals	Length of manger	Width of manger	Depth of manger	Height of inner wall of manger
Adult cattle	60 – 75	60	40	50
Adult buffalo	60 – 75	60	40	50
Calves	40 – 50	40	15	20

* Watering space should be provided @10% of the total manger length.

A comfortable dairy barn should provide

i) The air space per dairy cow : 22.5 m^3

ii) The maximum THI in the barn : 70

iii) The slope of roof (pitch)

i. Thatch roof : 35°

ii. Tile roof : 25 – 30°

iii. Asbestos sheet roof : 15 – 20°

iv. In heavy rainfall areas : 45°

Table 8. Feeding and watering space requirement

Type of animal	Space per animal (cm)	Width of manger/ water trough(cm)	Depth of manger/ water trough (cm)	Height of inner wall of manger/ water trough (cm)
Sheep and goat	40 - 50	50	30	35
Kid/lamb	30 - 35	50	20	25

Table 9. Floor space requirements for poultry

Type	Age (in weeks)	Deep-litter (ft²)	Cages (ft²)
Layers	0-8	0.60	0.20
	9-18	1.25	0.30
	>18	1.50	0.50
Broilers	0-4	0.30	-
	4-8	0.75	-

Table 10. Floor space, feeding space and watering space for chicks

Age weeks	Floor space Sq.ft./Chick	Feeding space inches/chick	Watering space inches/chick
1	0.2	1.5	0.5
2	0.2	2.0	0.7
3	0.3	2.0	0.7
4	0.4	2.5	0.8
5	0.6	2.5	0.8
6	0.8	3.0	1.0
7	0.9	3.0	1.0

(*Source:* Central Avian Research Institute)

Barn layout and arrangement

Cow sheds can be arranged in a single row for small herds (fewer than 10 animals) or in a double-row arrangement for larger herds. In double-row housing, animals may be aligned in either a tail-to-tail arrangement or a head-to-head arrangement.

A. Tail-to-Tail System

- Allows easy cleaning and milking due to a wider middle alley (Fig. 8)
- Ensures better ventilation as cows receive fresh air from outside.
- Reduces disease transmission.
- Facilitates quick identification of health issues in the hindquarters.

Fig. 8. Tail-to-tail barn arrangement

B. Head-to-Head System

- Improves the visual appeal of the shed.
- Facilitates easier stall entry for cows.
- Enhances sanitation, as sunlight can reach the gutter area.
- Simplifies feeding, as both rows can be fed simultaneously (Fig. 9)
- Suitable for narrow barns.

Although conventional barns offer better protection from extreme weather, they are expensive to construct and maintain. Consequently, this system is becoming less popular in favor of more cost-effective housing solutions.

Fig. 9. Head-to-head barn arrangement

2. Loose Housing System

In this system, dairy animals are kept in an open paddock with access to a covered shelter (Fig.10). They remain free to move except during milking. The open area is enclosed by walls approximately 5 feet high, and common feed and water troughs are provided. Milking is carried out in a separate milking parlor. This system is well-suited for Indian climatic conditions, except in temperate regions like the Himalayas.

A complete loose housing system includes the following structures

- Milking parlor
- Calf pens
- Maternity/calving pens
- Bull sheds
- Sick pens
- Feed and fodder storage
- Worker rest areas
- Manure pits/tanks
- Cattle crush facilities
- Water reservoirs

Fig. 10. Loose housing system

Advantages of Loose Housing

- Lower construction costs compared to conventional barns.
- Easier expansion to accommodate more animals.
- Simplified management with shared feeding and watering areas.
- Eliminates the need for additional exercise facilities.
- Facilitates easier heat detection, which is crucial for artificial insemination programs.
- Enhances milk quality due to structured milking practices.
- Requires minimal maintenance and efficient water usage.
- Reduces risk to animals in case of fire hazards.
- Existing non-dairy structures can be modified into loose housing setups.
- Provides freedom of movement, enhancing animal comfort.
- Group-based handling of animals leads to labor efficiency.

- Both housing systems offer distinct benefits, and their selection depends on factors such as climate, economic feasibility, and farm management goals. While conventional barns provide better environmental control, loose housing promotes animal welfare and cost-effectiveness.

Layout of a dairy farm

A well-designed dairy farm layout improves animal comfort, operational efficiency, and management. The layout must accommodate current needs and future expansion, considering animal numbers, feeding systems, waste disposal, and ease of cleaning. Structures like feed stores, manure pits, and hay stacks should be located to minimize transportation.In hot, humid regions, double-row sheds are preferred, especially for large herds. For smaller units (up to 12 cows), single-row housing is sufficient. A typical dairy cow barn in double-row arrangement includes a front feeding passage, standing space, a dung channel, and a rear milking passage. This tail-to-tail design simplifies milking and manure removal.

Its components

1. **Milking Shed** – Equipped with a milking parlor and storage for milking equipment.
2. **Loose Housing Shed** – Open area with partitions for cows to move freely.
3. **Calving Pen** – A separate area for cows in the last stage of pregnancy.
4. **Sick Animal Shed** – Isolation area to prevent disease spread.
5. **Bull Shed** – Housing for breeding bulls.
6. **Calf Rearing Shed** – Separate sections for young calves.
7. **Feed and Fodder Storage** – Protected area for storing feed materials.
8. **Manure Pit** – For proper waste disposal and composting.
9. **Water Troughs and Drainage System** – Proper access to clean water and efficient waste drainage.
10. **Office and Utility Room** – For record-keeping and farm operations.

Layout of Dairy farm for 25 cows

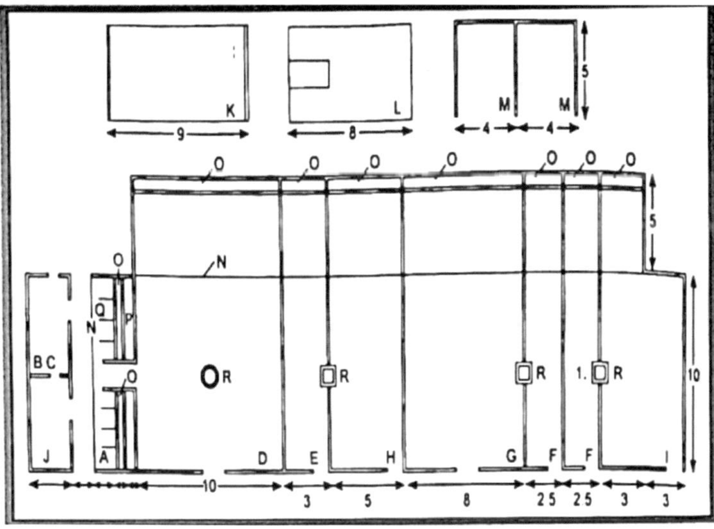

Fig. 11. Layout of Dairy farm for 25 cows

Layout showing the functional details of buildings required for a 25 cow dairy farm

A: Milking barn	H: Older calf pen	N: Shallow gutter
B/C: Milk house-cwn-record room.	I: Bull pen	O: Feed manger
D: Milch cow pen	J: Store	P: Feed passage
E:Young calf pen	K: Hay/Straw/Cart/Tractor shed	Q: Stanchion
F; Maternity pens.	L: Chaffing shed	R: Water tanks
G: Dry cow/Bullock pen	M: Silopits	All dimensions in metres.

Ref: Dairy Bovine Production (Thomas & Sastry)

Layout of Dairy farm for 100 cows

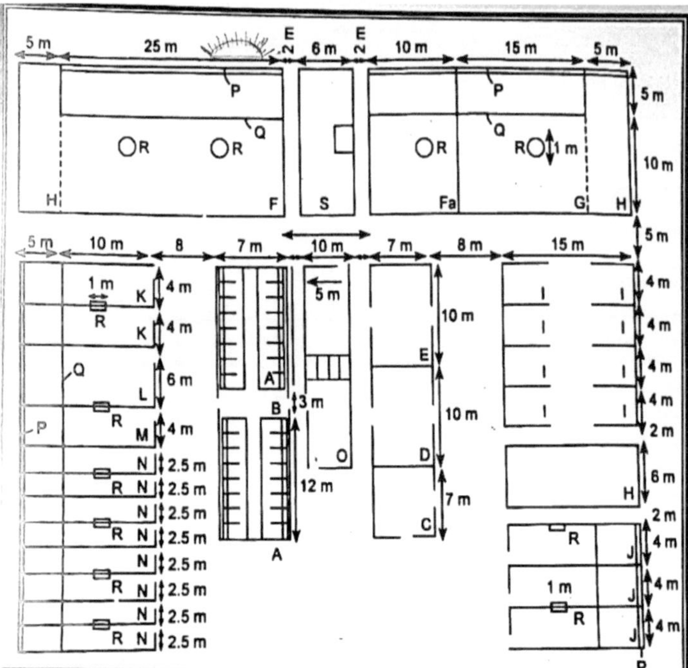

Fig. 12. Layout of Dairy farm for 100 cows

Labels

A. Milking barn	G. Dry cow shed	M. Pen for older heifers.
B. Milk house	H. Additional hay/straw store	N. Maternity pens
C. Office/record room	I. Trench silos	Handling yard
D. Al laboratory	J. Bull pens	P. Feed manger
E. Stores	K. Young calf pens	Q. Gutter or drain
F. Milch cowshed	L Older calf pen	R. Water tank
		S. Hay/straw/cart/tractor shed

Ref: Dairy Bovine Production (Thomas & Sastry)

Ancillary Structures in a Dairy Farm

Ancillary structures play a vital role in the effective functioning of a dairy farm. While the primary sheds are meant for animal housing, ancillary structures support various essential farm operations like feeding, milking, record-keeping, storage, and animal handling. Their inclusion in the farm layout ensures hygiene, efficiency, proper resource utilization, and animal welfare.These include:

- **Milk House:** Used for storing milk, weighing, and recording data. It must be well-ventilated and fly-proof. Floor space depends on daily milk output.(Table.11)
- **Shed for equipment and utensils:** Designed to house tools like milk cans, buckets, etc. It should be kept clean and protected from rodents.
- **Dry fodder shed:** Required for storing hay and straw. It must be moisture-proof and of appropriate size depending on the farm's feed requirements.
- **Store for concentrate feeds:** Used to store grains and feed supplements. It should be dry, damp-proof, and accessible.
- **Silage pit:** Required for silage-making and storage, especially for large farms. Silage pits must be constructed away from the main sheds and in dry, elevated areas.

Table 11. Floor Space Requirements for Milk House

Daily Milk Output (litres)	**Floor Space (m^2)**
Under 100	3.7 × 3.7
100 to 200	3.7 × 4.3
200 to 400	3.7 × 5.1
400 to 700	3.7 × 6.4
Above 700	Add 3.7 m^2 for every 700 L additional

Adjacent to the milk house, a shed is required for storing equipment and utensils such as milk cans and buckets. This structure must be kept dry, clean, and protected from rodents and pests. It generally requires a floor space of around 4 to 6 square meters, depending on the number of animals and the equipment used.

Dry fodder such as hay and straw must be stored in a dedicated shed that is moisture-proof, rainproof, and rodent-proof. Constructed on a raised platform to prevent dampness, this shed ensures the fodder remains suitable for animal consumption. Its size depends on the daily feed requirement of the farm but typically ranges between 5 to 10 square meters and should have sufficient capacity to store at least two to three days' worth of dry fodder. For concentrate feeds like grains, oil cakes, and mineral mixtures, a separate store is required. This storage area must be dry, damp-proof, and well-ventilated, with adequate arrangements such as bins or containers to prevent spoilage. The area for this store typically falls in the range of 6 to 10 square meters, proportionate to the herd size and daily feed usage.

In addition to these, a silage pit is essential for preserving green fodder, especially in large farms where continuous availability of quality fodder is necessary throughout the year. The silage pit should be located in an elevated, dry area to prevent water logging and ensure proper fermentation. Pits are usually constructed with dimensions of about 10 to 15 meters in length, 3 to 4 meters in width, and 2 to 2.5 meters in depth, depending on the size of the herd and the volume of green fodder to be stored.

Furthermore, a handling yard is necessary for routine operations such as vaccination, treatment, artificial insemination, and hoof trimming. It should be spacious, well-fenced, and designed to ensure animal movement without stress or injury. The yard must include facilities like a crush or race for safe animal handling and be constructed with a non-slippery floor and proper drainage to maintain hygiene. These ancillary structures, though often overlooked, are fundamental for the smooth and efficient management of a dairy farm.

References

Banerjee, G.C. (2014).A Textbook of Animal Husbandry. Oxford & IBH Publishing Co. Pvt. Ltd.

Banerjee, G.C. (2016). A Textbook of Animal Husbandry. Oxford & IBH Publishing.

Devendra, C., & Burns, M. (1983). Goat Production in the Tropics. Commonwealth Agricultural Bureau.

ICAR (Indian Council of Agricultural Research) (2022).Handbook of Animal Husbandry. 4th Edition, Directorate of Information and Publications of Agriculture, New Delhi.

Singh, B., Maurya, V.K. et al. (2020). “Performance of goats under different floor types.” J. Livestock Research, 10(2), 100–107.

Sastry, N.S.R., Thomas, C.K., & Singh, R.A. (2015).Livestock Production Management. 9th Ed., Kalyani Publishers.

Tamang, B. et al. (2021). “Evaluation of housing systems for goats in hilly regions.” Indian Journal of Animal Sciences, 91(3), 221–225.

Thomas, C.K. & Sastry, N.S.R. (2012).Dairy Bovine Production. Kalyani Publishers, Ludhiana

6

Management of Calves, Growing Heifers and Milch Animals

Utkarsh Kumar Tripathi, Anuradha Kumari, Manish Kumar and Ajeet Singh

Faculty of Veterinary and Animal Sciences, I. Ag.Sc., RGSC-Banaras Hindu University (BHU), Barkachha, Mirzapur, Uttar Pradesh-231001

India's livestock population has shown notable growth and trends according to the 2019 Livestock Census, which reported a total of 536.76 million livestock, reflecting a 4.8% increase since 2012. Of this, 95.78% (514.11 million) are in rural areas, while urban areas contribute 4.22% (22.65 million). The total bovine population, which includes cattle, buffalo, mithun, and yak, reached 303.76 million, marking a 1.3% increase. Specifically, the cattle population stood at 193.46 million, also showing a 1.3% rise; however, there was a decline of 6% in indigenous cattle populations, despite a significant increase of 29.3% in exotic/crossbred cattle, which totalled 51.36 million.

The buffalo population experienced a slight increase to 109.85 million (up by 1.1%), while the total number of milch animals (cows and buffaloes) rose to 125.75 million, representing a robust growth of 6%. Notably, the female indigenous cattle population increased by 10%, contrasting with a sharp decline in male buffaloes exceeding 42%. These trends reflect various economic influences such as mechanization and advancements in breeding practices that are shaping the dynamics of livestock demographics in India.

Table 1. Taxonomy and introduction of cattle and buffalo

Category	Cattle (Bos)	Buffalo (Bubalus)
Class	Mammalia	Mammalia
Subclass	Ungulata	Ungulata
Order	Artiodactyla	Artiodactyla
Suborder	Ruminantia	Ruminantia
Family	Bovidae	Bovidae
Subfamily	Bovinae	Bovinae

Tribe	Bovini	Bovini
Genus	Bos	Bubalus
Ancestor	Aurochs (*Bos primigenius*)	Wild Indian buffalo (*Bubalus arnee*)
Subspecies Origin	*Bos primigenius madicus* (Zebu cattle), *Bos primigenius primigenius* (Modern European cattle)	Water buffalo *(Bubalus bubalis bubalis),*swamp buffalo *(Bubalus bubalis kerebau)*
Chromosome	2N = 60 (Bos taurus and Bos indicus)	Water buffalo- 2N = 50 Swamp buffalo-2N = 48 African buffalo-2N = 52
Related Examples	Gaur Cattle (*Bos gaurus*), Banteng (*Bos banteng*), Kouprey (*Bos sauveli*), American Bison (*Bison bison*), European Bison (*Bison bonasus*), Yak (*Poephagusmutus*)	Anoa (*Bubalus depressicornis*), Mindoro Dwarf Buffalo (*Bubalus mindorensis*), African Buffalo (*Synceruscaffer*)
Domestication	Approximately 8000 to 10,000 years ago	Approximately 5000 to 6000 years ago
Groups	Indigenous (*Bos indicus*) Exotic (*Bos taurus*)	Bubalina (Asian buffalo) and Syncerina (African Buffalo)

In a dairy herd, adult cows or buffaloes produce milk, the primary income source for rural Indian dairymen. Calf rearing is essential for replacing older stock, as 20-30% of the herd is culled annually due to old age, breeding issues, udder infections, low production, vices, or chronic diseases. Herd replacement can be achieved through purchasing animals or raising farm-bred heifers. However, purchasing replacement animals often introduces lower-quality animals with defects or diseases, hindering consistent production improvement compared to raising one's own stock.

Calf care after birth

- **Clear Airways Immediately:** Remove membranes/mucus from mouth, nostrils, eyes, and ears for normal breathing.
- **Dry the Calf:** If the mother doesn't lick, rub the calf dry with clean towels/bedding to aid thermoregulation (especially important if mother has Johne's disease).
- **Ensure Rapid Breathing:** Calf should breathe within 30 seconds after birth (especially after difficult births); stimulate by tickling the nose with straw.

Immediate Actions Post-Birth

- Use a finger or instrument to pinch the nasal septum or insert a 20-gauge 1.5" needle at the acupuncture site in the center of the rostrum.

- Pour water over the head to trigger the grasping reflex in the calf.
- Induce artificial respiration by alternating pressing and relaxing the calf's limbs, holding it by the hind limbs if necessary.

Navel Cord Management

- Tie the navel cord 2.5 cm from the body and cut it about 1 cm below the ligature.
- Dip the navel in a 7% tincture of iodine and reapply antiseptics for 2-3 days.

Recording Birth Weight

- Document the birth weight of calves immediately after birth.

Colostrum: The First Milk

- Colostrum is the thick, sticky, yellow milk produced by a cow immediately after calving.
- Essential for both naturally suckled and weaned calves for the first 3-4 days.

Importance/Advantages of Colostrum

- **Rich in Nutrients:** Contains 4-5 times more protein than normal milk (around 17%), and is a rich source of minerals (copper, iron, magnesium, manganese), carotene, Vitamin A (10-15 times more), and B complex vitamins. These are crucial for early growth.
- **Laxative Properties:** Helps clear meconium (the calf's first faeces), a solid metabolic waste product.
- **Immunoglobulins:** Contains antibodies (IgM, IgG, IgA) that provide protection against infections.
- **Growth Factors:** Rich in insulin-like growth factors (IGF-I, II), prolactin, other growth factors, peptide cytokinines, lactoferrin, T_3, T_4, polyamine, enzymes, nucleotides, essential fatty acids (EFA), non-essential fatty acids (NEFA), and lactate.
- **Digestive Health:** Creates an acidic environment in the digestive system, which helps prevent diarrhoea.

Feeding Guidelines

- Feed colostrum preferably within 30 minutes of birth, but no later than 90 minutes, to maximize antibody absorption.
- Feed at 10% of the calf's body weight per day.
- Divide into 2-3 feedings per day, maintaining consistent intervals to avoid digestive issues.

Colostrum Quantity

- First feeding should be about 5% of the calf's body weight; total intake in the first 24 hours should be around 12-15% of body weight.
- Colostrum from second milking contains about 70% of antibodies found in the first.

ColostrumQuality

- Good-quality colostrum should be thick and creamy; avoid thin, watery colostrum.
- Collect colostrum from healthy cows, ensuring udder cleanliness before collection.
- Storage of Colostrum:
- Can be stored at room temperature for up to 1-2 days, refrigerated for up to 7 days, or frozen for up to one year.
- Thaw frozen colostrum using warm water baths or microwave on low power.

Colostrum Substitutes

In absence of colostrum, a mixture of eggs and oil can be used as a substitute.

Assistance for Weaker Calves

- Weaker calves may need help standing to nurse; use a nipple bottle or stomach tube if necessary.
- If no colostrum is available, supplement normal milk with 20 ml cod liver oil, 60 ml castor oil, and one egg yolk.

Alternatives

- If the mother's colostrum is unavailable, use colostrum from another cow.
- **Artificial Colostrum Recipe:**
  - 275 ml warm water
  - 1 egg (55g)
  - 3 ml castor oil
  - 10,000 IU Vitamin A
  - 525 ml warm milk
  - 80 mg Aureomycin
  - Give as a single meal.
- An intravenous injection of 50 ml of the dam's serum may also be recommended in the supervision of veterinarian.

Anaemia Treatment

- If signs of anaemia appear, administer an injection of a dextrin solution (150 mg) shortly after birth.

Meconium Passage

- Meconium should be passed within two hours of the first suckling.
- If not passed, administer an enema with 1 teaspoon of sodium bicarbonate in 1 liter of lukewarm water.

Identification

- Mark the calf with an appropriate method of identification before separating it from the dam. For example tattooing can be applied.

Housing

- Calves can be housed in individual calf boxes to prevent infection.

Calf Rearing Systems

- **Natural Suckling system:**
  - Common in rural India.
  - Calf initiates suckling or is allowed to suckle one udder completely.
  - Female calves often get more suckling opportunities.
  - Anti-suckling devices used to limit male calf suckling.
  - Challenges: Difficult if dam dies, risk of disease transmission, inconsistent milk intake, delayed ruminant development.
- **Weaning System:**
  - Calf is separated from the mother at birth or shortly after (e.g., 12 hours).
  - Advantages: Accurate milk production records, scientific calf feeding, no need for calf to stimulate milk let-down, prevention of teat injuries.

Feeding and Management Practices for Growing Calves (0-3 Months)

After the colostrum feeding period, calves are typically fed warm milk at approximately body temperature, amounting to $1/10^{th}$ of their body weight, divided into two equal feedings per day. Weaker calves may be benefited from three feedings daily, with the milk or milk replacer split into three equal portions.

Milk Feeding Techniques

- Calves can be taught to drink milk from a pail or nipple pail.

- To introduce pail feeding, bring the calf's nose to the milk, dip your fingers in the milk, and allow the calf to suckle your fingers. Gradually lower your fingers into the milk, repeating the process if the calf stops drinking.
- Nipple pails or bottles with nipples can be used, especially when time is limited.
- If calves are group-housed, tie them up during and for at least half an hour after feeding to prevent them from licking each other.
- After milk feeding, wipe the calf's mouth with a clean towel and rub some salt or mineral mixture in its mouth.

Nutritional Considerations

- Milk is deficient in iron (Fe), copper (Cu), manganese (Mn), zinc (Zn), and magnesium (Mg), so supplementation is necessary.
- Aim for a weight gain of 1 kg for every 1.39 kg of dry matter consumed in milk.
- Ensure clean water is always available.
- Consider milk replacers with higher protein levels and a protein-to-energy ratio that closely matches the calves' requirements. A good calf starter should have about 20% crude protein.

Solid Feed Introduction

- Solid feeds, including forages and concentrates, should be introduced from one week of age and be freely available before weaning.
- Introduce a high-quality calf starter early, around 3 days of age, to stimulate rumen development.
- A good calf starter includes coarsely ground corn, crushed or rolled oats, soybean meal, dried molasses, limestone, dicalcium phosphate, and trace mineralized salt.
- The calf starter should be coarse in texture, with few fine particles and a relatively high fibre concentration (12-15% NDF)

Milk feeding schedule of calf

The milk feeding schedule for calves is crucial in their early development. For the first four days (0-4), regardless of body weight up to 30 kg, calves should receive colostrum, specifically at a rate of $1/10^{th}$ of their body weight to ensure they receive vital nutrients and antibodies. Subsequently, from days 5 to 30, calves weighing up to 30 kg should be transitioned to milk at the same rate of $1/10^{th}$ of their body weight. As calves grow, those in the 31-60 kg weight range, from days 31 to 90, should have their milk intake adjusted to 1/20th of their body weight to accommodate their increasing consumption of solid feeds.

Phase 1: Colostrum (Days 0-5)

- **Weight:** Up to 30 kg
- **Colostrum:** 1/10th of body weight (provides essential antibodies and nutrients)

Phase 2: Milk and Introduction to Calf Starter (Days 6-30)

- **Weight:** Up to 30 kg
- **Milk:** 1/10th of body weight (gradual transition from colostrum to milk)
- **Calf Starter:** Introduce calf starter (quantity not specified, start with small amounts to encourage consumption)

Phase 3: Reduced Whole Milk, Introduction of Skimmed Milk, Increased Calf Starter (Days 31-60)

- **Weight:** 25-30 kg
- **Milk:** 1/15th of body weight (reduce whole milk as rumen develops)
- **Skimmed Milk:** 1/25th of body weight (introduce skimmed milk as a cost-effective alternative)
- **Calf Starter:** 125g (increase calf starter to promote rumen development)

Phase 4: Reduced Whole Milk, Increased Skimmed Milk, Higher Calf Starter Intake (Days 61-90)

- **Weight:** 31-70 kg
- **Milk:** 1/25th of body weight (further reduction of whole milk)
- **Skimmed Milk:** 1/15th of body weight (increase skimmed milk to meet nutritional needs)
- **Calf Starter:** 250g (continue increasing calf starter to encourage solid feed consumption)

Calf Starter

- **Definition:** A specialized concentrate mixture designed to meet the nutritional needs of pre-ruminant calves.
- **Timing:** Start feeding as early as 3 days old, but no later than 10-12 days.
- **Quality:** High-quality, palatable calf starter is essential.
- **Nutrient Content:**
    - 75-80% Total Digestible Nutrients (TDN)
    - 16-20% Crude Protein (CP)

- **Introduction:**
    - Rub starter on the calf's tongue and muzzle to familiarize them with the taste.
    - Place a small amount in the milk pail after feeding.
- **Ingredients:** Can vary based on regional feed availability.

Example Composition (per 100 kg)

- **Crushed Maize:** 35 kg
- **Crushed Barley:** 15 kg
- **Groundnut Cake:** 30 kg
- **Wheat Bran:** 10 kg
- **Fish Meal:** 7 kg
- **Mineral Mixture:** 2 kg
- **Common Salt:** 1 kg
- **Antibiotic:** 100 gm
- **Vitamin blend (A, B_2, D_3):** 15 gm

Milk Replacers and Calf Starters

Milk replacers and calf starters are essential components in modern calf nutrition, particularly for dairy operations aiming for optimal growth, early rumen development, and cost-effective rearing. Drawing from authoritative sources, the following elaborates on their composition, quality criteria, and feeding protocols.

Milk Replacers

Milk replacers are powdered feeds formulated to serve as substitutes for whole milk after the initial colostrum feeding period. They are especially valuable for large-scale operations or where whole milk is economically or logistically impractical.

High-quality milk replacers generally contain

- **Milk-derived proteins:** At least 50% spray-dried skim milk powder or whey products (such as whey protein concentrate, dried whey, or delactosed whey).
- **Fat:** 10–22% (ideally 10–15%) of stabilized, high-quality animal or vegetable fat (e.g., lard, tallow, or vegetable oils), often homogenized with the milk base before drying.
- **Carbohydrates:** Small amounts of glucose, soybean flour, or cereal flour may be added, but high starch or fiber content should be avoided (less than 0.5% crude fiber, ideally below 0.1%).

- **Vitamins and Minerals:** Supplementation with vitamins A, D, E, B-complex, and essential minerals (such as copper, zinc, manganese, cobalt, iron, and iodine).
- **Other Additives:** Lecithin for emulsification, and sometimes antibiotic feed additives (though the use of antibiotics is increasingly scrutinized due to resistance concerns).

Quality Criteria

- Contain a minimum of 18–22% crude protein, with no poorly digested (non-milk) protein sources.
- Have a fat content of 10–22%, with higher fat (up to 20%) recommended in cold weather.
- Be low in fiber and starch to avoid digestive issues.
- Be free-flowing as a powder, suitable for automatic feeders, and readily dispersible in warm water.
- Be mixed precisely according to manufacturer's instructions to avoid digestive upsets caused by incorrect dilution.

Mixing and Feeding Guidelines

- **Mixing Ratio:** Commonly, 1 part milk replacer powder to 8 parts warm water by weight (e.g., 150 g powder per 1 liter water for calves).
- **Water Temperature:** Mix with water below 60°C to prevent protein denaturation.
- **Feeding Frequency:** Feed twice daily in small quantities; over-frequent or excessive feeding can cause digestive disturbances like abomasal bloat.
- **Daily Intake:** Calves typically require around 500 g of milk solids per day.

Williams and Knodt (1949) Milk Replacer Composition

- 50 parts spray-dried skimmed milk powder
- 10 parts dried whey
- 40 parts non-milk sources (energy and nutrients)
- Resulted in calf growth and condition similar to whole milk feeding

Ohio Researchers' Suggested Milk Replacer Formula

- 70 kg dried skim milk
- 18 kg dried whey

- 2 kg lecithin (helps fat digestion)
- 10 kg animal fat (provides energy)
- 1.7 kg dicalcium phosphate (mineral source)
- Trace minerals included: copper ($CuSO_4$), iron ($FeSO_4$), manganese ($MnSO_4$), cobalt ($CoSO_4$)
- Antibiotics added (to prevent infections)

Key Points

- High-quality milk proteins from skim milk and whey are essential
- Added fats improve energy supply and growth
- Minerals and vitamins support healthy development
- Properly formulated milk replacers can match whole milk in supporting calf growth and health

Calf Starters

Calf starter is a high-quality, balanced concentrate mixture introduced to calves alongside milk or milk replacer, usually from the second week of life. Its primary role is to promote early rumen development, enabling the calf to efficiently utilize fibrous feeds and transition off milk by around 90 days of age.

Nutritional Composition

- **Crude Protein (CP):** About 22% CP or 18% digestible crude protein (DCP).
- **Total Digestible Nutrients (TDN):** 72–75% to support rapid growth.
- **Ingredients:** Typically includes ground maize, groundnut cake, fish meal, wheat bran, rice bran, and a mineral-vitamin premix.
- **Minerals and Vitamins:** Inclusion of a mineral mixture and vitamins is essential for balanced growth.

Feeding Protocol

Calf starters can be offered free-choice until the calf consumes approximately 1 to 1½ kg of the starter daily. Once this intake level is reached, the amount of starter feed may be gradually restricted. Typically, calves achieve this stage between 2½ to 3 months of age. Milk feeding can be safely discontinued at the earliest once the calf consistently consumes this quantity of starter feed.

- **Introduction:** Begin offering calf starter from 2 weeks of age, alongside milk or milk replacer.

- **Transition:** Gradually increase the amount; by 90 days, milk or milk replacer can be stopped, and the calf should be consuming 1 kg of calf starter per day.
- **Roughage:** Good quality hay and green fodder should also be introduced gradually to support rumen development.

Roughage feeding to calves

- **What:** Provide high-quality legume roughage (like lucerne) to promote rumen development.
- **When:** Start around 2 weeks old. Avoid pasture/silage before 3 months due to high moisture content.
- **Why:** Aids rumen function, but too much moisture hinders growth.
- **Water:** Always provide fresh water.
- **Transition:** After 3 months, stop milk and replace starter with a standard ration of good fodder and hay.

Weaning

Weaning refers to the process of making the calf independent from its mother. In an early weaning system, the calf is not allowed to suckle directly from the cow. Instead, the cow is fully milked, and the calf is fed a measured amount of whole milk, skim milk, or milk replacer.

Advantages of Early Weaning

1. **Controlled Feeding:** Calves receive an adequate and precisely measured quantity of milk, preventing both overfeeding and underfeeding.
2. **Accurate Milk Production Records:** The exact amount of milk produced by the cow is known, which is crucial for maintaining accurate production records that aid in breeding decisions and herd improvement. It also helps in providing the cow with appropriate nutrition.
3. **Milk Conservation:** Feeding calves with milk replacers saves whole milk for human consumption, improving farm efficiency.
4. **Cleaner Milk Production:** Since calves do not suckle directly, the risk of milk contamination is reduced, ensuring cleaner milk.
5. **Reduced Teat Injury and Mastitis Risk:** Preventing suckling protects the cow's teats from injuries and lowers the chances of infections entering through wounds, thereby reducing the incidence of mastitis.

Management of Heifer

Heifer

- A heifer is a female bovine from one year of age until her first calving.
- **Importance:** Heifers are the future of the herd; their care significantly impacts future milk production and reproductive success. Although they are actively growing but unproductive making them subject of neglection for their feeding and care.

Rearing Stages

1. Weaning to First Service

a) **Early Post-Weaning Period (4-8/10 months):**

- **Concentrates:** Feed 1-2 kg of high-quality concentrates daily. If possible, include some animal-origin protein.
- **Protein Content:** 14-16% protein if legume roughage is fed; 16-18% if not.
- **Minerals:** Supplement daily with 20-25g each of mineral mixture and common salt.
- **Roughage:** Feed good quality roughage, preferably mixed (legume/non-legume) and succulent. If only non-leguminous roughage is available, add about 25g of steam-sterilized bone meal or another calcium supplement daily.
- Liquid milk feeding is limited to 10% of body weight, with a max of 5-6 liters daily, for 6-8 weeks.

b) **Late Post-Weaning Period (After 8-10 months):**

- **Roughage Focus:** Heifers can adjust to a high-roughage, low-concentrate diet. The rumen is now fully developed.
- **Concentrates:** If good-quality leguminous or mixed roughage is available, concentrate feeding may not be necessary. If roughage is all cereal-based and of poor quality, feed 1-2 kg of concentrates.
- **Minerals:** Supplement daily with 25-35g each of mineral mixture and common salt. For both phases, the goal is to ensure proper growth and development, setting the stage for future productivity

2. First Service to Calving

Breedable heifers should be managed to achieve optimal body size (60% of mature weight) for breeding, as growth impacts puberty more than age. Avoid stunted growth to ensure timely calving and high milk production. Breed and species influence ideal weight/age targets.

- **Goal:** Maximize growth and development at the lowest cost to achieve early maturity and optimal breeding weight.
- **Growth Rate:** Influenced by breed, genetics, environment, and body weight. Nutrient needs vary based on desired growth rate. Underfeeding delays maturity.
- **Housing Systems:**
  - **Outdoor (Grazing):** Rotate grazing paddocks (no more than 4-6 days in one area). Supplement with concentrates and minerals if forage quality is poor. Provide shade and clean water.
  - **Indoor:** Provide shade (east-west orientation, ~3m high), forage, clean water, and minerals. Offer ~2.5-3 square meters of covered floor space per animal. Supplement of good quality of roughages/ crop byproducts with high-quality forage and concentrate.

Monitoring & Management

- Weigh heifers regularly and compare to breed-specific growth curves to identify slow growers.
- Groom heifers regularly and handle them gently to encourage a calm temperament.
- Introduce heifers to the milking herd routine (parlour, feeding during milking) 1-2 weeks before calving.
- Deworm heifers every 4-6 months.

Care and Management of Pregnant heifer and Down-Calvers

Gestation in cows lasts approximately 280-285 days (9 month 9 days approximately). Proper care during pregnancy is crucial for a healthy calf and mother, as well as better milk production.

Care of Pregnant Cows

- Avoid strenuous activities like long walks, running, or being chased or frightened.
- Ensure the ration contains adequate nutrients. From the sixth month of pregnancy, provide an extra 0.5 to 1 kg of concentrate mixture, depending on the animal's condition, to support the developing foetus and build body reserves for lactation. Heifers require additional feed for their own growth.
- **"Steaming up,"** or increasing feed in preparation for calving, is common. Start with 1.5 kg of concentrate per day two months before calving, gradually increasing to 4-5 kg per day until 2-3 days before parturition. This practice increases milk yield, lengthens the lactation period, and can improve butterfat percentage.

Table 2. Rationing strategy of cows in different stages of pregnancy

Stage of Pregnancy	Indigenous Cows	Crossbred Cows	Additional Notes
First 7 Months (3/4 of period)	Maintenance + 1.25 kg/day concentrate mixture	Maintenance + 1.50 kg/day concentrate mixture	Provide light, easily digestible feed. Ensure free access to good pasture grazing
Last Quarter (1/4 of period)	Increase concentrate by 1.25 kg/day (over maintenance)	Increase concentrate by 1.75 kg/day (over maintenance)	Steaming up is recommended, providing extra concentrate for foetal growth, mammary gland development, and lactation preparation. Consider supplementary feeding heifers, especially if pasture is low quality, to ensure a weight gain of 0.6kg/day to reach target weights.
General Recommendations	Maintain adequate nutrition from joining to calving to prevent growth restrictions. Aim for a body condition score of 3.5 for heifers. Ensure sufficient quantities of Vitamins A and E. Regional mineral formulas can provide necessary minerals. Trace mineral levels are often dictated by the soil		

Care of Down-Calvers (cows close to calving)

Separate pregnant animals from the herd and move them to a clean, advanced maternity pen about 1-2 weeks before the expected calving date.Advanced pregnant animals should be housed in a calving pen or calving box about two weeks before their expected calving date, with the number of pens sufficient to accommodate approximately 10% of the cows on the farm. Each pen should be thoroughly cleaned, disinfected, and provided with at least one foot of bedding. The pen should include a covered area measuring 3x4 meters and an open paddock of 4x5 meters. As calving approaches, animals should remain in the pen day and night, with monitoring every 2-3 hours for signs of calving. Additionally, they should be given a laxative ration to aid in the process.

Care during third trimester

The good care and management practices given to pregnant animal will give good calf and also high milk yield during the successive lactation.

- Extra concentrate mix of 1.25 to 1.75 kgs should be provided for pregnant animal as pregnancy allowance. Feed good quality of leguminous fodder.
- The animal should not be too lean and too fat in condition.
- Provide clean drinking water and protection from thermal stress.
- Do not allow them to mix with other animals that have aborted or that are suffering from or carriers of diseases like brucellosis.
- Sometime udder edema occurs before calving. This can be avoided by moderate exercise for a half an hour, two to three times per day. Do not tire them by making long distances especially on uneven surfaces.

- Do not allow them to fight with other animals and take care that they are not chased by dogs and other animals.
- Avoid slippery conditions, which causes the animal to fall receiving fractures, dislocation etc.
- If accurate breeding records are available, calculate the expected date of calving. Separate it one or 2 weeks before and shifted to individual parturition pens.
- Their pens are thoroughly cleaned and fresh bedding may be provided.
- Feed one kg extra concentrate during last 8 weeks of gestation.
- During the last few weeks of pregnancy there is a tendency of prolapse of vagina which may be caused by constipation, mineral deficiency and debility. Balanced and laxative rations should be fed to maintain the normal tone of the reproductive tract. Feed laxative about 3 - 5 days before and after calving (Wheat bran 3 kgs + 0.5 gms of Groundnut cake + 100 gms of mineral mixture of salt).
- Symptoms of delivery may be observed i.e. swelling of external genitalia, swelling of udder, usually majority of animals will deliver without any help.
- Take care, of the animal, if at all any abortion.
- **Guarding Against Milk Fever**
- In advanced pregnancy stage high yielding & first calves are susceptible to Milk fever. To avoid it, provide enough minerals especially calcium by bone meal in daily diet. Give large doses of Vitamin D about a week period to calving.
- **Avoid Milking :** Prior to parturition this is likely to delay parturition by few hours

Signs of approaching parturition

- Cow will leave the herd and seek isolation. Loss of appetite and distress.
- Distention of teat and udder, considerable milk appears in the udder and there may be dripping of milk.
- Relaxation of pelvic ligament one day before calving, the ligament on the sides of the tail head is loosened so that hollows appear on either side of the backbone and the tail head is raised and the quarters are dropped.
- The vulva become enlarged and flabby. Animal will be restless and will pace about often trying to kick or scratch the flank region.
- The parturition process has three stages a. preparatory stage (uterine contraction and dilatation of cervix) b. active expulsive stage c. expulsion of foetal membrane.

- Cow will deliver the calf within 12 hours after commencement of first stage and lapse in this vaginal examination of assistance is required.
- **Expulsion of placenta / after birth:** The placenta is discharged within 5-6 hrs. After calving in normal case while if not discharged within 12 hrs get the help of veterinarian and treat as per requirement.
- When placenta expelled, prevent cow from eating. The placenta should be properly disposed off by burying in ground.

Care at the time of calving

- Pregnant cow may be transferred to calving pen 1 to 2 weeks before the expected calving date.
- The number of calving pens required on a farm depends on the number of breedable cows and heifer, generally 5 per cent of the number.
- Ample amount of drinking water, laxative feed and generous supply of bedding may be provided.
- The calving pen should be scrupulously cleaned and sterilized before brining in the cow.
- After removal of calf, milk animal it will help in removal of placenta. Placenta is normally expelled within 2 to 6 hours after calving. If placenta fails to be expelled with 12 hours it is considered retained placenta. In case of retained placenta veterinarian should be called for its removal.

Calving pen

- It is a individual loose box or stall used for calving, which should be 3 m x 4 m size (12 m^2) and well ventilated. Sufficient lighting is essential.
- It provided better protection to the cow and calf and avoid disturbances from other cows.
- Attendant quarters may be nearer to calving pen to monitor calving process during night time.

Stages of parturition

1. **First Stage:** This preparatory stage involves regular contractions of the myometrium (12-24 contractions per hour), loosening of the cotyledonary attachments of the placenta, and dilation of the cervix. Signs of distress may include bellowing, kicking, and arching of the back. This stage lasts between 6 to 24 hours, generally shorter in older cows, and ends with the breaking of the water bag and appearance of the calf's hooves.

2. **Second Stage:** Characterized by increased contractions of abdominal muscles and myometrial contractions, this stage involves the expulsion of the fetus through the birth canal. Ferguson's reflex—stimulated by pressure from the fetus against the cervix—promotes further contractions. The cow typically lies down to give birth, while buffalo may remain half-standing. The second stage lasts between 0.5 to 4 hours; if complications arise, professional help should be sought.
3. **Third Stage:** Following foetal expulsion, myometrial contractions continue to separate and expel foetal membranes, which may take 4-8 hours. If placenta retention occurs beyond 12-18 hours postpartum, it can be removed manually. Cows should not be allowed to eat the placenta due to potential indigestion. Successful parturition is marked by the calf standing and membranes expelled. Uterine involution begins rapidly after calving due to oxytocin release and is typically complete within 23-35 days for cows and 15-60 days for buffaloes, influenced by factors such as calving month and parity.

Table Signs of Approaching Calving

Sign	Description	Approximate Timing
Isolation	Cow leaves herd, seeks solitude	Hours to 1 day before calving
Loss of Appetite & Distress	Reduced feed intake, withdrawal, restlessness	Hours before calving
Udder & Teat Distention	Enlarged udder, milk secretion or dripping	1–3 days before calving
Pelvic Ligament Relaxation	Tail-head raised, hollows near tail, dropped hindquarters	About 1 day before calving
Vulva Enlargement	Swollen, flabby vulva	1 day before calving
Labour Pain	Restlessness, pacing, kicking, flank scratching	Immediately before and during calving

Care and Management of Milking Animals

To achieve optimal milk production during any lactation, it is essential to provide milch animals with proper care and management. This includes ensuring a balanced diet, adequate hydration, and appropriate living conditions. Effective management practices in feeding, housing, disease control, exercise, milking, and breeding all play a significant role in maintaining the health and productivity of dairy cows. Proper attention to these areas not only supports high milk yield but also ensures the overall well-being of the animals, enhancing both their performance and longevity in the herd.

Watering

Adequate clean and fresh water should be provided at all time. An adult dry cow typically drinks 30-32 liters of water daily, in addition to 4 liters of water for every liter of milk produced. Water consumption increases as air temperature rises, especially in warmer climates like India, to help regulate body temperature and support milk production.

Housing

Good housing is essential for protecting animals from extreme weather conditions such as heat, rain, and strong winds. It should include proper drainage, ventilation, and exposure to natural sunlight to maintain a healthy environment. Loose housing systems with shelters during the hottest part of the day provide the animals with a comfortable and safe space. This setup also allows cows to get maximum exercise, promoting better physical health and reducing stress. Additionally, proper housing ensures better air circulation and minimizes the risk of disease transmission.

Cleaning and grooming

- Grooming or brushing of body hair coat is an important daily farm operation to make and keep the animals' body clean and fit. For grooming blunted type brush is used, if not available then use coarse rope made from paddy straw, coconut coir or dried grass. In India, grooming generally practiced before milking along with washing to improve the clean milk production.
- Daily brushing will remove loose hair and dirt from the coat
- Grooming will also keep the animal hide pliable
- Regulargroominghelpstokeepskin clean, helpsforbloodcirculation

Control of Disease in Milch Animals

Effective disease control is crucial for maintaining the health and productivity of milch animals. Key components include:

- **Detection of Diseases:** Regular observation for signs of illness (e.g., changes in appetite, behaviour, and milk production) is essential for early detection and prompt action, especially for common diseases like mastitis and hoof infections.
- **Timely Vaccination:** Vaccinating against diseases like Foot and Mouth Disease, Brucellosis, and Bovine Tuberculosis is essential to prevent outbreaks. Follow veterinary vaccination schedules to ensure immunity and reduce transmission.

- **Deworming:** Regular deworming is necessary to control internal parasites, which can affect nutrition and milk yield.
- **Treatment:** Immediate treatment of ill animals with prescribed medications, such as antibiotics or anti-inflammatory drugs, is crucial for preventing complications and ensuring recovery.
- **Sanitation:** Regular cleaning and disinfection of animal houses, feeding equipment, and water sources are vital for disease prevention. Proper waste management and maintaining dry, clean, and well-ventilated spaces reduce infection risks.

Vices

Common vices should be properly detected and care should be taken. Eg. Kicking, licking, suckling etc

Exercise

Exercise in dairy animals is an important daily activity that enhances the health and productivity of the animals. In loose housing management of dairy animals exercise is not much important because the animals move freely. However, in conventional housing management where animals are tied throughout the day and night, exercise is compulsory. Exercise should be recommended daily atleast once during morning just after milking for ½ to 1 hour. Exercise reduces the chance of deposition of fat, leg problems, bloat, calving difficulties etc

Milking

Milking is the most important daily routine activity in dairy farm. Milking is done commonly twice in most dairy farm morning and evening; however, if milk productivity of animal and labour availability is more then go for three times milking per day. Milking should be conducted gently, quietly, quickly, cleanly, completely and at regular intervals. Milk let down/milk ejection from secretory alveolar cell and small duct into cistern and major duct starts within ½-1 min of udder massaging under the influence of hormone oxytocin, so milking should be completed within 5-7 minutes, beyond that the effect of oxytocin is reduced.

Milking may be hand milking or machine milking. In hand milking, full hand milking or fisting is considered as best method of milking and commonly followed in high yielder dairy cows and buffaloes. In full hand milking the whole teat is hold in the fist, grasping the teat with all five fingers and pressing it against the palm. Then the teat base is closed by thumb and fore finger so that milk is trapped in the teat sinus and teat is compressed and thus forcing the milk out. The teat is alternatively compressed and relaxed in quick succession and removes milk quicker than stripping. Stripping is another method of hand milking that is followed in low yielder animals and animals with smaller teat size. Stripping consists firm holding of the teat between thumb and for finger and then drawing it down the entire length of teat and at the same time pressing it simultaneously, so that milk flow down in a stream. However, sometimes milker follow a defective method of milking called as knuckling, where thumb is bent against the teat during milking. Machine milking is practiced in dairy farm with high yielder exotic or crossbred cows and high yielder buffaloes. For machine milking minimum number of high yielder cows or buffaloes should be at least 10. The working principle of machine milking is under negative pressure created by the vacuum pump and the vacuum pressure for cows 280-320 mm Hg and for buffaloes 320-360 mm Hg. Although, machine milking is time and labour saving but improper cleaning leads to mastitis and reduces milk quality. Thus, regular cleaning of milking machine will maintain the cow health as well as ilk quality.

Breeding

Cow should be bred at 60 days after date of parturition which helps good reproductive health of cow. The shorter the interval between calving, the more efficient the animal is as a milk producer

Feeding of milch cows

- A cow's good genotype is ineffective without proper feeding and management to maximize economical milk production. Feed costs make up about 60% of milk production expenses. Dairy cows require feed for maintenance, foetal development, and milk production. In the first lactation, allowances for growth should also be considered.
- To optimize feeding, the total requirements of the cow should be calculated based on common feeding standards. As the prices of grains and oil cakes rise, feeding more roughage can reduce feed costs significantly. Research shows that good-quality leguminous fodder, fed *ad libitum*, can eliminate the need for extra concentrates for cows producing up to 10 kg of milk and buffaloes producing up to 7 kg. Above these amounts, cows need 1 kg of concentrate for every 2.5 kg of milk produced, while buffaloes require 1 kg per 2 kg of milk.
- Roughage should be fed after milking, and concentrates before or during milking to avoid dust. Dry matter from roughage should be 1-2% of the cow's live weight. In loose housing systems, roughage is often fed to groups rather than individually. Feeding should align with calculated requirements based on feeding standards.

- In general, the dry matter (DM) from roughage should not exceed 2 % of cow's live weight nor should it be less than 1%. Water should be provided to drink at will or at frequent intervals. But this is only a thumb rule. Actual feeding may be done on the basis of the requirements calculated as per feeding standards.
- Under loose housing system and self feeding practices, it is not possible to fed roughages to animals individually. Group feeding of roughage combined with individual feeding of concentrates at the time of milking is a generally accepted practice.

Feeding during early lactation

The recently calved high producing cow is unable to eat enough feed to support her milk production. This means that the cow should have enough reserve to store nutrient to be drawn to tide over the period of heavy demand in early lactation, during which period the cow loses weight. For the same one week or two before freshening, cows with high milk production potential should receive increasing amounts of concentrates to "challenge" them, preparing their digestive system for early lactation and ensuring adequate nutrients for optimal milk production.

Challenge feeding

- Challenge feeding means the cow with high milk production potential are to be fed with increased quantity of concentrate to 'challenge' them to produce to the maximum level of milk production
- This starts two weeks before expected date of calving.
- This challenge feeding will condition her digestive system for the increased amount of concentrate and provide enough nutrients to initiate lactation on a higher plane
- Two weeks before the expected date of calving start feeding 500 g of concentrate mixture
- The quantity should be increased daily by 300-400 g until the cow is consuming 500-1000g concentrate for every 100 kg body weight

Feeding during mid and late lactation

- The nutrient deficit period of early lactation is followed by a relatively stable period during which the cow can consume enough feed to meet the various demands for nutrients and the body weight of the cow remains more or less stable

- During this period the cow may be fed a well balanced ration of good quality fodder and concentrate according to the milk yield and fat percentage of milk
- During the late lactation, intake ability of the cow exceeds nutrient needs. This is the time when the cow starts needing extra allowance for the growing foetus.
- This is also the period when the cow can readily replenish the already depleted body reserve and gain weight very fast
- From 7½ month to 10 months of lactation, cow may be fed 1-2 kg concentrate feed in addition to their nutrient requirement for maintenance and milk production to replenish the condition lost in early lactation

Guidelines to Feed High Yielding Dairy Cows

1. Forage Proportion and Quality

- At least 20–25% of the cow's total dry matter intake should come from forages to ensure proper rumen function and maintain milk fat levels.
- Forages provided to high yielders must be of superior quality, harvested at the right stage of maturity for optimal nutrient content. Ideally, 30–50% of the roughage should be leguminous crops, which are richer in protein and more digestible.

2. Forage Processing

- Avoid excessive processing or reducing forages to very small pieces, except when dealing with extremely coarse fodder. Maintaining adequate particle size supports rumen health and effective fiber digestion.

3. Feeding Concentrates

- When large quantities of concentrates are needed, feed them immediately after roughage or, preferably, mixed together. This practice helps buffer rumen pH and reduces the risk of digestive upsets like acidosis.
- The roughage-to-concentrate ratio should be managed carefully, typically 50:50 in early lactation, shifting to 60:40 or 75:25 as lactation progresses and milk yield declines.

4. Nutrient Density and Digestibility

- Maintain a density × digestibility value (a measure of energy and nutrient availability) of at least 35 to ensure the ration meets the high demands of milk production.
- The fiber content should not fall below 17–18% crude fiber, 21–22% acid detergent fiber (ADF), or 28% neutral detergent fiber (NDF) to prevent metabolic disorders and maintain milk fat.

5. Feeding Frequency and Rumen Health

- Feed high-yielding cows a minimum of four times per day, with each feeding including both concentrate and roughage. Frequent feeding helps maintain stable rumen fermentation, supports higher dry matter intake, and minimizes fluctuations in rumen pH.
- Fresh feed should be available after each milking, and cows should have access to feed for at least 22 hours per day.
- Avoid drastic changes in the ration to prevent digestive disturbances.

6. Additional Nutritional Considerations

- Include essential minerals and vitamins, such as magnesium, selenium, and vitamin E, particularly during early lactation and the transition period to support immune function and reproductive health.
- Limit concentrate per feeding to 2.5–3.5 kg to avoid digestive upsets.
- If feeding oilseeds or fats, limit total added fat to about 4% of the concentrate mixture or 0.5 kg per day to boost energy without depressing fiber digestion.

Period	Concentrate Allowance	Forage Allowance	Notes/Adjustment Criteria
Last 2 weeks before calving	Start with 500 g/day; increase by 300–400 g daily until cow is eating 500–1000 g per 13 kg body weight	Good quality forage ad lib	Gradually adapt rumen to higher concentrate levels; ensures smooth transition to lactation diet35.
First 2 weeks of lactation	Increase by 500 g/day to near free-choice level	Good quality forage ad lib	Rapidly increase concentrate to support rising milk yield and minimize negative energy balance35.
Second week to peak yield	Free choice (ad libitum)	Good quality forage ad lib	Allow cow to consume as much concentrate as she will eat safely; supports peak milk production35.
From peak yield (test day) onward	Adjust according to milk yield (e.g., 1 kg concentrate per 2.5 kg milk produced)	Good quality forage ad lib	Use monthly milk tests to fine-tune concentrate; prevents over- or underfeeding235.
Remaining lactation	Continue adjusting concentrate to match production and body condition	Good quality forage ad lib	Maintain cow health and body reserves as milk yield declines235.
All periods	As above; always provide adequate green and dry fodder	Green and dry fodder ad lib	Forage is the foundation of the diet; ensures rumen health and fiber intake

Feeding the Dry Cow

Feeding management during the dry period-the non-lactating phase, typically lasting 6–8 weeks before calving-is critical for the health of both the cow and her unborn calf, as well as for optimal milk production in the next lactation. Despite the absence of milk yield during this phase, neglecting nutrition can result in poor future performance, health issues, and impaired calf development.

Aim of dry cow feeding

1. Maintain the cow's health and body condition.
2. Build sufficient nutrient reserves for the next lactation.
3. Support proper fetal growth, especially in the last trimester.
4. Ensure high-quality colostrum production.
5. Core Principles from Authentic Sources

Body condition management

Cows should be dried off at a body condition score (BCS) of 3.0 (on a 1–5 scale) and maintained at this level until calving. Cows in poor condition (BCS <2.5) need additional concentrates to regain condition, while over-conditioned cows (BCS >3.5) are at higher risk for metabolic disorders.

Nutrient requirements

The nutrient needs of the cow increase in the last 3 weeks before calving due to rapid fetal growth and declining dry matter intake. Diets should be balanced for energy, protein (12–14% crude protein), and minerals, with close attention to controlling energy intake to avoid over-conditioning.

Forage and concentrate use

The diet should be based on high-fiber, moderate-energy forages (e.g., grass hay, straw) with low potassium to reduce the risk of milk fever. Only thin cows or those in poor condition should receive extra concentrates.

Disease prevention

Consistent feed intake and controlled energy diets help prevent metabolic disorders such as ketosis, fatty liver, displaced abomasum, and milk fever.

Condition at dry-off	**Forage allowance**	**Concentrate allowance**	**Additional notes**
Good condition (BCS 3.0)	Bulky, moderate-energy forage (ad libitum)	None or minimal	Maintain condition; avoid over-conditioning; keep rumen function optimal36.
Poor condition (BCS <2.5)	High-quality forage	1–2 kg/ day of concentrates	Support weight gain; aim for 20–40 kg gain before calving depending on cow size36.
Over-conditioned (BCS >3.5)	Low-energy, high-fiber forage	None	Avoid further weight gain; monitor for metabolic risks23.
Last 3 weeks before calving	Increase nutrient density; supplement with pre-calver minerals	As needed, based on condition	Support fetal growth and colostrum; manage declining intake

7

Management of Sheep, Goat and Swine

Utkarsh Kumar Tripathi, Anuradha Kumari
Santosh Marandi and Kaustubh Kishor Saraf

Faculty of Veterinary and Animal Sciences, I. Ag.Sc., RGSC-Banaras Hindu University (BHU), Barkachha, Mirzapur, Uttar Pradesh-231001

The total goat population in India reached 148.88 million in 2019, reflecting a growth of 10.14% compared to the 2012 Livestock Census. Goats contributed approximately 27.8% of the total livestock population, showcasing their vital role in India's rural economy. Among major states, West Bengal exhibited the highest growth in goat population, increasing from 11.51 million in 2012 to 16.28 million in 2019, a remarkable rise of 41.49%. In contrast, Rajasthan, which traditionally has a large goat population, saw a decline of 3.81%, with its numbers dropping from 21.67 million to 20.84 million during the same period. Uttar Pradesh also experienced a decrease of 7.09%, with its goat population falling from 15.59 million to 14.48 million. These trends underline the shifting dynamics of goat rearing across states and its significance in ensuring sustainable livelihoods, particularly through the production of milk, meat, and skins that support rural economies and food security across India.

The total sheep population in India reached 74.26 million in 2019, marking a 14.13% increase compared to the 2012 Livestock Census. Sheep accounted for approximately 13.8% of the total livestock population in the country. Among the states, Telangana recorded the highest growth, with its sheep population rising from 12.8 million in 2012 to 19.1 million in 2019, reflecting a remarkable increase of 48.51%. Andhra Pradesh followed with a population of 17.6 million, up by 30%, while Karnataka saw a growth of 15.31%, reaching 11.1 million sheep. These figures highlight the significant role of sheep in supporting the rural economy, providing meat, wool, and livelihoods to millions across India.

Table 1. Introduction of sheep and goat and their common information

Feature	**Goat *(Capra aegagrus hircus)***	**Sheep *(Ovis aries)***
Classification	Genus: Capra, Species: C. hircus	Genus: Ovis, Species: O. aries
Chromosomes	60	54
Body Size	Generally smaller	Relatively larger
Coat Type	Short, coarse hair	Thick wool (except hair breeds)
Tail Position	Stands upright	Hangs down, often docked
Horns	Straight and upright	Curved, if present; many breeds are polled
Foraging Behavior	Browsers; prefer leaves, twigs, and shrubs	Grazers; prefer grasses and broadleaf plants
Behavior	Curious and independent; often stray from the herd	Flocking instinct; prefer to stay together
Odor	Strong odor in males during breeding season	Less pronounced odor in males
Estrus Cycle	Longer estrus cycle	Shorter estrous cycle
Nutritional Needs	Require copper and other trace minerals	Sensitive to copper; excess can cause toxicity

Care and Management of Goats

Reproduction in the Male (Goat/Buck)

Breeding Age and Semen Quality

- Male goats (bucks) generally have exceptionally high libido (sex drive).
- A well-grown buck-kid may be bred to a doe as early as six months old.
    - However, at this age, semen volume and motility are low.
    - Semen quality (volume and sperm motility) is significantly higher in adult bucks compared to 5-month-old males.
    - Semen from 4–5-month-old kids is generally inadequate for breeding, despite satisfactory libido.
- By 18-24 months of age, bucks can serve 25-30 does per breeding season.
- Fully mature bucks can serve 50-60 does in a breeding season.

Factors Affecting Puberty and Semen Quality

- Age at puberty varies with breed.
- Climatic factors, nutrition, and male reproductive hormones can influence the onset of puberty.
- Bucks tend to have higher ejaculate volume but lower sperm density compared to other livestock.

Reproduction in the Female Goat

Estrous Cycle and Duration

- The estrous (heat) cycle in goats generally recurs every 18–21 days unless pregnancy occurs.
- Duration of estrus (standing heat) is usually 24–48 hours, but can be as short as 18 hours in certain breeds, such as the Indian Sirohi.
- Ovulation typically occurs toward the end of standing estrus, between 9–72 hours after onset.

Gestation and Kidding

- Gestation period ranges from 145 to 153 days (about 5 months).
- Indian goats commonly kid twice a year, and sometimes three times in two years.

Seasonality and Breeding Patterns

- Goats are seasonally polyestrous, with reproductive activity induced by shortening day length (more pronounced in temperate regions).
- In India, due to minimal variation in day length, estrous cycles can recur throughout the year, but the highest incidence is from May to October.
- Breeding can occur in other months, but winter heats are often shorter and less effective.

Puberty and Factors Affecting Onset

- Sexual desire in female kids can be observed as early as 14 weeks.
- Age at puberty varies by breed, body weight, climate, nutrition, and presence of a male.
- Barbari does reach puberty at 9–10 months, while Anglo-Nubians take about 15 months.
- For optimal results, breeding is recommended at 14–18 months, with first kidding around two years of age.

Estrus Detection and Breeding Management

- Some does are shy breeders and may show only subtle signs of heat, such as slight redness of the genital opening.
- Bringing a buck near the females each morning during breeding season helps identify those in heat.
- Behavioural signs of oestrous include bleating, tail wagging, seeking out males, vulvar swelling, and mucus discharge.

Special Observations

- Occasionally, does may accept the male even when pregnant; the reason for this species-specific behaviour is unclear, and it is not known if ovulation accompanies such oestrus.

Breeding Recommendations for Milch Breeds

- For dairy goats, one kidding per year is preferable to ensure better kid health and prolonged lactation.

Care and Management of Pregnant Does

Pregnancy should be confirmed, and does should be dried off 6-8 weeks before kidding to support foetal development and future milk yield. Provide a laxative, leguminous, and nutritious diet with concentrate mixture, adjusting for body weight. Avoid sudden feed changes and ensure adequate protein, minerals, and vitamins. Water should be available at all times. Shift pregnant does to a separate kidding pen one week before kidding, providing clean, soft bedding. Reduce concentrate mixture in the last week of gestation to 100 gm/day/doe. Signs of approaching kidding include an enlarged udder, sunken look at the tail and hip, heavy breathing, restlessness, and vaginal discharge.

Management of Does After Kidding

Clean and disinfect the kidding area, disposing of the placenta immediately; consult a vet if it's not expelled within 30 minutes to 4 hours. Disinfect the doe's hindquarters with antiseptic solution. Provide warm bran mash with oat meal, ginger, minerals, salt, and jaggery. Protect against cold weather. Provide laxative feed for two days, then resume normal feeding. Limit concentrate to 300 gm/day for small breeds and 500 gm/day for large breeds. Does may return to heat after a month, but breeding should wait until 40 days post-kidding.

Management of Lactating Does

Provide high-quality green fodder and a balanced concentrate mixture (240 gm per liter of milk). Wash the udder with antiseptic solution, milk completely with dry hands, and clip hair from hindquarters. Ensure clean milk production and groom daily to maintain cleanliness, docility, and health.

Care and Management of Kids

Immediately after birth, clear the kid's nose of mucus to prevent suffocation and clean its body with a dry towel. Cut the navel cord 3 cm from the body and dip it in tincture of iodine. Provide artificial respiration if needed. Ensure colostrum intake within make it 30 minutes of kidding, feeding 10% of body weight, and avoiding over or underfeeding. Start a starter ration after three weeks, gradually increasing it. Discontinue milk after 3-4 months as the kid consumes hay, greens, and concentrate. Identify kids by tattooing and record weights. Deworm at 4 weeks of age.

Care and Management of Young Stock

Keep young stock in a well-ventilated place, protecting them from other animals. Provide clean drinking water and adequate exercise. Feed good-quality green fodder and concentrate to develop the rumen early. Deworm every three months, wean at three months, and cull or castrate male kids not needed for breeding.

Care and Management of Bucks

Select a purebred buck with good breeding ability. House bucks in individual stalls (2.5m x 2m) with food and water, avoiding keeping two together to prevent fighting. Ensure proper nutrition during the breeding season with green pasture, concentrate, minerals, and vitamins. Only introduce the buck to the female pen for breeding. Provide ample exercise and daily grooming to keep them clean and docile. Trim hooves periodically to prevent lameness. Generally, one mature buck is sufficient for 25-30 does, but for a high conception rate, follow a 10:1 ratio.

Feeding management of goat

Most village goats do not get enough grain or good fodder, leading to low milk production. However, when given proper care and feeding, dairy goats can produce much more milk. They should be treated with the same attention as other dairy animals.

Feeding Tips

Goats are picky eaters and care about cleanliness. They prefer fresh and clean food and will not eat anything dirty or smelly. They also dislike wet or old fodder. To keep their feed clean, it's best to use hay racks or hang feed in bundles. Portable hay-racks that have two sides are best for feeding goats in stalls. It's better to give them small amounts of food at a time, because they tend to waste large quantities by stepping on it. Goats are ruminants, and they especially enjoy legume fodders. They don't like fodders such as sorghum, maize, silage, or straw. They also dislike hay made from forest grasses, even if cut early, but they really enjoy hay made from legume crops.

Feeding Guide for Goats

Birth to 3 days

- Feed colostrum ad libitum for essential antibodies and nutrients.

3 days to 3 weeks

- Whole milk or milk replacer: 450 ml daily.
- Provide water and salt ad libitum.

3 weeks to 4 months

- Gradually minimize milk feeding and completely stop by 4 months.
- Whole milk: 450 ml daily up to 8 weeks.
- Creep feed: 450 g daily.
- Lucerne hay and water with salt: ad libitum.

4 months to freshening

- Concentrate mixture with 15-16% crude protein: 450 g daily.
- Lucerne hay and water with salt: ad libitum.

Dry pregnant does

- Concentrate mixture with 15% crude protein: 400-500 g daily.
- Lucerne hay and water with salt: ad libitum.

Milking does

- **Concentrate mixture:** 350 g for each litre of milk produced.
- **Trace mineralized salt:** 1% of concentrate mixture.
- **Molasses:** 5-7% of concentrate mixture.

Bucks

- **Pasture only:** ad libitum.
- **Concentrate mixture:** 400 g daily during non-breeding season; adjust during breeding season.

Table 2. Some green roughage for goat feeding

Type of Fodder	Common Varieties
Green Roughages	Lucerne (*Medicago sativa* L.), Berseem (*Trifolium alexandrinum*), Napier grass (*Pennisetum purpureum*), Green arhar (*Cajanus cajan*), Cowpea (*Vigna sinesis*), Soybean (*Glycine max*), Cabbage leaves, Cauliflower leaves, Shaftal, Senji, Methi, Various shrubs and weeds, Babul leaves (*Acacia arabica*), Neem leaves (*Azadirachta indica*), Pipal leaves (*Ficus religiosa*)
Dry Roughages	Straws of arhar, Urid (*Phaseolus mungo*), Mung (*Phaseolus aureus*), Gram (*Cicer arietinum*), Dry leaves of trees, Lucerne hay, Berseem hay

Table 3. Nutrient Requirement for the goats

Nutrient Requirement	Details
Maintenance Ration	Goats have higher maintenance needs than cattle due to a higher basal metabolic rate.
	Additional feed of 25-30% is recommended for maintenance.
	Maintenance requirement: 0.09% DCP and 0.09% TDN.
	Goats can consume 6.5-11% of their body weight in dry matter, compared to 2.5-3% for cattle and sheep.
Production Ration	***Starch equivalent (SE)** is a measure of the energy value of animal feed, specifically its ability to produce fat compared to pure starch. It's expressed as the amount of starch (in kg) that would produce the same amount of fat as 100 kg of the feed being evaluated. **For Example:** SE compares the fat-producing potential of a feed to that of pure starch. For example, if a feed has a starch equivalent of 70, it means 100 kg of that feed can produce the same amount of fat as 70 kg of pure starch.
	For 1 liter of milk with 3.0% fat: Requires 43g DCP and 200g SE.
	For 1 liter of milk with 4.5% fat: Requires 60g DCP and 285g SE.
	A 50 kg goat producing 2 liters of milk with 4% fat needs 400g concentrate and 5kg of berseem or lucerne hay.
	Ration should contain 12-15% protein, depending on hay & milk protein content.

Table 4. Common concentrate mixture, their ingredients and percentage composition

Concentrate Mixture Ingredients	Percentage Composition
Wheat Bran, Maize Grain, Linseed Cake	Wheat Bran: 25%
	Maize Grain: 50%
	Linseed Cake: 25%
Maize Grain, Barley, Mustard Cake, Gram Husk	Maize Grain: 40%
	Barley: 20%
	Mustard Cake: 20%
	Gram Husk: 20%
Wheat Bran, Barley Grain, Groundnut Cake	Wheat Bran: 25%
	Barley Grain: 50%
	Groundnut Cake: 25%
Gram Grain, Wheat Bran	Gram Grain: 67%
	Wheat Bran: 33%
Mineral Mixture: 2% & Salt: 2%	

Care and management of sheep

Reproduction in Sheep

Age of Maturity

- Sheep generally reach full growth at about 2 years of age, though this can range from 18 months to 3 years depending on breed and locality.
- Ewes are typically mated between 9 to 14 months, but it is preferable to wait until they have attained proper growth to ensure healthy lambs.
- Rams can be used for breeding as early as one year, but it is recommended to use rams for mating from 2 years up to about 7 years of age for best results.

Mating Season and Estrous Cycle

Sheep are seasonally polyestrous, meaning ewes have multiple cycles during certain seasons.

In India, the main breeding seasons are

- **Summer:** March–April
- **Autumn:** June–July
- **Post-monsoon:** September–October
- Fertility is generally highest in autumn in the plains and in summer in hilly areas.
- The estrous (heat) cycle averages 17 days, with a range of 14 to 19 days.
- The heat period (estrous) lasts about 24 to 36 hours, with an average of 27–30 hours.
- Ewes show few external signs of estrus except standing to be mounted, so heat detection is usually done with the help of a teaser or ram.

Ovulation and Gestation

- Ovulation occurs in mid to late estrous, typically about 12–30 hours after the onset of heat.
- Conception is most likely if breeding occurs late in the heat period.
- The average gestation period for sheep is approximately 147 days, or about 5 months.
- The gestation period can vary slightly depending on breed, age, and number of fetuses.
- Ewes usually come into heat about 2 months after lambing.
- Only about 2–3% of rams are used for mating with ewes in a flock.
- Environmental factors such as photoperiod (day length), breed, and nutrition influence the reproductive cycle and seasonality.
- Practices like "flushing"-feeding extra grain or lush pasture 2–3 weeks before breeding-can increase ovulation rates and the incidence of twinning

Preparing the Ewe and Ram for Breeding

Flushing

Flushing is the practice of feeding extra grain or lush pasture to ewes two to three weeks before the breeding season. This nutritional boost increases the number of ova shed from the ovary, leading to higher chances of twins and an increase in lamb crop by 10–20%. Typically, about 250 grams of grain is given daily per ewe during this period.

Tagging

Tagging involves shearing wool and removing dirt from the dock (the area around the tail). This prevents wool or tags from obstructing the ram during mating, ensuring more successful breeding. Rams are also trimmed around the sheath for the same reason.

Eyeing

Eyeing refers to clipping excess wool around the eyes to prevent wool blindness, especially in breeds prone to this condition. Regular eyeing helps maintain good vision and overall health.

Ringing

Before breeding season, rams should be shorn completely or at least clipped around the neck and belly, particularly near the penis. This process, called ringing, makes mating easier and more effective.

Marking the Ram

To identify which ewes have been bred, rams are marked with paint or chalk on their brisket. When a ram mounts a ewe, the paint leaves a mark on her rump, allowing for easy identification of bred ewes and helping to detect any sterile rams if ewes are repeat breeders.

Additional Preparations

Evaluate and manage the body condition score (BCS) of ewes, aiming for a BCS of 2.5 to 3 at breeding. Thin ewes should be supplemented with grain or quality pasture, while over-conditioned animals should not be flushed.

Check and trim hooves to prevent lameness and hoof diseases.

Treat for internal parasites before breeding season to ensure optimal health.

Ensure all vaccinations are up to date, particularly those recommended for breeding

Care and Management of Pregnant Ewes

Pregnant ewes should be kept in separate enclosures and provided with leguminous fodder *ad libitum*, along with a balanced concentrate mixture daily. Avoid long-distance grazing and ensure a well-ventilated, clean house with access to clean drinking water. Dipping should be avoided in the advanced stages of pregnancy, and regular deworming is

Care and Management of Newborn Lambs

Immediately clear mucus from the lamb's mouth and nose, and clean its body with a clean cloth to dry it. Allow the ewe to lick the lamb for cleaning and bonding. Cut the naval cord 3 cm from the body with sterilized scissors and treat it with tincture of iodine. Help the lamb reach the ewe's teat to suckle, weigh the lamb, and feed colostrum as soon as possible. Identify the lamb according to age, sex, and breed, and weigh it periodically for better management.

Care and Management of Ewes After Lambing

Clean the posterior part of the ewe, including genitalia, with lukewarm water. Dispose of the placenta immediately. Provide a laxative diet to avoid constipation, along with fresh forage and a small amount of concentrate ration. Ensure *ad libitum* access to water throughout the day and night.

Care and Management of Rams

Select purebred male animals from recognized sources, providing regular exercise to keep them active and fertile. Change the breeding male every 2 years to avoid inbreeding depression. One adult male can successfully breed 25-30 ewes. Follow criss-crossing for better breeding, provide a well-balanced ration to maximize fertility, and house rams individually to avoid fighting. Vaccination and deworming should be done periodically, and wool should be clipped from the sheath and penis region to facilitate breeding.

Dipping

Dipping is crucial for eradicating ectoparasites, preventing sheep scab, warding off sheep blowflies, and obtaining clean wool. It is typically performed before winter shearing or after autumn shearing using either a hand bath (for small flocks) or a swim bath (for large flocks). Precautions include avoiding dipping pregnant ewes, providing drinking water before dipping, avoiding dipping sick or wounded animals and young lambs, and avoiding dipping rams during the breeding season. Dipping should be done on sunny days, avoiding rainy days. Hold sheep in a drain pen after dipping to recover the dip solution.

Table 5. Feeding strategy for the Sheep at different stages

Stage	Nutritional Needs & Practices
General Grazing	Sheep have small muzzles and split upper lips for grazing on pastures.
	In India, they primarily rely on roughages, reducing production costs.
	They graze on crop stubbles, weeds, and grasses on fallow and rangelands.
	Supplement diet with cultivated fodders, grains, and oil cakes during critical production periods.

Flushing (Pre-Breeding)	Increase ewe nutrition 2-3 weeks before breeding to boost body weight and encourage early heat.
	Promotes uniform lamb crops and higher lambing rates.
	Flushing Rations:
	• Good mixed pasture of legumes and grasses.
	• Grass pastures + 250g grains + 450g oil cakes.
	• Legume hay (full fed) + 100g wheat bran + 150-200g grain.
	• Green fodder (10% of body weight) + 150-200g concentrate per head per day.
Early & Mid-Pregnancy	Good feeding is essential for healthy lambs.
	Inadequate feeding results in weak or dead lambs.
	Can be met by grazing on stubbles and harvested fields, supplemented with 100g of oil cakes per head per day.
Late Pregnancy	Most critical period nutritionally.
	Graze on crop aftermaths, wild grasses, and weeds, supplemented with 5 kg of available green fodder per head per day.
	Last Month: Rapid fetal growth requires energy.
	Option 1: Molasses or grains (barley, maize, oats) at 225g per head per day + 7 kg green fodder per head per day.
	Option 2: 600g quality legume hay or 300g concentrate (12-14% DCP, 65-70% TDN) during the last 45 days.
Lambing Time	Reduce grain allowance; provide good quality dry roughage.
	Gradually increase ewe's ration post-partum, dividing it into 6-7 daily doses.
	Include bulky, laxative feedstuffs.
	Wheat bran and barley/oats/maize (1:1) mixture is excellent.
	Provide slightly warm water after lambing.
	Formulate creep feeders with lamb 'starter' ration (16 parts groundnut cake, 84 parts barley/maize grain) and available fodder.
Lactating Ewes	Supplement ration to maintain milk production for lamb growth.
	Good pasture can meet requirements.
	Supplementary Feeding:
	Replace 50% of pasture with 450g good hay, 1.4 kg silage, or 250g grain.
	10 kg cultivated green fodder per head.
	400g concentrate mixture or 800g quality legume hay per day for 75 days after lambing, plus 8 hours of grazing.
Rams	Need additional nutrients during breeding season.
	Over-fat rams need to be thinned through feed reduction and exercise.
	Can graze with ewes on the same ration.
	If separate feeding: 300-500g concentrate mixture (3 parts oats/barley, 1 part maize, 1 part wheat) per day

Concentrate Mixture for Pregnant Ewes

The nutritional needs of pregnant ewes are critical for the health of both the mother and her developing lambs. Below are several recommended grain mixtures designed to meet these needs, with varying compositions:

- **Mixture 1:** This blend consists of 60% whole barley, corn, or wheat, 30% whole oats, and 10% wheat bran.
- **Mixture 2:** Comprising 75% whole barley, corn, or wheat and 25% dried beet pulp, this mixture is designed to provide additional fiber and energy.
- **Mixture 3:** This formulation includes 75% whole barley, corn, or wheat and 25% whole oats, offering a high-energy diet.
- **Mixture 4:** With a composition of 50% whole barley, corn, or wheat and 50% whole oats, this mixture provides a balanced approach to energy and fiber intake.

Lamb feeding (important point of consideration)

- From approximately 2 weeks of age, lambs should have unrestricted access to creep feed.
- If pasture is limited, lambs should be creep-fed for 1–2 months until adequate forages are available.
- In cases where pasture is not available until lambs are 3–4 months old, they can be finished in a dry lot.
- The grain used for feeding should initially be ground coarse or rolled; whole grains can be introduced as the feeding period progresses.
- Fresh, clean grain should be gradually introduced to the lambs' diet in small amounts.
- The amount of grain should be increased gradually until the lambs are on full feed.
- Feeding lambs in a dry lot, combined with early weaning at 2–3 months of age, is becoming more popular nowadays.
- A complete diet consists of hay, grain, and vitamin-mineral supplements, which can be ground and mixed or pressed into pellets (5-10mm size).
- Lambs typically reach market weight in 3.5–4 months when fed this way.

In case of foster feeding for orphan lambs

- Orphaned lambs, extras, triplets, or those from poor-milking ewes can be raised on milk replacers to improve productivity.
- Lambs should receive 10%–20% of their body weight in colostrum, divided into multiple feedings within 18–24 hours of birth.

- If ewe colostrum is unavailable, use available fresh cow colostrum.
- Milk replacers designed for lambs are available and contain approximately 30% fat, 25% protein, and a high level of antibiotic.
- Consider injecting orphaned lambs with vitamins A, D, and E and selenium.
- Ewe milk replacers are preferable; good quality calf replacers may be used.
- Mix milk replacers carefully to ensure proper suspension.
- Feed small quantities frequently to reduce bloat and/or diarrhoea.
- Feed milk replacers at 10%–20% of the lamb's body weight, divided into 4–6 feedings/day during the first week, reducing to twice a day by 3–4 weeks of age.
- Use multiple-nipple pails or containers.
- Cold milk replacer can be used by older lambs who nurse more often.
- By 9–10 days of age, provide water in addition to milk if a creep ration is offered.
- Lambs can be weaned abruptly at 4–5 weeks of age if creep feed and water intake is reasonable.

Table 6. Feeding strategy for the lambs at different stages up to market age

Feeding Stage	Key Points
Suckling Lambs	Depend heavily on mother's milk for nutrition.
	Grazing on good pasture is essential for sustaining milk production.
	Supplementation with grains and oil cakes may be necessary if pasture quality is poor.
Early-Weaned and Orphan Lambs	Usually weaned at three months; orphaned lambs need special care.
	Grains should be cracked for lambs up to six weeks; thereafter, whole grains can be used.
	High-quality legume hay or pellets should be provided.
Rations for Creep Feeders	Recommended mixtures include:
	1. Maize 40%, oats 30%, barley 30%, plus lucerne hay.
	2. Oats 20%, maize 40%, barley 20%, groundnut cake 20%.
	3. Maize 25%, oats 40%, wheat bran 20%, groundnut cake 15%.
From Weaning to Market	Feeding methods vary based on economic and climatic conditions.
	Utilize grazing lands and supplement with quality fodder as needed.
	Average concentrate intake ranges from 225g to 450g depending on grazing conditions.

General Management of Sheep and Goat

Familiarize yourself with the animal's body parts to properly handle them for shearing, administering medicine, castration, etc.

Housing

- Sheep and goats have similar building requirements, with additional structures needed for goats raised for milk 8.
- They don't need elaborate buildings; simple, dry, and clean shelters suffice.
- Flocks can be penned in the open during good weather, with temporary shelters used in monsoon and winter.
- Sheep can be economically reared under a ranch system.

Buildings for Sheep and Goat Farms

- **General Flock Shed:** Houses about 60 adult ewes or nannies, with the shed being 3 meters high and having a moorum or brick floor. Floors should be elevated in low-lying and heavy rainfall areas and made of strong wood in temperate Himalayan regions.
- **Shed for Ram or Buck:** Rams or bucks are housed individually or in wooden-partitioned stalls.
- **Lambing or Kidding Shed:** Maternity pens for pregnant ewes or nannies, made draught-free. In colder regions, warming devices should be used to protect newborns.
- **Lamb or Kid Shed:** Houses lambs or kids from weaning to maturity, at about 75 animals per shed8. Partitions can separate unweaned, weaned, and nearing-maturity lambs, or separate sheds can be constructed for each category
- **Sick Animal Shed:** Constructed away from other sheds, sized approximately 3 x 2 or 3 m, to isolate ailing animals.
- **Shearing Hall:** A specialized room for shearing sheep, required only on wool-producing farms. It should be well-lit with large glass windows, have smooth cement floors, and include fenced lots on either side for collecting sheep before and after shearing. The entrance features a 1.5 m wide passage.

Breeding Season in Sheep and Goats

The breeding season in sheep and goats is influenced by factors such as breed, photoperiod, and environmental conditions. Tropical sheep breeds typically cycle throughout the year, whereas temperate breeds are seasonal breeders, cycling with declining daylight hours in autumn (August to November) and sometimes extending to February. Rams of indigenous breeds produce good-quality semen year-round under proper management, while temperate rams may experience reduced semen quality during hot and humid seasons if not protected. Breeding should be planned to align lambing with favorable environmental conditions, such as mild weather and abundant vegetation, to ensure higher lamb survival. For indigenous breeds, the major breeding season occurs from July to October, often after the onset of the monsoon, while a minor season is observed in March-April when grazing on stubble is supplemented with pods like Acacia and Prosopis.

Flushing and Nutritional Management

Flushing-feeding ewes extra concentrate 3–4 weeks before the breeding season-can enhance ovulation rates, increasing the likelihood of twins or triplets. However, in tropical flocks or extensive systems, twinning is less desirable due to management challenges. Flushing proves most effective for ewes in poor body condition by improving estrous incidence, ovulation rates, and reducing early embryonic mortality through better fetal membrane integrity. This can be achieved by supplementing 250 g of concentrate daily or 500 g of good-quality legume hay. Proper timing of breeding also ensures lambing coincides with optimal feed availability and favourable weather conditions, which are critical for lamb survival and overall productivity.

Selection of Ewes and Rams for Breeding

When selecting ewes for breeding, desirable traits include a long, low-set body, roomy hindquarters, well-formed pliable udders, active foraging habits, and strong mothering instincts. Ewes with poor milking capacity, jaw deformities, blind teats, or meaty udders should be excluded from the breeding program. Ideally, the body weight of a ewe at breeding should be less than the adult weight typical for her breed. The ram's libido is also crucial for successful breeding; poor libido can stem from inadequate nutrition, extreme heat stress, or health issues. In situations where multiple rams are present in a flock, larger and older rams tend to dominate younger ones. Aggression among unfamiliar rams can lead to serious injuries unless they are housed in confined spaces that prevent charging.

System of Mating in Sheep and Goats

Hand Mating

In the hand mating system, a teaser ram or buck is used to detect ewes or does in heat, which are then mated individually with a selected ram or buck. This method allows for careful monitoring of breeding, as each female is mated one at a time, ensuring that no more than three ewes or does are bred per day. Hand mating not only helps predict the timing of lambing but also reduces the risk of injuries among animals and is particularly beneficial when mating older males with younger females. Additionally, this approach enhances the breeding efficiency of the male, allowing for a greater number of females to be bred in a shorter time frame.

Pen Mating and Flock Mating

In pen mating, ewes or does are grouped into batches of 20-25 and males are introduced only during the night while being separated during the day. This arrangement allows ewes to rest and be properly fed without disturbance from males during grazing hours. Conversely, flock mating involves allowing rams to run with females continuously throughout the day and night. While this method is commonly practiced by farmers, it has significant drawbacks; rams may exhaust their body reserves chasing ewes, leading to reduced body condition. Additionally, if a ram develops a preference for a particular ewe or doe, others may remain unbred, resulting in empty females and lower fertility rates. Furthermore, rams may serve multiple times overnight, which can lead to suboptimal sperm delivery to later-served females, increasing the chances of non-conception.

Care and Management of Pigs

The total pig population in India stood at 9.06 million in 2019, marking a decline of 12.03% compared to the 2012 Livestock Census. Pigs contributed approximately 1.7% to the total livestock population. Among the categories, exotic/crossbred pigs saw a significant decrease of 22.76%, with their population dropping from 2.46 million in 2012 to 1.90 million in 2019. Indigenous/non-descript pigs also declined by 8.66%, from 7.84 million to 7.16 million during the same period. The northeastern states dominate pig rearing in India, with Assam leading at 2.10 million pigs (a growth of 28.30%), followed by Jharkhand with 1.28 million (a rise of 32.69%), and Meghalaya with 0.71 million (an increase of 29.99%). However, some states like West Bengal experienced a decline, with its pig population dropping by 16.63% to 0.54 million in 2019. These trends highlight the shifting dynamics of pig farming across India, which remains vital for meat production and rural livelihoods, particularly in the northeastern region where pigs are integral to local culture and economy.

Table 7. Taxonomical classification of Pigs

Kingdom	Animalia
Phylum	Chordata
Class	Mammalia
Order	Artiodactyla
Family	Suidae
Genus	Sus
Species	*Domesticus* and *vittatus*
Scientific name of domestic pig	*Sus domesticus*
Scientific name of wild pig	*Sus scrofa*

Advantages of Pig Farming

- **Efficient Feed Conversion:** Pigs are highly efficient at converting feed into meat, requiring only 3-3.5 kg of feed to produce 1 kg of pork.
- **Versatile Feed Usage:** They can convert inedible feed, forage, grain by-products, meat by-products, garbage, garden waste, vegetable market waste, and dairy by-products into valuable meat.
- **High Productivity:** Swine are prolific producers, with an average litter size of 8 and the potential for 2 or more litters per year. Gilts can be bred as early as 8 months old.
- **Rapid Growth and Short Generation Interval:** Pigs grow quickly and are ready for market by 6 months of age. Their short generation intervals allow for rapid genetic improvement.
- **Multiple Products:** Pigs provide pork, lard, bristles, and manure, and are also valuable in medical research.
- **High Dressing Percentage:** They have a high dressing percentage of 65-80% because the skin is also used for consumption, and a higher percentage of edible meat due to smaller bones.
- **Nutritional Value:** Pork contains less cholesterol and has a high energy value compared to other meats. Pork fat is a good source of polyunsaturated fatty acids.
- **Adaptable to Various Farming Systems:** Pigs thrive in both intensified and diversified agriculture, offering profit in commercial and mixed farming systems.
- **Low Initial Investment and Quick Returns:** Starting a swine enterprise requires a relatively small initial investment and offers quick returns due to the pig's short life span. They adapt to self-feeding, reducing labor costs.

Disadvantages of Pig Farming

- **Religious and Social Taboos:** Pork consumption is restricted by religious taboos, such as in Islam, and may be seen as less dignified.
- **High Grain Costs:** Production costs can be high when grains are the primary energy source, as pigs compete directly with humans for food. Raising pigs on non-cereal diets can be limiting.
- **Susceptibility to Diseases and Parasites:** Pigs are highly susceptible to diseases and parasites, requiring careful management to maintain a disease-free herd. Well-established farms to supply stocks are limited in number.
- **Sensitivity to Unfavorable Conditions:** They are sensitive to unfavorable rations and require skilled management to thrive. There are limitations to daily weight gain as compared to developed countries.
- **Skilled Farrowing Management:** Skilled attention is required at the time of farrowing to minimize piglet loss

Care and Management of Boars

Select purebred boars with pedigree records, signs of masculinity, and both testicles in the scrotum. Keep boars in a separate, clean, dry, and well-ventilated pen, avoiding both overfeeding and underfeeding to maintain breeding capacity. Boars should not be used for breeding before eight months of age and limited to one service per day during the breeding season. Trimming the boar's feet prevents lameness, and clipping tusks avoids injury to sows and attendants. Boars should be culled after 5-6 years of age, and slippery floors should be avoided in breeding areas.

Care and Management of Pregnant Sows

Pregnant sows should be kept in separate pens at least two weeks before farrowing in a clean, dry, and non-slippery sty. Avoid rough treatment and provide a well-balanced ration according to body weight, ensuring constant access to clean drinking water. Provide bedding material like chopped straw, and consider calcium supplements 10 days before farrowing. Withdraw feed just before farrowing. Signs of approaching farrowing include restlessness, abdominal contraction, enlarged udder, twitching of the tail, frequent drinking/urination/defecation, expulsion of blood-stained fluid from the vulva, a slight increase in body temperature, and increased grunting.

Care and Management of Newborn Piglets

Remove piglets immediately after farrowing, clearing mucus from the mouth and nostrils and transferring them to a creep area for warmth. Dispose of the placenta quickly. Assist piglets in getting colostrum as soon as possible. Cut the navel cord with sterilized scissors, leaving 2.5 cm, and disinfect with tincture of iodine. Clip needle teeth to prevent irritation to the sow while suckling. Allow piglets to nurse soon after clipping the needle teeth. Check the udder for agalactia, mastitis, and abnormal teats. Ear notch piglets for identification. Administer an iron injection or paint the sow's udder to prevent piglet anemia. Raise orphan piglets with foster mothers or milk replacers. Wean piglets at 7-8 weeks of age.

Care and Management of Sows After Farrowing

Clean the sow, especially the posterior part, with lukewarm water and allow her to lick her piglets. Feed the sow 12 hours after farrowing. Provide clean and fresh water on a free-choice basis. Spray the sow's udder with a saturated solution of ferrous sulfate to prevent piglet anemia. Provide a balanced ration to lactating sows for sufficient milk production. If the litter size is large, cross-foster piglets of similar sizes.

Care and Management of Growers

Group grower piglets by weight rather than age and provide optimum floor space to minimize tail biting. Deworm growers for internal parasites two weeks after weaning. Transition growers from creep feed to a grower protein ration according to age, and vaccinate them for swine fever according to the vaccination schedule

Swine Feeding

Feeding Management of Sows and Litters

Sows should receive a nutrient-rich gestation diet to ensure healthy piglets, maintaining good body condition without being too fat or thin. Thin sows tend to produce smaller piglets with lower survival rates. After farrowing, sows should quickly return to full feed. If constipation occurs, adding 5%–10% wheat bran or dried beet pulp can help, or chemical laxatives like potassium chloride or magnesium sulphate can be included.

Newly farrowed piglets should be monitored to ensure nursing. If necessary, oxytocin can stimulate milk flow. Weak piglets may require artificial milk if the sow's milk is slow to come in. To prevent nutritional anemia, piglets should receive an iron injection before three days of age. Transfers from larger to smaller litters should occur within the first 24 hours after birth, and a palatable starter diet should be introduced if weaning occurs after three weeks.

Pre-starter feed serves as an initial diet for piglets weaned before three weeks, acting as a milk substitute. However, as many farmers wean pigs at around five weeks, they might skip this type of feed, administering it in small quantities until the piglets reach 5–7 kg due to the high cost. This feed is heavily supplemented with proteins, minerals, vitamins, and antibiotics, with the latter included at high levels to prevent diarrhoea.

Starter, or creep feed, can be introduced around the second week after birth while the piglets are still suckling. It is particularly useful when the sow's milk production is insufficient, or when she has a large litter to support. Starter feed can be used after a prestarter feed or as a general creep feed for piglets weaned at five weeks. Starter feed is less fortified, and therefore less expensive, than prestarter feed.

Both pre-starter and starter feeds must be highly palatable. Ingredients like cane sugar, beet sugar, and molasses can improve the feed's palatability. Young pigs tend to accept diets containing cane sugar, stabilized lard, or beef tallow better than those containing only maize-soy. Dried skim milk is another highly palatable ingredient for pig starter diets.

Feeding Management of Weanling Pigs

Pigs weaned at 3–4 weeks benefit from a complex starter diet for 1–2 weeks, which includes dried whey, lactose, and high lysine levels. Some producers implement early weaning programs at 10–16 days with enhanced nutritional management and higher lysine and lactose levels.

Growing-finishing pigs thrive on full-feeding programs, while limit-feeding may improve carcass quality but reduce growth efficiency. Proper self-feeder design is essential to minimize feed wastage and ensure optimal growth.

After reaching 10-15 kg, pigs transition to a **grower diet**, which is less fortified and less expensive than creep feed. This diet continues until they weigh 35 kg.

Pigs weighing 35-60 kg are fed a **developer diet**.

From 60 kg to market weight, pigs receive a **finisher diet**, which is the least fortified and least costly of all the diets.

Alternatives for Feeding Pigs from 10 kg to Market Weight Include

1. **Pasture vs. Confinement Feeding:** Confinement feeding is increasingly common. High-quality pasture can replace 10-15% of concentrates and 25-50% of protein supplement. Protein levels can be reduced by 2% when pigs are on pasture.
2. **Complete Mixed Feed vs. Maize and Supplement Free Choice:** Both methods have pros and cons. In the latter, maize and supplement should have similar palatability to ensure balanced consumption.

3. **Pelleted vs. Non-Pelleted Feeds:** Pelleting is most beneficial for young pigs, while the cost of pelleting may outweigh the benefits for older pigs.

Feeding Herd-Replacement Gilts During Growth

Reproduction and lactation benefit from unidentified factors found in lucerne meal, high-quality pasture, and animal protein concentrates. These factors, stored during growth, impact a pig's ability to conceive, reproduce, and lactate later in life. With maize and soybean meal, all essential amino acid requirements of pigs are met.

Summary of Pig Feeding

- During the **pre-starter stage** (from 2 to 5 kg), which lasts from days 1 to 21, pigs require a diet with a crude protein (CP) content of 24% and digestible energy (DE) of 3500 kcal per kg. The daily feed intake for this stage is approximately 1.5 to 2.0 kg, with an expected daily weight gain of around 150 grams.
- As pigs transition to the **starter stage** (from 5 to 15 kg), typically from days 22 to 56, the CP level decreases to 22%, while the DE remains at 3500 kcal per kg. During this period, daily feed intake increases to about 9.0 to 11.0 kg, and the expected daily gain rises to approximately 300 grams.
- In the **grower stage** (from 15 to 35 kg), which spans days 57 to 90, the CP content further decreases to 18%, and the DE is adjusted to 3300 kcal per kg. Pigs in this stage consume between 50 to 60 kg of feed daily, with an anticipated daily weight gain of about 580 grams.
- Finally, in the **growing and finishing stage** (from 35 kg onwards), starting from day 91, the CP level drops to 14%, while the DE remains at 3300 kcal per kg. Daily feed intake during this stage is significantly higher, ranging from 140 to 150 kg, with expected daily gains reaching up to 600 grams over a period that may extend up to 180 days.

Selection of Breeding Stock

When developing a successful piggery unit, several key characteristics must be taken into account. These include the size of litters, the strength and vigor of the offspring, milking ability, temperament, feed efficiency, weight gain of the progeny, longevity, fertility, and freedom from defects. Each of these traits plays a crucial role in ensuring that the breeding stock contributes positively to the overall productivity and health of the piggery.

Selection of Gilts and Boars

Selecting the right gilts and boars is essential for optimal breeding outcomes. Ideally, gilts should be chosen when they reach a market weight of approximately

90 kg. The mother of the gilt should have produced larger litters, ideally eight piglets or more. Additionally, the minimum weaning weights should be 120 kg for gilts and 150 kg for sows. Both gilts and boars should achieve a body weight of around 90 kg within six months and exhibit desirable physical traits such as adequate body length and depth, well-muscled hams, and a prominent neck. It is crucial that they are free from leg defects and have at least 12 functional teats. A negative blood test for brucellosis is also necessary to ensure their health, along with being free from other diseases and physical defects.

Management for Optimal Production

Ensuring maximum fertility and large litter sizes is critical for the success of swine projects. The cost of raising pigs from birth to weaning remains similar regardless of litter size, making it essential to maximize productivity. While breeding and selection are important, improving management practices, feed efficiency, and disease control can significantly enhance herd productivity. Key measures include breeding gilts at 12–14 months of age when they weigh at least 100 kg and ensuring they come into heat shortly after weaning. Additionally, hand breeding involves mating sows twice during their heat period-once at the onset and again after 24 hours-to improve conception rates. In pen mating, sows are kept with a boar until the heat subsides, while commercial pigs may be bred to multiple boars during the same heat period to increase conception chances.

Strategies to Improve Litter Size

Although the heritability of litter size is low, continuous selection for this trait over generations can gradually increase litter size. Females selected for breeding must have at least 12 evenly spaced functional teats and good body length to support larger litters. Proper heat detection is essential for timing insemination accurately, whether through natural mating or artificial insemination (AI). AI programs require precise timing based on the onset of standing heat to maximize reproductive success. Additionally, maintaining good health by preventing diseases like brucellosis and ensuring proper body condition are vital. These strategies, combined with effective record-keeping and consistent evaluation of herd performance, contribute significantly to optimal production outcomes.

Selection of Breeds for Swine Enterprises

When selecting a breed for a swine enterprise, litter size is a critical factor to consider. Yorkshire pigs are particularly notable, producing 10–11 piglets per litter, which is higher than the average litter size of 9–10 in most breeds. Some breeds, however, may have smaller litters of only 6–8 piglets. If older boars are used for breeding, it is essential to select those with a proven record of high conception rates and large litters. Additionally, gilts and sows should be in medium condition before breeding; overly fat animals should have their condition reduced, while thin ones should be improved. During hot summer months, breeding and gestating sows must be kept cool through water sprinkling or wallowing to maintain reproductive health.

Breeding and Farrowing Management

To optimize breeding outcomes, intense inbreeding must be avoided as it significantly reduces litter size and weaning success. Females should be checked for diseases such as brucellosis and leptospirosis before breeding, and sows with flu or high body temperatures during early gestation should not be bred to prevent complications like abortion. Gilts and sows should be moved to farrowing stalls or pens 5–7 days before their due date. Proper farrowing management is crucial, as most losses occur within the first three days postpartum due to causes like starvation or digestive issues. A well-trained attendant is necessary to ensure newborn pigs are dried, breathing properly, and not overlaid by the sow. Creep feeding from three weeks of age until weaning is recommended to supplement milk production, which peaks around three weeks of lactation. Facilities must also be cleaned and left vacant for at least a week between groups of farrowing females to maintain hygiene and prevent disease transmission.

Management of Sow at Farrowing

Farrowing is a critical period in swine production, requiring meticulous care to ensure the health and safety of both the sow and her piglets. The sow should be moved to a farrowing pen one week before the expected farrowing date, determined through accurate mating records. Before transfer, the sow must be cleaned thoroughly with soap and warm water, especially around her sides and underside, and treated for external parasites. The farrowing pen should be sanitized using phenyl water and sprayed with a 4% disinfectant solution to eliminate pathogens. Adequate bedding, such as clean chopped straw, must be provided for nest preparation. The sow should have access to ad-lib clean fresh water, and her ration should be adjusted by replacing one-third with wheat bran to make it bulky while reducing the quantity by one-third until farrowing. Feed should be withheld 12 hours before farrowing to prevent complications.

Piglet Care and Environmental Management

To protect newborn piglets from crushing by the sow, the farrowing pen must be equipped with guard rails or a farrowing crate positioned 9 inches above the floor and 12 inches from the wall. Since piglets have underdeveloped temperature regulation mechanisms, maintaining an ambient temperature of 24–28°C for the first three to four days is crucial. This can be achieved using artificial heating sources like infrared bulbs in the creep feeding area. An attendant should be present during farrowing to assist in drying piglets, ensuring they breathe properly, and preventing overlying by the sow. Piglets should begin suckling immediately after birth, typically nursing 8–10 times per day initially. Close monitoring is essential during the first three days postpartum to prevent losses due to starvation, digestive issues, or unknown causes. Proper hygiene and environmental control are vital for successful farrowing outcomes

Farrowing Troubles and Management

Farrowing complications, such as mastitis and agalactia, require careful attention to ensure the health of both sows and piglets. Mastitis, characterized by inflammation of the udder, can be treated with antibiotics, while agalactia often results from insufficient oxytocin secretion or constipation. Intravenous administration of oxytocin (5–10 IU) can stimulate milk production, and laxatives like castor oil can address constipation-related issues. Injuries or inflammation of the vulva during farrowing may lead to fever, reduced feed intake, and poor milk secretion. In such cases, piglets may need to be reared by a foster mother or through hand-feeding to ensure their survival.

Management of Lactating Sows and Piglets

Proper care and management of lactating sows are essential for the growth and vigor of weaning piglets. During lactation, sows produce approximately 180–185 kg of milk over 56 days, with peak production occurring at three weeks postpartum. To support this demand, sows require a high-nutrient diet and access to clean drinking water at all times. Maintaining a clean, cool, and dry farrowing environment helps prevent conditions like MMA (Mastitis-Metritis-Agalactia) syndrome. For large litters, split suckling-allowing smaller piglets to nurse first-can improve survival rates. Additionally, newborn piglets should be kept in a clean environment to minimize disease exposure as they explore their surroundings and establish teat order within the first few days of life.

Challenges in Large Litters and Immediate Post-farrowing Care

In larger litters, weaker piglets often struggle to compete for teats, risking starvation despite frequent nursing intervals. Piglets typically nurse over 24 times daily, with sessions lasting only a few minutes, though frequency decreases as lactation progresses. To prevent infections like "naval ill," the umbilical cord should be tied off, cut 3–5 cm from the ligation point, and dipped in 2% iodine or 70% ethyl alcohol. Additionally, **needle teeth**-four pairs of sharp, nonfunctional teeth-must be clipped shortly after birth using side-cutting pliers to avoid injuring the sow's udder or littermates. Care is taken to prevent jagged edges or gum damage during clipping.

Health management for the piglets

Piglets should be ear-tattooed at birth and ear-punched at six weeks for identification. Male piglets not intended for breeding require castration 7–10 days before weaning to ease management and prevent uncontrolled mating, allowing recovery before weaning-related growth checks. Tail docking is recommended for pigs raised in confinement, performed alongside teeth clipping using sterilized tools. These procedures, combined with strict hygiene protocols, minimize stress and health risks, ensuring optimal growth and welfare during the critical weaning phase.

Piglet Mortality: Causes and Mitigation Strategies

Piglet mortality, averaging 16–20% pre-weaning, primarily stems from stillbirths, crushing, and starvation, with over 50% of live-born deaths occurring within the first three days. Stillbirths (17.4%) often result from intra-partum anoxia due to uterine contractions or premature umbilical cord rupture, particularly in aged sows, while pre-partum deaths may arise from maternal iron deficiency. Crushing by sows accounts for ~66% of deaths, especially in loose farrowing systems, whereas conventional crates see higher starvation rates. Immediate interventions include farrowing supervision to assist dystocic sows, ensuring colostrum intake for weak piglets, and maintaining a warm microenvironment (24–28°C) to prevent hypothermia. Hygiene protocols, such as disinfecting pens and clipping needle teeth, reduce disease risks, while split suckling and early fostering improve survival in large litters.

Long-Term Solutions and Considerations

Reducing mortality requires coordinated genetic, nutritional, and management strategies. Selecting sows for litter viability, providing high-nutrient diets during lactation, and optimizing pen design (e.g., guard rails) mitigate crushing risks. Economic analyses highlight cost-effective interventions like improving ventilation and enriching sow environments to enhance welfare and productivity. Addressing congenital abnormalities (e.g., atresia ani) and ensuring sow health through iron supplementation further lowers pre-weaning losses. Combined with stockperson training to minimize sow stress, these measures not only curb mortality but also enhance farm profitability, offsetting labour costs through improved survival rates

Minimizing Agalactia

Agalactia is a complex condition that is part of the MMA syndrome, which includes mastitis, metritis, and agalactia. This syndrome can be influenced by metabolic, bacterial, and hormonal factors, with stress also playing a significant role. The primary consequence of agalactia is the loss of milk production in the first three days after farrowing, leading to increased piglet mortality due to starvation. To control MMA, it is essential to regularly monitor the rectal temperature of sows, as elevated temperatures are often associated with this condition. Prompt treatment with antibiotics and oxytocin can help mitigate the issue. If not detected early, signs such as weight loss in piglets may indicate a problem that is difficult to rectify quickly. In severe cases, providing an alternative food source for piglets through foster sows or artificial feeding can help minimize losses. Additionally, heat stress from artificially brooding piglets can exacerbate the situation; therefore, maintaining cooler conditions for the sow is beneficial. Ensuring that the udder and teats are kept clean and dry will also help prevent agalactia.

Preventing Piglet Anemia

Newborn piglets have limited iron reserves in their livers for haemoglobin synthesis due to poor placental transfer of iron. Typically, they contain only about 50 mg of iron at birth, while their daily requirement is around 5 to 10 mg. Since sow's milk is low in iron, piglets should be supplemented during their first few days to prevent anaemia. Access to fresh, clean soil can be a source of iron; however, indoor rearing on concrete flooring often restricts this access. Symptoms of piglet anaemia include paleness around the ears and belly, weakness, rapid breathing, and diarrhoea. To control anaemia, fresh earth can be placed in the piglet pen daily or soil drenched with a solution made from 500 grams of ferrous sulphate, 75 grams of copper sulphate, and 3 liters of water may be used. Daily administration of a 4 ml dose of 1.8% ferrous sulphate solution or painting the mother's udder with a ferrous sulphate solution mixed with sugar (0.5 kg of ferrous sulphate in 10 liters of water) are also effective methods. However, these approaches can be labour-intensive; therefore, the safest and most efficient method is to inject piglets with 100–150 mg of iron in the form of iron dextran three days after birth, with a second smaller injection given around three weeks later if necessary.

8

Incubation and Hatching

Anuradha Kumari, Utkarsh Kumar Tripathi
Manish Kumar and Vipin Maurya

Faculty of Veterinary and Animal Sciences, I. Ag.Sc., RGSC-Banaras Hindu University (BHU), Barkachha, Mirzapur, Uttar Pradesh-231001

Incubation: The process of subjecting selected eggs from mated flocks to proper conditions outside the bird's body for the embryo to develop and hatch into a chick. Incubation is also defined as a process in which a microscopic germ cell is being transformed into a chick capable of moving, feeding, and drinking etc.

Incubator room

- Optimum results can be expected if the temperature can be maintained at between 24 and 27°C, with uniform humidity below the level that is required in the incubator.
- Size depends upon the type of incubator
- Should not be less than 5.5 m^3 for 1000 egg capacity
- The ceiling height should not be less than 12 feet.
- Keep distance of 0.3 m between wall and incubator.

Methods of incubation

1. **Natural Method:** Eggs are incubated with the help of broody hens (does not lay egg, dull or inactive, no lustre on body, drink less to keep her body warm, lack appetite, shank and comb pale and reduced size). Broodiness decreases egg production; hatching is seasonal with a hen; one hen can incubate only 15 eggs at a time; the hen requires more space per egg than the artificial incubator; and there is a greater risk of disease transmission from hen to egg under natural incubation.
2. **Artificial method**: eggs are incubated in "egg incubators". The artificial method of incubation of eggs is known for more than 2000 years and the earliest records of artificial incubation are from China and Egypt.

Incubator: An apparatus for maintaining optimal conditions (temperature, humidity, etc.) for growth and development and for hatching eggs. Machine into which eggs are usually set for hatching.

Parts of incubator

Setter: This is the part where eggs are set for first 18 days. It has multiple trays for setting of eggs. Eggs are set with broader end up. Temperature of hatcher part of incubator is 99.5°F

Hatcher: Eggs remain in hatcher for last three days. i.e days. Eggs lie flat in hatchers. Temperature of setter part of incubator is 99°F .

- During incubation, the hatching eggs are set vertically, with the large ends up in trays or flats in a setter and turned mechanically until about three days prior to hatching (setting period).
- The eggs are then transferred to a Hatcher (hatching period) in a horizontal position and not turned during the hatching process.

Incubator types

Few years back, eggs used to be kept in a machine till they are hatched. After hatching and removal of chicks and other materials, they used to be cleaned and disinfected before next use. Even now, such machines are used on small-scale and they are referred to as "Setter-cum-hatcher" As the chick hatches out, part of small hair-like feathers (down feathers) on the chick also falls. After coming out of the shell, chick passes its first faecal matter called "Meconium". Therefore, a settercum- hatcher will become dirty during hatching process. When many batches of eggs are set in the machine, it can infect other eggs and reduce hatchability. In addition to the above, temperature and humidity requirements of eggs during hatching process are different. Therefore, it is better to have separate machine called **"Setter"** to keep the eggs during the first 18 days (in case of chicken) and transferred to another machine for the last 3 days exclusively for hatching purposes called **"Hatcher".**

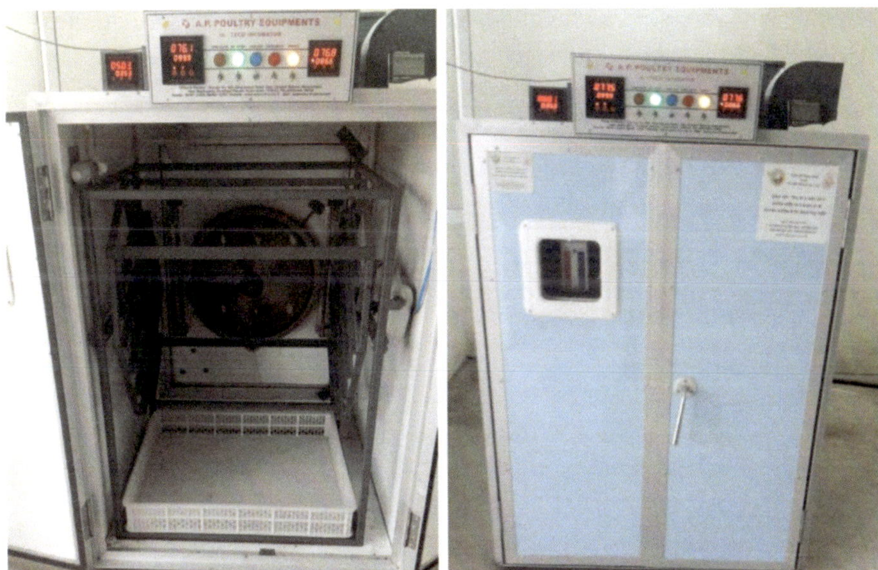

Fig. 1. Setter cum Hatcher incubator (1000 egg capacity)

The two popular types of incubators are the flat (table) or still air type incubators and the cabinet type incubators (force drought).

Still air type incubators: Heat is supply to the top of machine and as air cool it passes down to the bottom through the egg. There can be as much as 7.7 ^{0}C temperature difference between top and bottom of machine.

- The flat type incubators are usually of small capacities; of between 50-500 eggs with only single layer incubation on which the eggs are laid horizontally i.e. lying flat.
- It has no fans, so the air is allowed to stratify. Difficult to maintain proper temperature and humidity. Temperature: 100-101 °F, Humidity 60-65% during incubation and 70- 75% at hatching time. Ventilation provided through natural means.
- The heating system is either by hot water or hot air supplied by electricity. Although early incubators were oil (kerosene) heated, most of these types of incubators now rely on electricity to heat the air or ın some cases, the water of the system

Cabinet Incubators

These Incubators are of very large capacities, which can contain very large number of eggs, from about 3,000 to 30,000 and even more. The incubators are highly specialized type of equipment which are electrically heated and operated. They are sold as separate setter and hatcher units. Ventilation is by fans with automatic heat/humidity controls and egg turning devices during the incubation period.

Fig. 2. Incubator setter

Fig. 3. Incubator Hatcher

Incubation period

Number of days required for an egg to hatch is called "Incubation period" and it differs between species.

Incubation period for some common avian species

Chicken	**21 days**
Turkey	28 days
Japanese quail	17 days
Duck	28 days
Pheasant, Partridge	24 days
Guinea fowl	26 days
Muscovy duck	35 days
Goose	1-32 days

Factor affecting incubation period

- Breed – light breed hatch few hours earlier than heavy breeds
- Egg size – larger egg takes more time then smaller eggs
- Storage temperature – longer storage period of egg more will be time of incubation
- Stock – eggs of inbred stock takes longer time than out breed stock

Principal factors in incubation

1. Temperature
2. Humidity
3. Ventilation
4. Position
5. Turning of eggs

1. Temperature

- Temperature is the most critical part and it plays the key role in the incubation system. Hatching eggs may be warmed to a temperature of 25 to 30ºC, prior to setting in incubator.
- In small still-air incubators, the temperature of the upper surface of the egg is higher than on the lower surface, while in large incubators, the air movement maintains the same temperature over the entire surface.
- For this reason, a still-air incubator must be operated at a higher temperature than a forced draft incubator.
- In small still-air incubators, a constant temperature of 39ºC is considered satisfactory (between 37.5 and 39.5ºC) (99.5-103.10ºF)
- In large incubators, the temperature, humidity, and speed of air movement are very closely dependent on each other, and since air speed varies in different incubators , generally it is around 37.5ºC (99.5ºF) for a setter and 37ºC (98.6ºF) for a hatcher.
- High temperatures even for a very short period of time during any part of the incubation period will cause more harm than low temperatures.
  - Temperature prior to egg laying: 40.6ºC and 41.7 0C (105-107ºF)
  - During the first 18 days of incubation: 37.7ºC (99.86ºF)
  - During the last 3 days of incubation: 36.7ºC -37.2ºC (98.06-98.96ºF)
- Adverse effect of high temperature: smaller chicks, lack of alertness, crooked toes, straddled legs, croocked neck. High temperature as 46ºC for 3hr and 49ºC for 1 hr kills all embryos.
- Sub normal temperature causes late hatching and poor hatchability.

2. Humidity

- The maintenance of consistent relative humidity is more difficult during incubation and can only be constantly maintained by using adjustable ventilation apertures and by surface water sprays during incubation. During the incubation period, eggs should lose 11 to 12% of their weight (another 3 to 4% in the hatcher, after day 18), due mainly to a loss of moisture.

- The amount of moisture (humidity) in the incubator controls the rate of evaporation from the egg. The evaporation rate is also related to temperature, air speed, shell thickness, and size of eggs; the smaller the eggs, the greater percentage of moisture loss.
- Rarely is the humidity too high in properly ventilated still-air incubators. The water pan area should be equivalent to one half the floor surface area or more. Increased ventilation during the last few days of incubation and hatching may necessitate the addition of another pan of water or a wet sponge. Humidity is maintained by increasing the exposed water surface area.
- Too great a moisture loss from the egg in the early stage of incubation will cause the embryo to adhere to the shell, and small, hard chicks.
- Insufficient evaporation may cause large, soggy chicks. The incubation period is shortened, and the chicks are wet at hatch and residual albumen may be present. The chicks will have poorly healed navels.
- The best guides to the correct amount of humidity in an incubator is the weight loss and the size and enlargement of the air cell during incubation, or the position at which the chick pips the shell.
- In the first 17 days the relative humidity level should be maintained at 65% and the last four days the relative humidity level should be around 70% to 75%. (Niranjan *et al.,* 2021)
- The degree of enlargement of the air cell should be determined by candling several eggs and estimating the average evaporation.

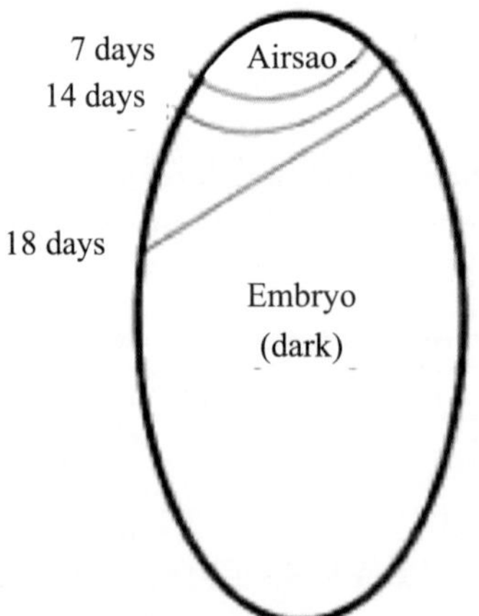

Incubation day	Depth of air cell
First day	0.31 cm
Second day	0.562 cm
Eight day	0.952 cm
Fifteen day	1.587 cm
Nineteenth day	1.903 cm

3. Ventilation

The free movement of oxygen, carbon dioxide and water vapor through the pores of the shell should be maintained in the incubation process. To do so the air in and air out exhaust fan s should be used to regulate the oxygen level in the incubator. This is controlled by the proposed method using the air vent openings. As embryos grow, the air vent openings are gradually opened to satisfy increased embryonic oxygen demand

- Oxygen content of 21% (present in air at sea level) and a carbon dioxide content not exceeding 0.5% in the air are considered optimum for good hatching results.
- Room temperature, room humidity, the number of eggs set, the period of incubation, and the air movement in the incubator all influence ventilation requirements.
- Ventilation problems are not the same in small incubators as they are in large incubators, where a large number of eggs are set in a very small space.
- Ventilation is very important in any incubator at hatching time.
- The appearance of chicks panting in a hatcher at normal temperature is an indication of a rise in the carbon dioxide content of the hatcher air. Under such conditions chicks must breathe faster to obtain the required amount of oxygen and to eliminate the excesses carbon dioxide. If excessive panting occurs, increase the air flow in the hatcher.
- Oxygen-21%
  - Down 1% O_2 = down 5% Hatchability
  - Up 1% O_2 = down 1% Hatchability
- Carbon dioxide is given off by the developing embryos particularly at the later stage of incubation.
- Carbon dioxide (CO_2): 0.3-0.5% for maximum hatchability
  - Up 1% CO_2 = Increases embryonic mortality
  - Up 5% CO_2 (Prolonged period) All embryos will die

4. Position of Eggs

- In small incubators, the eggs are maintained in a horizontal position (horizontal or section type incubator) during the entire incubation period.
- In large incubators eggs should be placed in a vertical position, large end up, for the first 18 days (in setter) and in horizontal position for the last 3 days (in hatcher). This position also permits the embryos to develop head towards the large end of egg near air cell. The chicks ready to hatch are able to break in air cell to intimate pulmonary respiration. Reverse setting of eggs reduces the hatchability by 10 % and quality of chicks is also reduced.

5. Turning of Eggs

The most essential part in the incubation is the egg turning, in the 1st week the embryo has no circulation system hence egg rotation is critical, this helps proper development of additional embryonic membrane which in turn helps the chick to pep out without any problem. It is necessary that the embryos are gently but frequently moved within the eggs to prevent their setting and adhering to other structure of eggs.

- In small incubators, the eggs are moved when turned, while in large incubators they remain in a stationary position on the incubator tray and tilting the egg trays at 45^{0}c in both directions from the perpendicular or a total of 90^{0}c gives the best result with frequent turning.
- The objective is the same in both types of incubators; namely, to prevent the embryo from sticking to the shell membranes. Turning also ensures a complete contact of the embryonic membranes with the food material in the egg, especially in early stages of incubation.
- Hand Turning - In a still-air incubator, where the eggs are turned by hand, it may be helpful to place an "X" on one side of each egg and an "O" on the other side, using a pencil. This serves as an aide to determine whether all eggs are turned.
- Turning of eggs should be done for the first 18 days of incubation once in every 3 hours (6-8 times turning/day). There is no need to turn in last three days.
- Eggs should be transferred to hatcher when about one per cent eggs are slightly piped. At on average 19 days and 12 hours of incubation, the embryo pierces the inner membrane and starts lung ventilation in the air cell. This is called internal pipping.
- About 12 hours after it has pierced the inner membrane, the embryo starts tapping the eggshell repeatedly with its egg tooth, a sharp and

strong structure that can temporarily be found on the top of the beak of the embryo. Repeatedly tapping the eggshell in the same spot causes the shell to weaken and eventually break. This is called external pipping.

- After external pipping, the embryo takes a rest. It has one last challenging task ahead: breaking free from the eggshell – and it normally takes another 12 hours after external pipping to hatch.

Hatching

Fertility: Refers to the capacity to reproduce young ones**.** It is the factor, which determines the successful offspring that may be obtained from a given number of eggs. It may be defined as percentage of eggs those have been fertilized.

Factors affecting fertility

- **Age of the parent stock:** There is an increase in fertility in a breeder flock between the ages of 25 to 40 weeks after which fertility gradually diminishes
- **Breed:** Lighter breeds like white leghorn is more fertile than heavier breeds like the broiler breeders
- **Genetic factors:** Many genes influence fertility eg: in wyandotte the gene responsible for rose comb (RR) lowers fertility in males
- **Environmental factors:** excessive high and low temperature reduces fertility due to poor mating frequency because of the inactiveness of the birds. It is high during spring.
- **Disease conditions:** Many diseases like Ranikhet disease, Mycoplasmosis, Salmonella etc., affect fertility
- **Rate of lay:** Those flocks laying at a high rate usually have a high rate of fertility and hatchability. Eggs laid in longer clutch hatch better.
- **Age of males:** Although the presence of viable sperms has been reported in the semen of 10 week old chicken, but the breeding males should be at least six or seven months old for resulting in high fertility.
- **Laying pattern:** Fertility and hatchability are higher in first year of laying than in subsequent year. This is again higher in the first twelve to fifteen weeks of laying then starts gradually declining.
- **Inheritance:** Fertility is poorly heritable traits
- **Sex ratio:** Both higher and lower males to female ratio will reduce fertility. The recommended ratio in lighter breeder is 1:10-12,. In broiler breeder 1:8-10 and in J.quails 1:1-2. The semen volume, sperm concentration and number of successful mating also alter fertility. Inseminating the birds during the afternoon can lower fertility

- **Nutritional factors:** Some deficiencies like vitamin A, E , Biotin, Pantothenic acid and B2 and minerals like calcium, phosphorous, sodium, Magnesium, Manganese, Zinc and Iodine lower fertility.
- **Photo period:** A photo period of 16 hrs per day will give optimum fertility. By either lowering the length of period to 12 hrs or increasing it to 18 hours lowers the fertility
- **Male nutrition:** Male breeders should be fed with lower protein levels of 12-14% for optimum fertility

Hatchability

Hatching of eggs refers to the production of baby chicks. Hatchability is defined as the number of chicks produced from eggs.

Methods of hatching

Natural hatching: Desi hens used for this purpose. Only 10 to 12 eggs can be put under 1 hen. This method of hatching is highly unsatisfactory for large-scale production of baby chicks.

Artificial hatching: Artificial hatching of chicken eggs involves using an incubator to replicate the natural conditions as that of broody hens needed for eggs to hatch. This method is more efficiently, are used at present for hatching of eggs.

Measurement of hatchability: it is measured by two means (two ways)

1. Based on the fertile eggs set (FES) = calculation is made usually on research farm to pinpoint cause of unusual hatchability.
2. Based on the total eggs set (TES) = Refers to the % of chicks hatched to the eggs set. From the hatchery point of view, this calculation is more important as it affect the profitability directly.

Factors affecting hatchability

- Size, shape and condition of egg shell are important for hatching. Eggs having abnormal shape, too small or extra large eggs, thin shelled eggs with poor internal quality do not hatch well.
- Incubation condition: Condition and duration of storage of eggs prior to incubation as well as setter and hatcher environment affect hatchability. Faulty pre-incubation storage conditions for eggs reduce the hatchability considerably. Optimum temperature and humidity in incubator is most essential for desired hatchability. Abrupt and frequent variation in these factors alters hatchability seriously. The desire levels of oxygen and carbon dioxide in incubator play major role in obtaining optimum

hatchability seriously. The reverse setting of eggs with narrow end up lower the hatchability seriously. Inadequate and faulty turning of eggs during incubation lower the hatchability. Use of separate hatcher with slight decrease in temperature and increase in humidity as compared to that setter improves hatchability

- Hatchability is positively correlated with the rate of eggs production. Therefore, when birds are selected for high egg production, hatchability is automatically improved. Hatchability is higher in egg from younger flocks and vice versa. Eggs from birds between the age of 21 to 40 weeks hatch well.
- Internal quality of egg – egg with balance nutrient in it, gives more hatchability.
- Bacterial contamination have adverse effect on hatchability. Diseases which are vertically transmitted lowers hatchability
- Fertility of eggs – fertility of parent has a positive correlation with hatchability.
- Breed, strain and individual variation: Presence of lethal and semi lethal genes like creeper, crooked toe, crooked beak, polydactyl conditions. Intense Inbreeding leads to reduced fertility while out-breeding improves it
- Management and nutritional status of breeding stock with special reference to minerals and vitamins alter hatchability considerably. Vitamins A, B_2 and E are critical vitamins which affect hatchability. In practical rations only few nutrients like riboflavin, pantothenic acid, manganese and B_{12} may become low enough to cause the problem of low hatchability.
- Too high and very low temperature in breeder houses also lower hatchability.

Selection and Care of Hatching Eggs

Proper selection and care significantly improve the chances of a successful hatch, leading to healthier chicks.

Selection of Hatching Eggs

1. Select eggs from breeders that are:
   a) Well developed, mature and healthy and compatible with their mates and produce a high percentage of fertile eggs.
   b) Reared under stress free environment.
   c) Fed a balanced breeder diet

2. Egg size. Select eggs for hatching that are normal in size. For optimum results of hatchability the optimum size of egg should be 56.7 gm. The extra large or small eggs are often infertile also too small or too large eggs as they create hinderance in setting in incubation trays. Large eggs hatch poorly and small eggs produce small chicks. Duck eggs should vary between 65 to 70 g and turkey eggs should be between 80 to 85 g.
3. Egg shape. Do not incubate eggs that are misshapen. Oval shaped eggs hatch better than spindle and round shape eggs.
4. Egg shell colour. Dark brown eggs tend to hatch better than light brown eggs due to the fact that dark brown eggs are thick shelled so water loss from these eggs is less than light brown eggs are thin shelled.
5. Egg shell thickness. For best hatching results eggs shells should be between 0.33 to 0.35 mm in thickness.
6. Specific Gravity. A positive relationship exists between fertility of eggs and their specific gravity.
7. Defects in eggs. Remove eggs with obvious defects, dirt, cracked or thin shells.
8. Age of eggs. Always set fresh eggs for incubation.
9. Shell texture. Eggs set for incubation must have smooth, thick and uniform shell texture.
10. Internal egg quality. Eggs should have good albumin yolk quality and should be free from blood and meat spots or any other defects.
11. Cleaning of egg shells: Set only clean eggs. Dry cleaning of soiled eggs with rough cloth or sand paper may be done before setting. If hatching eggs are washed, the temperature of the water must be 43 to 44°C. However, this removes not only the soil but also the cuticle from the egg shell surface. Dirty eggs can be reduced if the eggs are gathered four or five times daily from nests supplied with fresh, clean nest material and if the litter in the breeding pens is kept dry so the hens feet are clean. Collect eggs with clean hands onto clean flats.

Care of hatching eggs

- Temperature: There are two types of temperature to be considered, namely temperature during the whole storage period and pre warming of eggs. After hatching eggs are laid or brought to hatchery, they should be cooled to a temperature well below, Physiological zero temperature (20 to 21 ^{0}C) to arrest the embryonic development. On farm egg storage conditions: Temperature – 65 ^{0}F Humidity – 75%.

- º The hatching percentage will be the highest if eggs are held at a temperature of 16 to 17ºC for not more than one week before setting. Higher temperatures initiate embryo growth. Storage temperature should be reduced to 13ºC if eggs are being held for two weeks or longer.
- º Pre-warming of hatching eggs during storage improves hatchability particularly when the eggs are stored at low temperature for long periods. A pre warming temperature of 23°C for 18 hours before setting is suggested. Abrupt warming from 55 degrees to 100 degrees causes moisture condensation on the egg shell that leads to disease and reduced hatches.

- Humidity: relative humidity (RH) is maintained at or above 80% during storage to minimize water (moisture) loss from hatching eggs in the egg holding room
- Duration of storage: prolonged storage leads to reduced hatchability and also produces a longer intubation time. It is common practice in the hatchery to avoid the pre-setting storage period beyond the 7th day in fact many hatcheries aim to set eggs after only 3rd or 4th day of storage.
- Controlled Atmospheric: storage plastic packs improves hatchability by reducing moisture and carbon dioxide loss if storage is prolonged, particularly if a high holding temperature is maintained. Permeability of the plastic is an important factor; hatchability is higher in bags of low permeability. Greater chemical stability of eggs has been observed when nitrogen was flushed in plastic package of eggs during pre incubation period and as such was more desirable for preservation of hatchability.
- Position of egg during storage: Turning eggs during the holding period is not beneficial. It has been shown that eggs held for more than 2 weeks hatched better when stored small end up (contrary to the accepted traditional large end-up position). Eggs stored at small-end-up position had yolk positioned quite centrally compared to small-end-down position.
- Transportation of hatching eggs. Eggs which have been shaken or jarred in transit should be allowed to settle for 24 hours before setting in incubator. Truck drivers and their helpers must shower and change into clean clothes, headgear and foot wear before entering any truck involved with the egg transport or before entering a hatchery. All trucks, trollies etc., delivering hatching eggs must be disinfected and fumigated with formaldehyde gas before eggs are placed in them.

- Fumigation: eggs and equipments before storage or use are fumigated. The main objective of fumigation is to destroy bacteria inside to reduce infection in newly hatch chicks. A room or cabinet large enough to hold the eggs is required. It must be relatively air tight and equipped with a small fan to circulate the gas. Formaldehyde gas is produced by mixing 20 gram of potassium permanganate ($KMnO4$) with 40 ml of formalin (37.5 % formaldehyde) for each m3 of space (1 X concentration) in the fumigation chamber. The ingredients are mixed in an earthenware or enamelware container with a capacity at least 10 times the total volume of the ingredients. For effective fumigation, the temperature should be more than 21°C with 65% RH. Duration of fumigation should be 15-20 minutes only.
- Do not allow hatching eggs to be placed in front of the fans. Increasing the flow of air around the hatching eggs increase the rate of egg evaporation and thus, dries out the contents more rapidly.

Hatchery performances

a) % Fertility = No. of fertile eggs/ No. of eggs set x 100

b) % Hatchability = No of Chicks hatched/ No. of fertile eggs x 100

c) % Hatch = No. of Chicks hatched/ No of eggs set x 100

Hatchery Operations

The sequences of hatchery operations followed in commercial hatcheries are

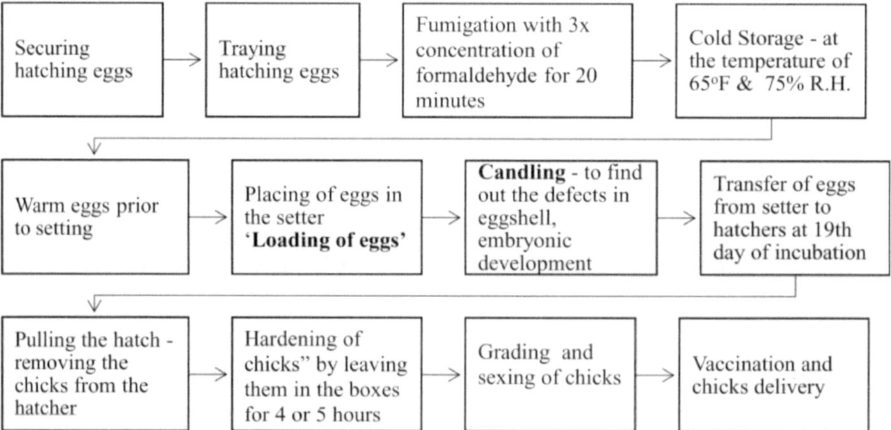

Fig. 4. Setting eggs in setter tray

Fig. 5. Hatching tray with hatched chicks

Reference

Niranjan, L., Venkatesa n, C., Suhas, A.R.,Satheeskumaran, S. and Nawaz, S.A. (2021) 'Design and implementation of chicken egg incubator for hatching using IoT', Int. J. Computational Science and Engineering, Vol. 24, No. 4,pp.363–372 (11) (PDF) Design and implementation of chicken egg incubator for hatching using IoT. Available from:

9

Brooding Management

Anuradha Kumari, Utkarsh Kumar Tripathi Manish Kumar and Ajeet Singh

Faculty of Veterinary and Animal Sciences, I. Ag.Sc., RGSC-Banaras Hindu University (BHU), Barkachha, Mirzapur, Uttar Pradesh-231001

- Brooding is the process of providing heat and other management services to the baby chicks at an early age i.e upto 4 to 6 week of age.
- **Brooding** consists mainly the process of providing chicks with natural or artificial heat to help maintain their body temperature. The mother hen supplies natural heat to its young while chicks hatched, in an incubator get heat from a brooder. Brooding is done immediately after newborn chicks are taken out from the incubator. The length of the brooding period may last from two to five weeks depending upon some factors like the rate of feathering of the chicks and the season of year when, brooding is done. Artificial heat supply related with environmental temperature.

What is Brooder?

The device or equipment used for providing artificial heat to baby chicks is known as brooder.

Objective of brooding

- Chicks do not possess a well-developed thermo- regulatory (hypothalamus) mechanism.
- The day old chicks don't possess the insulating feather coverage to protect them from chillness.
- The body temperature of chick is 107°F which is always more than the ambient temperature. It may result in the losing of body heat to the environment.
- So a source of heat is given by natural brooding or by artificial brooding up to 4 weeks of age.

- The general concept is to accelerate growth rate in the first week to enhance structural development and feed capacity intake, so add to maximize genetic potential.
- To improve overall health of the bird

Methods of brooding

1. Natural method of brooding

- This is the brooding of chicks with the mother hen or trained capon. Natural brooding is still the most common practiced in the rural areas. The hen after hatching the eggs rears her brood on a natural process. Depending upon size a hen will brood 15-20 chicken.

2. Artificial method of brooding

It is the process of providing the chicks with the required temperature to make them warm and comfortable. In artificial brooding large number of baby chicks are reared in the absence of broody hen.

If artificial method of incubation have been practised, then artificial method of brooding should follow.

Advantages of artificial method of brooding

- Chicks may be reared at any time of the seasons.
- Thousands of chicks may be brooded by a single person.
- Sanitary condition may be controlled.
- Temperature may be regulated.
- Feeding may be undertaken according to plan.

Systems of artificial brooding

- Cold room brooding
- Hot room brooding

System of brooding is the method of keeping chicks warm during the brooding period.

1. **Hot room brooding-** In this system entire brooder house is heated through central heating system (circulating hot water in pipe or under floor heating or blowing hot air)
    - Make ventilation easier
    - Keep litter dry

Warm Room Brooding

Some people heat their barns with space heaters or hot water pipes. Without a heat lamp or brooder stove to act as a hot point, the birds cannot move towards or away from the heat source to regulate their body temperature. As a result, it is more difficult to judge bird comfort when you warm room brood. Comfortable birds will be spread out and making full use of the pen. Cold birds will tend to bunch together, sometimes near the walls. Hot birds may also gather around cold, outside walls. Panting is a sign of heat stress.

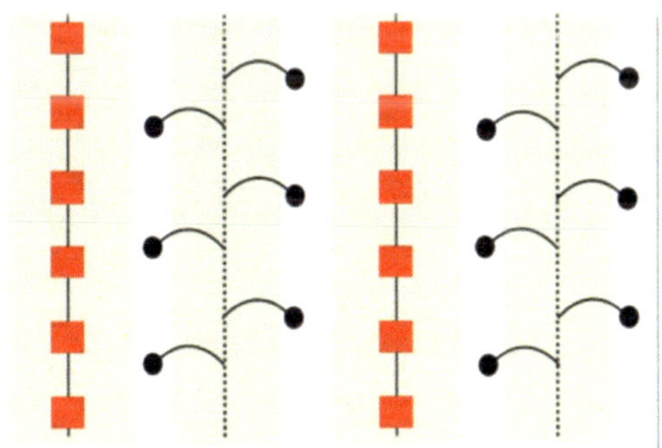

Centralised heating system of broading is followed in enviromentally controlled poultry houses; where the room temperature is maintained at an average of 32° C during first week of age

2. **Cold room brooding-** heating the area only under brooder canopy.
    - Used mostly in tropical countries for both layers and broilers.
    - Feathering appears to be better with this system of brooding

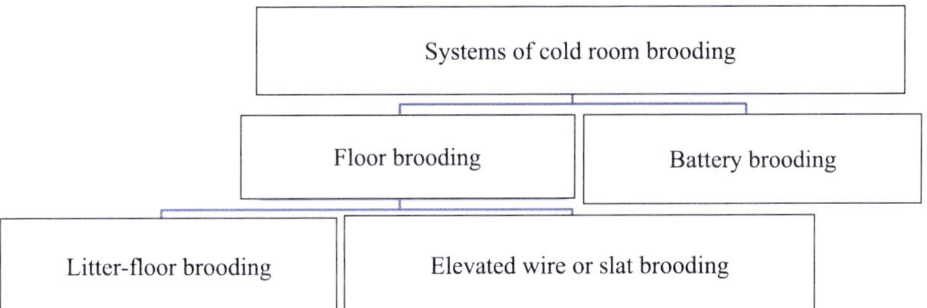

1. Litter-floor brooding

- It uses the floor as place for brooding
- Litter should be organic in nature, fine in texture, light in weight, soft, cheaper, free from dust and less danger of fire.
- Litter spread 5 cm deep in the begging increase the depth 5 cm every month till it reaches 15-20 cm.
- The litter or material used are rice hull, wood shavings and sawdust
- Wet litter causes the spread of disease. Prevention done by changing litter regularly
- Power 1-2 watt/chicks is required. A metal hover/inverted bamboo basket of 90 cm diameter is sufficient for 250 chicks

2. Elevated Wire or Slat Brooding

- Elevated pens (choices of) wire mesh/wooden/bamboo slats
- More economical/convenient to use than litter-floor types.
- Reduces problems of diseases and parasites.
- Promotes better growth
- Easier collection of manure and good supervision.

Battery Brooding

Battery brooders – Multi-tier cage brooding is also practised.

Bulbs or heaters with thermostats are used to provide warmth to chicks in the battery cages

- More than 80 % of all commercial layers are brooded and reared in cages.
- Coccidiosis is less frequent.
- It consists of 4-5 batteries each having a heating space comprising 1/4th of total area.
- Labour requirement is less.
- Mortality is less.
- But initial cost of battery is more and there is more problem of flies

Location of Brooder House

Distance between brooder house and other poultry houses should 100 m (45-100 meter).

Prevailing air movement must be from the brooder area to other poultry area and never in the opposite direction. The brooder house should separated by distance and management both.

Brooders

- Equipments used for brooding are called brooders. Brooder comprises of three elements:
- Heating source
- Reflectors
- Brooder guard

Heating source

- The broiler growers use many types of brooders. Name of brooder depends on either energy source or material they use to brood the chicks. Heating source may be electrical, gases like natural gas, LPG and methane, liquid fuel like kerosene and solid fuel like coal, wood can be used as a heating material.

- Charcoal stove / kerosene stove- where electricity is not available, ordinary charcoal / kerosene stoves are used to provide supplementary heat to chicks.
- These stoves are covered with plate / pans to dissipate the heat

Electrical brooder

- It is also thermostatically controlled heating system that spread required amount of heat uniformly above large area, this avoid crowding of chicks under brooder directly.
- One electrical brooder can be used for 300 to 400 chicks.

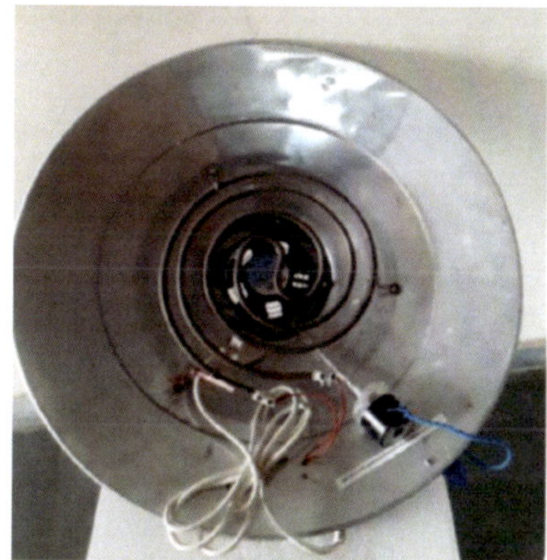

Electrical brooder

Gas brooder

Infrared Brooding

- No need of canopy as the infrared light heats any object that comes in contact with it.
- 150 and 250 watt infrared bulbs are available for 100-250 chicks.
- The infrared bulb should be hung 25-30 cm above the litter floor.
- Infrared light has germicidal effect.
- Infrared light helps in the synthesis of Vit-D.
- It reduces cannibalism.

Gas brooder

- Natural gas, LPG or methane is connected to heating element which is hanged 3 to 5 feet above the chick to provide heat. Now practised particularly in semi-environmentally controlled and environmentally controlled houses. These are costly but can take care of 1500 to 2000 chicks.

Reflectors

- These reflectors are called Hovers .
- Flat type hover – These hovers are provided with heating element, heating mechanism and pilot lamp and in some cases thermometer is also there in order to record the temperature.
- Canopy type hover – These reflectors are in concave shape consisting of ordinary electrical bulb, thermostat mechanism and in some cases thermometer.

Brooder guard / chick guard

- They are used to prevent chicks from straying too far away from heat supply until they learn the source of heat.
- 5 feet diameter chick guard can accommodate 200-250 chicks.
- Height of the brooder should not exceed 1.5 feet.
- For this purpose, we can use materials like cardboard sheet, GI sheet, wire mesh, and mat etc. depending upon the season of brooding.
- Chick guard placed 60-75 cm from edge of hover(at distance of 76cm in winter and 91cm in summer)
- Guard is not necessary after 1 weeks.

Fig. Chick Guard

Brooding management

Preparation of poultry shed/house before arrival of chicks

- Clean the equipments- Remove all the movable equipment (lightings, feed and water pans) from the shed. Soak in water and clean thoroughly in tap water and finally dip in disinfectant solutions. Finally wash in clean water, sun dry and store.

- Removal of old litter - Litter should be removed from the shed and transported away from the farm in closed containers or in gunny bags and disposed off properly.
- Clean and scrub the house - Accumulated dust and cob web formed on the wall, ceiling, mesh etc., should be removed.
- Insecticide is to be sprayed over the litter, walls, mesh, roof etc., Shed should be washed using a pressure washer.
- Disinfection of the room is carried out by spraying phenyl, lysol, etc at 5% concentration
- All the repair works of the shed including cages equipment and mesh should be carried out
- Fumigation in tight house, fumigation of house and equipments with 3X concentration of formaldehyde.
- *(Note-For the fumigation take a two part of Formalin and one part of potassium permanganate. When two compound mixed together the fume will be generated and that fume will destroy the microbes present in brooder house e.g. 35 ml of formalin and 17.5 gm potassium permanganate is sufficient to disinfect 2.83*m^3 *space which is known as 1X fumigation, but for fumigation we have to keep in our mind that always add potassium permanganate in formalin).*
- Water tank, pipeline and water channel in the shed be cleaned thoroughly. First drain the water, fill the water lines with de-scaling and disinfectant agent overnight. Flush with water for 2-3 times to remove all dirt and debris.
- Flame guns should be used inside and outside of the houses.
- Walls should be white washed and metal surfaces should be painted if needed.
- The equipment and fitting should be re-assembled.
- Hang the gunny bags around the brooder house to maintain the room temperature to maintain temperature.
- *Downtime* - The house may be prepared before hand for the purpose and kept vacant for considerable length of time (minimum two weeks downperiod)
- ***Setting of brooder*** - About 24 hours before the anticipated time of arrival of the chicks, the brooder arrangements are made and kept ready. Number of brooder depends on the total number of chicks. It is easy to handle 250 chicks per brooder. Number of chicks can be increased in case of gas brooder.

- *Litter materials* - Dry litter material such as saw dust, wood shavings, paddy husk etc. is spread to a height of 5 cm and newspaper has to be placed over this to prevent the chicks from consuming litter material
- The brooder guards with diameter of 5 feet to accommodate 150-200 chicks have to be arranged in a circular fashion on litter material with necessary heating arrangements.

- The chick linear feeders and chick waterers should be placed alternatively in a radiating way, to give a *"cart wheel"* appearanc
- Care should be taken to avoid placing them crowded at the centre under the source of heat.
- 4 hours before the scheduled arrival of chicks place the waterers with water in order to bring water to room temperature (65 °F)
- Depending on weather condition put curtains on all four sides of room to maintain room temperature.
- Set heating system (switch on brooders) at 29-32°C (85-90 °F) for cage brooding or at 32-35 (90-95) at chick level for floor brooding.
- When chicks are delivered, do not allow the delivery van into the farm premises. Take delivery at the entrance itself.

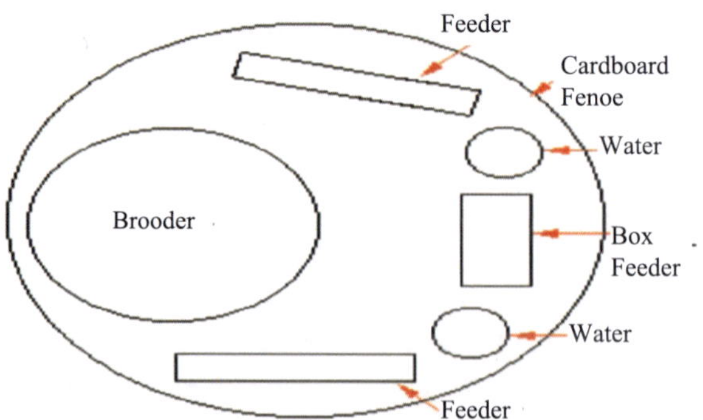

After arrival

- As the chicks arrive, check whether the chicks are healthy, of uniform weight count the chicks, and moist the beak of the chick by dipping it in the water containing vitamins, electrolytes and/or antibiotic and place it gently into the brooder arrangement
- Provide instant energy and restore normal body function.
- For first 3-4 days, feed may be spread over the news paper. Along with this feed should also be provided in chick feeders.

Note: *feed should be supply after 3 hrs on arrival to avoid early enteritis in chicks. For the best result chicks should given water 24 hours after they hatch.*

- Check that the chick's move actively scratching and taking feed and water. Return weak, inactive, unhealthy chicks with matted feathers at the back and the dead chicks and ask for replacement.
- In good management farm mortality upto the sexual maturity is about 10%. It is customary that private hatcheries supply 2 to 4 % extra chick to take care of sexing mistake and early mortality.
- Distribute chicks equally after counting under the brooders. Chick should be vaccinated against Marek's disease on the day of hatch at the hatchery.
- Heat source should be placed at the right height at the centre and the temperature has to be checked by observing the chick comfort zone
- Newspaper - removed after 3 or 4 days and must be burnt.
- Chick or brooder guards - removed after a week or 10 days.
- As age advances the initial chick feeders and waterers may be replaced with bigger ones to match the growth of the chicks.

Brooding temperature and duration

- Before receiving the chick, the heat source should be switched on
- Temperature required during the first week is 90-95°F, later temperature reduced by 3°C (5°F) per week, until the room temperature of 20°C (70°F) is reached. After six weeks of age, temperatures in the 18 to 21°C (65 to 70°F) range are desirable.
- As a thumb rule the temperature inside the brooder house should be approx 20°F(-6.7°C) below the brooder temperature.
- Temperature taken at a point of 15 cm outside of canopy and 5 cm above the top of the litter.
- The house temp 75°F for 1st 4-5 days and 65° -70°F for rest of period.

Age of chicks	Brooder Temperature	
	°F	°C
1st week	95	35
2nd week	90	32.2
3rd week	85	29.4
4th week	80	26.7
5th week	75	23.9
6th week	70	21.1

Effect of temperature on chick comfort

- It is necessary to verify whether the warmth given is sufficient to the chicks. A thermometer kept at the bird level will indicate the temperature
- More practical way of assessing the adequacy of warmth provided is by watching the distribution of the chicks within the brooder guard management.
- If they crowd under or near the source of heat, then the warmth given is not sufficient. Then, a bulb may be added to the hover or the height of the hover may be brought down.
- If chicks have moved to the periphery and are reluctant to come to the centre under heat source, then temperature in the environment is higher than required. The hover may be pushed up or a bulb removed.
- If the chicks feel comfortable at the given temperature, they walk actively throughout the area unmindful of the heat provided and some take rest setting their head down on the side, the posture being given the name as *"Chick comfort"*.

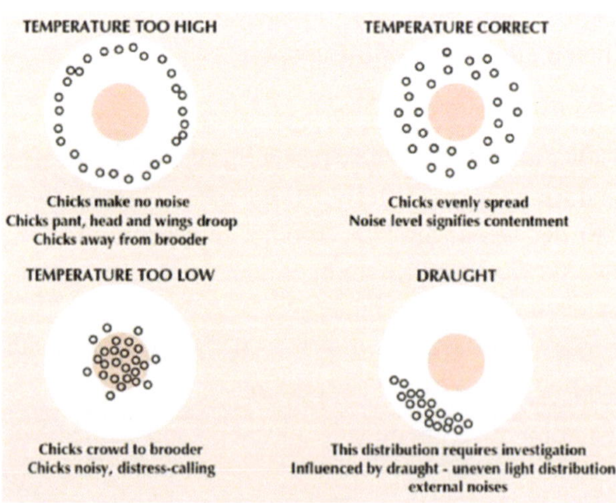

Birds distribution under brooder

Humidity

- Relative Humidity (RH) in the hatcher, at the end of the incubation process will be high (approx.80 percent).
- To limit the shock to the chicks of transfer from the incubator, RH levels in the first three days should be maintained near 70 percent.
- The goal is to maintain in-house relative humidity at 50-65% during brooding as long as possible.
- If it falls below 50 percent in the first week, chicks will begin to dehydrate, dusty litter resulting in respiratory trouble and poor growth of feathers. In such cases, action should be taken to increase RH.If the house is fitted with high-pressure spray nozzles (i.e., foggers) for cooling in high temperatures, then these can be used to increase RH during brooding
- As the chick grows, ideal RH falls.
- High RH can cause wet litter which encourage the development of coccidiosis.
- As broilers increase in live weight, RH levels can be controlled using ventilation and heating systems.

Light

- Light for 1st 48 hrs – 23 hours (1 hr darkness) at 3.5 fc of illumination at floor label, for this 3.5 watt bulb for 4ft square of floor space when bulb hang at 8 feet above floor label.
- Light after the 1st 48 hrs – 1fc of illumination at floor label, for this 1 watt bulb for 1 ft square of floor space when bulb hang at 8 feet above floor label. Light enters to the eye of the bird and sends a message to the brain
- Through optical nerve which stimulates pituitary gland to release hormone. It stimulates broiler chicks to eat more to gain muscle.

Ventilation

- The aim of ventilation is to supply of fresh air and removal of heat and toxic gas. Excess ammonia or carbon dioxide, along with too high or too low relative humidity, can become serious problems, especially during winter flocks. Obnoxious odor of ammonia irritates the eyes of chicks and retards the body growth. Wet litter due to poor ventilation is often predisposing factor for outbreak of coccidiosis. The only way to solve or reduce air quality problems once they have occurred is to increase the ventilation rate.

- Three to five air exchange per hour.
- Concentration of carbon dioxide >0.01 % becomes poisonous to chick.
- CO_2 levels should be kept below 3,000 ppm.
- Ammonia levels should be kept below 25 ppm.

Space requirements for chicks

- Two types of brooder space should be considered in this context-space under the hovers and floor space in brooding house.
- Brooder capacity – in a canopy type 500 chicks
- Space requirement under the hover depends upon type of hover being used and source of heat employed.
- 10 sq inch (65.5 sq.cm/chick)- electric hover
- 7 sq inch (45 sq.cm/chick)- kerosene or coal

Floor space

- 0.05 sq.m/chick (500 sq.cm or 0.53 sq.ft) for start, which is increased by 0.05 sq.m after every 4 week until pullet are about 20 week.
- For broiler at least 0.1m sq (1000 cm. sq or 1.07 ft. sq) of floor for female and 0.15 m sq (1500 cm.sq or 1.61 ft.sq) for male till 8 week of age.

Feeder and waterer space for chicks

- Feeder space for 0-2 week = 2.5 cm & for 2-6 weeks = 4.5 cm or (3 hanging feeder of 36 c.m diameter with 12 kg capacity for 100 chicks)
- Waterers space 0-2 week = 0.5 c.m/chick (50 cm for 100 chick) and at 6-8 week- 1.52 to 1.90 cm (152-190 cm/100 chick)
- Two fountain type waterers (capacity 5 lit) for 100 chicks.
- The waterer placed on litter so that the water level will be convenient to chicks for 1st 2 days. After that at 1'' (2.5 cm) high.
- Feeder space should be enough so that at least two-third of all the birds can eat at one time. Drinker space will be half of the feeder space.
- At 24 to 36 hours of age when chick given their first feed.
- The lip of feeder and waterers should be only 5c.m in height from the level of the litter so that the chick can readily reach the feed and water.
- During first week, feeder may be filled up to the brim there after to avoid feed loss feeders not be fill more than one third of the capacityfull at any time.

Feeding of chicks

- Brooder mash with 22% crude protein and 2700 Kcal/kg of metabolisable energy has to be prepared and provided (As per the recommendation of BIS). Feeding of chicks intially on news paper.

Litter

- Organic materials like rice husk, sawdust, chopping straw, dry leaves, wood shavings, dried cane fiber and corncobs etc can be used as litter materials for brooding purpose. Inorganic materials such as calcium oxide powder, ash, sand etc are used with organic litter. Litter materials should be inexpensive and nontoxic. It should have strong moisture and urine absorbing capacity from the poultry droppings. At first 2 days of brooding newspaper is used on litter. Calcium oxide powder inhabits the growth of parasite and microorganism. Fresh and shallow litter is used in the brooder house. It should be 4-5 cm in depth.

Debeaking/Beak trimming

- To avoid feed wastage, the chicks are to be debeaked.
- Layer chicks are comparatively more active, they tend to peck at each other's back (vent pecking) cause injury and death also. Debeaking helps to prevent such instances.
- Beak trimming ensure low mortality, less feather pulling (feather pecking), better feed conversion.
- Debeaking if carried between one day to 6 weeks of age. Sometime repeated and done in pullet at 16 week when pullet transfer to layer house.
- The upper beak has to be cut 2/3rd and the lower beak 1/3rd portion. The cut portion has to be cauterized (2 Sec) (Destroying the tissue responsible for beak re growth) by touching on the hot plate (1500^{0} F/815°C). The tongue should be carefully held back.
- Undertake debeaking during cooler parts of the day.
- Provide anti-stress B-complex vitamins and vitamin K in drinking water before, during and after the day of debeaking; adjust the feeder and waterer height to be lower than before to suit the shortened beak.

Advantages of debeaking

- Toe picking reduced
- There is less stress in the flock
- It help to prevent feather picking and cannibalism

- Feed efficiency is improved as a result of less wastage
- Liveability is better, with fewer culls.
- Thee is more uniformity of the birds in the flock.

Disadvantages of debeaking

- Birds loose weight
- Growth rate is reduced

Dubbing

- Removal of comb
- Done in day old chicks.
- Done in breeds having larger combs
- Prevent comb injuries during fight and picking, provide better vision, and lower damage by frostbite.
- **Toe clipping** - clipped to prevent tearing the back of the female during Mating.

Vaccination Schedule - Broiler Chicken

To prevent diseases every bird should be vaccinated

SI. No.	Age in days	Type of vaccine	Remarks
1.	4-5th day	RD VF/LaSota or F1	Drops at the nostril/ eyes/drinkingwater
2.	12-14th day	IBD (Gumboro) (livevaccine)	Drinking water
3.	28-30 days	RD (Lasota/F)	Occulonasal /Drinking water

Problems occur during brooding operation

1. **Coccidiosis control** – It is the most common disease of poultry at young age,coccidiostats are added to feed in sufficient quantity to suppress the multiplication of oocytes

2. **Stress** – Majority of stress is occur when birds are handling during the vaccination and due to that bird are huddle together. To overcome the problem we may increase the brooder temperature to fill birds comfortable or we may add anti stressor compound in water/feed
3. **Inclement Weather** – Environmental heat may create a severe stress although young chicks can tolerate higher temperatures than older birds. When the temperature is more the birds will eat less and drink more water. To overcome this problem increase the feed and water along with an increase in the floor space allowance.
4. **Unabsorbed Yolk** – High temperature of chicks during the first two days under the brooder also lowers the yolk absorption. Diseases that raise the body temperature prevent utilization of the yolk material in young chicks. Feeding chicks soon after hatching also causes a slower absorption of yolk materials in young chicks.
5. **Mortality Standards** – Chick mortality during the first week in the brooder house is higher than any week. Losses during the second week should be slightly less.

10

Management of Growers and Layers

Anuradha Kumari, Utkarsh Kumar Tripathi and Manish Kumar

Faculty of Veterinary and Animal Sciences, I. Ag.Sc., RGSC-Banaras Hindu University (BHU), Barkachha, Mirzapur, Uttar Pradesh-231001

Introduction

Period after brooding till sexual maturity is referred as growing period and from one sexual maturity (when hen housed production is 5 per cent) till the hen is sold-out as of laying season or retired for moulting is referred as laying period or laying cycle.

Grower management

The growing stage of chicks starts from 9 weeks of age and ends at 18-20 weeks of age. During this stage, the optimum body weight and growth of reproductive organs are critical to produce eggs in its egg laying phase. Growers may be reared in separate grower house or continue to be reared in brooder-cum-grower house. Proper cleaning and disinfection of grower house is needed before introducing grower birds. Floor space allowance is 1.5 sq.ft./ bird in deep litter houses. Spread litter material to a height of 4" in case of deep-litter system. In cages, each grower is provided with 48 sq.in of cage space only. During this period grower diet should be provided with adequate protein and energy levels to support growth of the birds.

Rearing Sexes separately: Males should be separated from females at least the first several weeks of the growing period. Separation reduces the stress and allows the males and females to attain the target weight.

Floor space requirement for growing chicken on deep-litter

Type of birds	Floor space/bird (sq. m.)
Egg type pullets (small strain)	
to 18 weeks	0.13
to 21 weeks	0.18

Egg type pullets (medium and large strains)	
to 18 weeks	0.18
to 22 weeks	0.20
Egg type breeder pullets	0.22
Egg type breeder cockerels	0.25
Meat type pullets	0.3

Feeding of growing birds

Feeding and watering space should be increased as the birds grow older. If restricted feeding programme is followed in growing birds, it is necessary to provide more feeding space than usually recommended in order to obtain uniform growth and avoid introduction of bad vices in birds.

Restricted feeding

The important feature of grower phase feeding is restricted feeding. It is not advisable to provide ad lib. feeding at this age as the birds may tend to put on more fat and thus, their future egg laying ability may be affected. When full fed during the growing period they gain excessive weight and mature earlier. This will lead to little persistency of peak production and decrease the number of hatching eggs. So, in recent years emphasis has been placed on the merit of restricted feeding of poultry during the growing period.

Advantage of feed restriction

A. Uniformity in the flock
B. Produces better egg size
C. Reduces body weight at sexual maturity
D. Offers better livability during egg production
E. Avoid problems of fertility and hatchability
F. Leg disorders due to over feeding can be controlled.
G. Mortality due to over feeding will be prevented.
H. Avoid mating difficulties due to their large body size in males

The higher net returns from restricted birds (income over feed and chick coasts) are due to both higher income from eggs (egg income is higher in restricted birds due to increased egg production, higher percentage of large and medium size eggs, increased livability in laying house) and lower feed costs.

A few suggestions and cautions may be helpful in practicing restricted feeding in growers

1. 20 to 30 per cent restriction of feed on weight basis is recommended. Eighty per cent of ad lib intake is perhaps the best level under most of the circumstances.

2. Feed restriction should never be practised during brooding period. It should be started after 6 or 8 weeks of age and may be terminated by 20th week.
3. Birds may be fed twice daily.
4. Pullets on restricted feeding programme need enough feeder space so that all can eat at the same time.
5. Pullets on restricted feeding programme adopt themselves to a feeding schedule. At feeding time they are very hungry, ready and anxious for feed. Failure to feed on scheduled time may lead to feather picking and other vices.
6. Feed restriction will also restrict the drug intake given in feed which may not be sufficient to prevent or cure particular ailment unless the concentration of drug is enhanced proportionately in feed.
7. Full feeding should be practised during the disease period. Restricted feeding should be started after recovery.

There are two types of restricted feeding

Quantitative feed restriction

- In which the amount of feed is reduced below the normal requirement of birds.
- This can be done on day-to-day basis or skip-a-day programme or skip-two days in a week programme.
- But this restriction depends on the matching of the flock average body weight with standard body weight provided by the breeder.
- Quantitative feed restriction is usually followed in commercial breeders.

Qualitative feed restriction

- In which the quality of the feed is reduced below the standard requirement of the bird.
- This can be done by including unconventional feeds or lesser nutrient feed ingredients in place of high protein or high energy diet.
- Here the quantity of allotment to the bird is not restricted.
- During restricted feeding programme, provide more number of feeders and see that all the birds are taking feed simultaneously or otherwise dominant birds will take more amount of feed and the weaker will be subjected to feed deprivation and hence the uniformity will be affected.

Flock uniformity

The purpose of feed restriction is to maintain uniformity in body weight of grower phase, thereby attaining a desirable standard body weight at the end of the growing phase. Flock uniformity is measured by comparing body weight of individual birds with flock average. Flock uniformity of 90% is desirable. It means 90% of the birds body weight lies in the range of flock average body weight ± 10%.

Water restriction

Birds under restricted feeding consume excessive quantities of water to satisfy physical hunger. This will lead to distended crop, sudden death syndrome and wet litter problems. It is advisable to provide water to the birds 30 min to 1hr prior to feed delivery and an additional hour of water supply can be given in the afternoon.

Light during growing period

The effect of light on growth rate is due to the pattern of activities of birds leading to food intake induced by the period of lighting. There is no evidence of any direct physiological effect of light on growth. The recommended lighting programme for growing period of 9-16 weeks is 12 h. It means natural photoperiod of 12 h is enough during this period.

De-beaking or beak trimming: Beak trimming is performed early in the life of commercial hens to decrease injuries caused by the behavioural bad habits (vices) like pecking and eating , to avoid feed wastage and to avoid egg-eating vice. For this , pullets must be trimmed at 7-10 days of age and then re-trimmed at 8-10 weeks of age. For the majority of birds, it involves the partial removal of the upper and lower beak using an electrically heated blade. For birds beaks should be trimmed 6 to 7 mm beyond the nostril with 2 seconds of cauterization. After beak-trimming feeders must be kept full with feed to help birds eat easily. Also provide vitamins B-complex, C and K can be through water to help reduce stress.

Transfer of pullets to laying house

One of the most profitable practices is to follow "All-in, all-out" practice. This simply means that the farm has birds of only one age group. It helps in disease control of the flock. With decreased stress from disease and reduced mortality, rate of lay is higher. Over and above these, birds of one age group greatly simplify the management.

Placing of Laying Nests

Nest-boxes should be placed at least one month before the start of lay in order to encourage the birds to lay eggs in the boxes (multi-tiered egg boxes can be used for chickens but not for ducks. All nest-boxes should be at ground level for ducks.

Management of Layers

The layer chicken refers to the hybrid chickens reared exclusively for the production of table eggs. These chickens are selected for high egg production and free from broodiness. In their production process, at the hatchery itself, day-old chicks are sexed. Only female chicks are sent to commercial layer chicken farming for further rearing and table (infertile) egg production for food purpose.

Laying phase of layer chicken starts essentially from 18-20 weeks of age and lasts up to 72 weeks and above. The extension of production beyond 72 weeks is based on the flock's egg production performance and market rate of egg. Growers should be transferred to layer cages on 17th week of age to adapt them for the new environment and lighting schedule.

The mean mortality during laying period is 10%. The layers will be culled after 72 weeks and above, and sent for table purpose as culled bird. Decision on culling age is not rigid but flexible, which depends on factors like hen day egg, replacement pullet cost etc

Light during laying period

Even distribution of light in poultry house is important for proper utilization of all space available in the shed. There are evidences to suggest that light affects growth and reproduction by different physiological actions. Light entering the eyes of birds induces response in the hypothalamus which through releasing factors in turn affects the rate of secretion of gonadotropic hormones from the anterior pituitory gland. The gonadotropic hormones affect the activity of gonads and in turn ape reproductive behaviour of birds. Age of sexual maturity and rate of egg production are affected by pattern of lighting in laying house but this does not have the power to either stop or start these reproductive processes.

The ideal photoperiod during laying period is 16 h. However, from the 12 h lighting period of growers, increase in lighting period should be smooth and incremental. From 17th week onwards increase the photoperiod to the tune of half an hour per week and end up in 24th week with a photoperiod of 16 h. The recommended photoperiod is 22-24 h during brooding period gradually reduced to 12 h at the end of eight weeks of age. The photoperiod should be kept constant for 12 h up to 16 weeks of age. From 17th week onwards, the photoperiod of 12 h/day is gradually increased to 16 h/day with the weekly increment of half an hour per week up to 24th week. Then throughout the laying period, a constant photoperiod of 16 h should be provided. The 16 h photoperiod includes 12 h of natural and 4 h of artificial photoperiods. The 4 h artificial photoperiod can be split into two phases for morning and evening

Intensity of light is not very important factor. It may vary between 0.1 and 100 F.C. without having significant effect on growth, and it may vary between 0.5 and 38 F.C. without having effect on egg production. For practical purposes there should be read in the shed. This will provide enough light to stimulate the process of reproduction.

Layer chickens are reared in following two general systems

Deep litter system of rearing: In this system, layers are reared on floors of the deep litter houses, and it is almost not in practice in commercial layer farm operations in India. The design of buildings for rearing layer type birds on deep litter is similar to that of broilers. The specifications for width, length, sidewalls, floor and roof of broiler house will also apply to layer house.

Space Requirement

During laying period, hens require 1400 cm^2 of floor space, 8.75 cm of feeder space and 2.5 cm of drinker space.

Cage system of rearing: It is the most popular system of rearing in commercial layer production. The cage rearing of chicken layers is of two types, viz. floor level and elevated cage houses. Floor level cage houses are mainly used in brooding phase of layer chicken. Elevated cage houses are mainly used for growers and layers. Considering the different phases in layer chicken, poultry houses may be classified as follows:

Fig. 1. Layer cages

Brooder house: It is a house, in which chickens are reared from day-old to eight weeks of age. These are mostly floor level cage houses.

Grower house: In this house, chickens are reared from 9 to 20 weeks period and they are elevated cage houses. (Brooder-cum-grower house models are also available in which chickens are reared from day-old to 20 weeks in one house. This system is not common now in commercial operations).

Layer house: Layer chickens are reared in this house during their egg laying phase (21-72 weeks and above) and they are also elevated cage houses.

To ensure continuous supply of eggs, proper planing is required in terms of number of different houses. The number of various houses required varies according to the batch interval of chicks. Based on this, the layer farm may be established as follows:

1+1+4 pattern- One brooder house + one grower house + four layer houses and batch interval for chick placement will be 15 weeks.

1+1+5 pattern- One brooder house + one grower house + five layer houses and batch interval for chick placement will be 12 weeks.

1+1+6 pattern- One brooder house + one grower house + six layer houses and batch interval for chick placement will be 10 weeks.

Feeding of layers

Layer ration which is usually in mash form is offered for layers. Layers are never feed-restricted; they are given feed *ad libitum* meaning as much as they want.

The age at sexual maturity is 18-20 weeks, also called as point of lay. During this time, birds lay small size eggs called as pullet eggs. At the point of lay, prelayer ration shall be offered with incremental additional calcium level to condition calcium reserve of birds for supporting initial shell egg production. There will be upward swing in egg production, attaining a peak egg production of more than 90% on hen day basis and remain there for up to 45-50 weeks of age. The production will start a downward swing to 85-90% production on hen day basis. Based on the production performance and age, laying phase can be (20-45 weeks of age) and phase-2 (45 weeks of age and above). Mean hen day phase-1 egg number in 52 weeks production period will be 310 eggs. At an ideal temperature of 21.1°C, White Leghorns (most of the commercial layers in market are strains of this breed) consume 110 g feed and 220 ml water per day. The total feed intake during the laying period will be 42 kg. As the age advances the egg weight will also increase, for which little bit of additional protein in terms of essential amino acids is required. Because of these variations in production in laying, phase feeding is followed in commercial layers. The egg laying time is between 10 AM to 1.00 PM and eggs should be collected twice a day and in summer thrice a day to maintain the quality of eggs.

Pre lay diets: Pre lay diets are designed to make the bird store adequate calcium in medullary bone reserves, which is necessary for the birds for calcification of the first egg. Generally, these diets are fed two weeks prior to egg production and up to 5% hen day egg production. Pre lay diets generally contain 2-2.5% calcium. Lack of good pre lay diets may result in lameness, cage layer fatigue and drop in egg production.

Phases of a commercial layer

The system of rearing, feeding and management specifications with different phases of commercial layers. The layers are generally reared in three phases:

a) **Starter (0 to 8 weeks):** Young chicks need more of body building nutrients in their ration. There is a greater need of protein for growth of tissues and feathers.

b) **Grower (9 to 20 weeks):** The rate of growth is slowed down during this period, but chicks need enough nutrients for proper body development and reserves for the subsequent laying period.

c) **Layer (21 weeks & above):** During the early period of laying, a young pullet must consume sufficient amount of all nutrients to allow her to produce: (a) eggs at a maximum rate; (b) to maintain her body weight; (c) eggs of bigger/ optimum size; (d) reserve nutrients in the body against stress and diseases; and (e) requirements of Ca increases significantly with increased rate of egg laying.

BIS specification for poultry feed

Characteristics	chick feed	Growing Chicken feed	Laying chicken feed	Breeder laying feed
Moisture max %	11	11	11	11
Crude protein min %	20	16	18	18
Crude fibre max %	7	8	8	8
Acid insoluble ash max %	4	4	4	4
Salt (as (NaCl) Max %	0.6	0.6	0.6	0.6
Calcium min %	1.0	1.0	3.0	3.0
Available phosphorus min %	0.5	0.5	0.5	0.5
Lysine min %	0.9	0.6	0.65	0.65
Methionine min %	0.3	0.25	0.3	0.3

Metabolizable energy (ME) Kcal/ kg	2600	2500	2600	2600
Manganese, mg/kg	90	90	90	90
Vitamin A, IU/kg	6000	6000	6000	6000
Vitamin D_3, IU/kg	600	600	1200	1200
Vitamin E, mg/kg	15	10	10	10
Vitamin K, mg/kg	1	1	1	1
Riboflavin, mg/kg	6	3	3	3
Biotin, mg/kg	0.2	0.15	0.15	0.15
Choline	1300	900	800	800
Pyridoxine, mg/kg	5	5	5	8
Aflatoxin, max, P.P.b.	50	50	50	50

Phase Feeding

The production cycle is divided into three stages and phase feeding is applied for energy re-striction. Thus, phase feeding refers to:

a) Changes in the laying hen's diet;

b) Adjust to age and state of production;

c) Adjust for season of the year and for temperature and climatic changes;

d) Account for difference in body weight and nutrient requirement of different strains of birds; and

e) Adjust for one or more nutrients as other nutrients are changed for economic and availability reasons.

In phase feeding, high protein feed (usually 18-19%) is given from onset of egg production to peak production period. Therefore, a low level of protein (about 16%) is fed for the next 5 or 6 months, followed by still lower level (usually 15%) until the laying period is completed. Phase feeding thus helps to reach higher peak production and sustain it longer. It is known that the requirement of calcium increases while that of phosphorus decreases with each succeeding phase.

i) Phase-I

From the age of 20 to 40 weeks, the birds are expected to reach from zero to peak egg production of 85 %. This further is accompanied by increase in body weight by 500 g and increase in size of eggs from 40 to 60 g. Therefore, the first phase of reproduction is critical for maximum egg production and tissue develop-ment. Thus, the energy content of poultry ration should be adjusted to supply required quantity of protein.

ii) Phase-II

It is the period from 41 to 60 weeks when hens have attained mature body weight and egg production has not gone below 60 %. The eggs produced are larger and efficiency of protein utilization is approximately 56 % during this phase.

iii) Phase-III

It ranges from 60 weeks and above when the production declines but the body weight and egg weight slightly increases or till the spent birds are discarded. During this phase, egg production is less than 60 %. Requirements of proteins are reduced particularly in this phase.

The special care of calcium nutrition needs is considered, as increasing calcium level in the feed, increases vitamin D level, supplementation of shell grit in the feeder during the noon @ 3 g/bird and supplementation of organic acids increase calcium absorption. The above measures are needed as the calcium absorption decline as the age of the birds advances leading to egg shell breakage problem

Fig. 2. Shell grit box

Recycling of the layer flock: In commercial chicken egg production process, replacement pullets are the second costly item after feed. Considering that, to economize the production cost, recycling of layer birds is preferred. Recycling is the process of keeping birds for a longer period of egg production with a pause in the midway, which is by induced-molting programs.

Moulting

It is natural, physiological process to renew old feathers at the end of first year of laying. Natural moulting represents the renewal of old feathers few times at different stages of life cycle including the end of the egg laying period. If this process is artificially induced, it is called as artificial molting. The time and duration of the first annual moulting are important points in distinguishing the poor and good layers.

During moulting process, the feathers on different parts of the body are shed in the following manners: head, neck, breast, body, wings, and tails. It takes about six-weeks for a primary to grow to its full size-in three weeks a primary feather grows 2/3 of the size and for remaining 1/3 size the rest three weeks are utilized.

Early moulters are usually poor layers. Such birds start moulting early and take unusually long time (24 weeks or so) to complete moulting. Often they stop laying during the moulting period. They shed only one primary at a time at the intervals of two weeks, during moulting period. In contrast to this, good layers start moulting late during their first year of laying. Some exceptionally good layers do not shed their old primaries at all and continue with them to the second year of lay. Good layers complete the moulting quickly by sheding two or more primaries together. They may sometime continue to lay during moulting period. Process of moulting is completed only in 1 to 2 months period whereas poor layers may take 6 months.

Forced moulting in laying hens

Normally hens live for more than six years. Egg laying in commercial hens starts giving eggs at around 16-18 weeks of age, reach peak at 25 weeks of age and continue up to 72 weeks. After that flocks are economically not viable and slaughtered as spent hens. However, instead of slaughtering, hens are force moulted to restart egg laying for one or two more laying cycles.

It is also termed as recycling or controlled moulting. Force moulting refers to moult which has been induced in laying hens by introducing planned stress usually during non-moulting periods. The objective is to give the laying hen rest or to rejuvenate her lost vigour during laying period, and make her lay eggs again after the rest.

Force moulting can be considered in one more of the following circumstances

1. Non-availability of funds to replace new pullets.
2. Slow-down in number of pullets raised because of loss from certain diseases.
3. When seasonal rate of egg is low, the hens can be force moulted during that period, i.e., summer in Indian condition.

4. When there is a market for extra large eggs or premium rates.
5. When quality replacement flock is not available and costly.
6. When there is low return from spent out birds.

There are several methods of moulting suggested using single or combinations of light, feed, water restriction and use of hormones.

Moulting through light restriction: It is a preferred procedure when light controlled house is available in which light infiltration is essentially eliminated. The common method is given below:

Day 1- Reduce light from 14 to 16 hours to 6 or 8 hours.

Day 22: Remove all feed except shell.

Day 26 to 28: When egg production stops, resume full feeding of regular ration.

Day 50: Increase light to 14 or 16 hours.

Moulting through restriction of feed and water: Turn off lights, withdraw water for 48 hours to 72 hours, depending upon season of the year. Feed is withdrawn completely upto seven days or until birds are out of production. Water is supplied during feed with drawal period. Feed is given only 1/6 to 1/4 of normal intake during the resting period of 30 to 45 days. Lighting period in the shed is increased after rest and when full feeding is restored.

The advantages of flock recycling are extended egg production period up to 100-110 weeks with improved egg weight on the other hand shell thickness and internal quality of eggs will be compromised.

Culling of non-layers and poor-layers

Culling is an important managemental procedure to be practiced in case of layers and breeder females. Even if one starts with good stock, invariably there will be some which will not grow well and ultimately will form poor layers. Such birds in laying house will reduce the efficiency of flock and often will remain as continuous source of problem. Therefore, usually monthly culling is followed to improve economy by removing unproductive birds

For culling bird should be held in natural position while examining. Its body should rest comfortably on the palm of left hand leaving right hand free for examination and handling. Birds should be removed from and replaced to the crate or cage with head first,

Night culling is desirable when not more than 10 to 15 per cent birds are to be removed as it avoids undue stress on birds due to excitement and handling. Flash light may be used to catch the birds showing traces of yellow pigment at the base of beak, shrunken and scaly combs and loss of primary, secondary or tail feathers. The birds captured in night are examined in detail during the day time.

Table 1. Distinguish poor layers from good layers

Body part	Layers	Non-layers
Health	Healthy, vigorous, well fleshed but moderately fatty	Appearance sick, sluggish, thin or debilitated or extremely fatty
Feathering	Tight and close feathering, tail and wings well carried up	Loose and scattered feathering, tail and wings are droopy
Plumage	Initially bright and dark coloured but faint and dull in latter stage	Always bright and dark coloured
Comb	Large, red, glossy and warm	Small, pale, dry, shrunken
Wattles	Full	Scaly and cold.
Eye	Bright	Dull
Vent	Large, dilated, oval and moist	Small, contracted, round and dry.
Abdomen	Enlarged, soft and pliable	Contracted, hard and fatty.
Skin	Soft, thin, loose, silky	Thick, dry, underlaid with fat.
Pubic bone	Soft and flexible, well spread, distance between two pubic bones 2 to 3 fingers	Hard, brittle, tightly placed distance between two pubic bones less than 2 fingers
Keel bone and distance	Soft, flexible, distance between keel and pubic bones is more than 4 to 5 fingers	Hard, brittle, distance between keel and pubic bones is less than 3 to 4 fingers

Depigmentation

Xanthophylls, the yellow pigment responsible for yolk colour will be present in various parts of the body viz. round the vent, earlobes, eyelids, beaks, shanks etc. Before the birds begin to lay (Non-layers), the pigment is replenished regularly by the feed the birds eat. But, when the birds begin to produce eggs, the xanthophylls will be taken for yolk colouration and hence, the tissue from where the xanthophyll is oxidized does not get the pigment; and therefore, they get discoloured or bleached or de-pigmented. It can also be noted that non-layers are characterized by presence of yellowish colour round the vent whereas both poor and good layers will definitely have bleached vent. The depigmentation of body parts follows a particular pattern depending upon the blood circulation in those parts.

Vaccination: The following vaccines are administered during the laying period:

Age	Disease	Route
Above 18 to 20 weeks	RD	Spray or drinking water (Every 3 months)

Evaluation of Performance of Layers

The following are some of the practical indicators of the performance of layer farm.

a) Hen-Day Egg Production (HDEP)

For a particular day

$$\text{HDEP} = \frac{\text{Total number of eggs produced on a day}}{\text{Total number of hens present on that day}} \times 100$$

For a longer period =

$$\text{HDEP} = \frac{\text{Average number of eggs produced over a period}}{\text{Average number of live hens}} \times 100$$

It is mostly used for the scientific studies and truly reflects the production capacity of the available birds in the house. A farm average of 85% or more per year is desirable.

b) Hen-Housed Egg Production (HHEP)

For a particular day

$$\text{HHEP} = \frac{\text{Total number of eggs laid on a day}}{\text{Total number of hens housed at the beginning of laying period}} \times 100$$

For a long period

$$\text{HHEP} = \frac{\text{Total number of eggs laid during the period}}{\text{Total number of hens housed at the beginning of laying period}}$$

For a given flock, HDEP and HHEP will be same only when all birds housed survive throughout the period.

c) Feed efficiency (FE) / Kg egg mass

- It is the ratio between the feed consumed and the egg mass in Kg.
- This takes into consideration the feed intake, egg weight and egg production.
- A value of 2.2 or less is advantageous to the farm.

d) Feed efficiency (FE) / dozen eggs

- It is the feed consumed per dozen eggs.
- It is the most commonly used index in layers.
- The value should be 1.5 or less in a well maintained farm.

e) Egg mass / Hen housed

- It is the kilograms of egg produced per hen housed. This will take into account the rate of egg production, mortality, as well as the egg weight.
- Average egg mass = Per cent HDEP X Average egg weight in grams (Per hen per day in grams)

f) Egg: Feed Price Ratio (EFPR)

- It is used to find out the ratio between the receipts from egg and expenditure on feed.
- An E.F.P.R. ratio of 1.4 and above is desirable.

g) Break Even Point

- It is the stage, at which the sum of prices of all inputs and outputs will be the same.
- The breakeven point depends on the following factors
- Rate of egg production.
- Prevailing egg price.
- Prevailing feed price
- Culled hen price
- Daily feed intake
- Predictable future trends in eggs feed and culled hen prices.

Table 2. Mean performance of commercial layer chicken (white egg shell)

Traits	Performance
Hatch weight	35-38 g
8th week body weight	625-650 g
Age at sexual maturity	18 weeks
Body weight at sexual maturity	1.2-1.3 kg
Age at 5% egg production	20th week
Period at peak production	26-45 weeks
Hen day egg number up to 72 weeks	310-320
Hen housed egg number up to 72 weeks	300-310
40th week egg weight	56-58 g
Feed intake up to 8th week	1.75 kg
Feed intake up to 18th week	6.0 kg
Feed intake during laying (19-72 weeks)	42 kg
Mean daily feed intake during laying	113 g

Feed efficiency/dozen eggs	1.6-1.7
Feed efficiency/kg egg mass	2.3-2.4
Livability during growing period	95%
Livability during laying period	90%
Age at culling	72 weeks and above
Body weight at culling	1.6-1.7 kg (72 weeks)
Egg shell thickness	0.32-0.34 mm

11

Digestion and Reproduction in Poultry

Manish Kumar, Mukesh Bharti, Anuradha Kumari and Utkarsh Kumar Tripathi

Faculty of Veterinary and Animal Sciences, I. Ag.Sc., RGSC-Banaras Hindu University (BHU), Barkachha, Mirzapur, Uttar Pradesh-231001

The digestive system of domestic fowl is simple yet highly efficient, lighter in weight compared to species like cattle. Due to this simplicity, high-quality and easily digestible feed is essential for optimal productivity. The system comprises the alimentary canal, liver, and pancreas.The alimentary canal is a long tube extending from the beak to the vent (cloaca), lined with mucous membranes and digestive glands. Unlike mammals, fowls lack lips, cheeks, and teeth. Instead, they have a horny beak over the mandible and incisive bones. The egg tooth, aiding hatching, disappears shortly after birth. The hard palate has a median slit connecting to the nasal cavity and five rows of backwardly pointing papillae. Salivary glands release secretions into the mouth cavity. The tongue, long and pointed, matches the beak's shape and has a thick, horny epithelium. It is supported by the hyoid bone, but birds have a weak sense of taste. The oesophagus is wide and highly expandable, connecting the mouth to the crop while running alongside the trachea. The crop, a dilated part of the oesophagus, temporarily stores food before digestion. Within the thoracic cavity, the oesophagus leads to the proventriculus, a glandular stomach that initiates digestion.

Crop structure is similar to that of oesophagus except there are no glands present in fowls. Ducks and geese have glands in crop mucous membranes. In pigeons surface cells of crop slough off during brooding to form pigeon's milk which is used to feed baby pigeons in nest.

Proventriculus: is relatively small and tubular, wall is very thick and is composed of five layers:Outer serous membrane, Muscle layer composed of three separate layers:Two thin longitudinal layers, Thick circular layer, Layer of areolar tissue containing blood and lymph vessels, Thick layer composed mainly of glandular tissue and Mucous membrane.

The gizzard is a muscular stomach located after the proventriculus, partially between and behind the liver lobes. It has a flattened, rounded shape, resembling a convex lens with one side slightly larger. Beneath its outer layer are powerful red muscle masses. The inner surface is lined with thick, horny tissue forming ridges. The gizzard's layers include straight tubular glands and a strong, flexible inner lining, which protects against the grinding action of food, often aided by stones or insoluble materials.

Small Intestine

The small intestine extends from the gizzard to the junction of the caeca and colon. It is long and uniform in diameter, with only the duodenum being easily distinguishable. The jejunum and ileum lack clear separation, appearing as a single tube. Most digestion and all nutrient absorption occur here.

Duodenum

The duodenum forms a 20 cm loop starting at the gizzard, with the pancreas nestled between its arms. The jejunum and ileum extend 120 cm, beginning at the bile and pancreatic duct papilla and ending at the ileo-caecal-colic junction, where the small intestine, caeca, and colon meet. Structurally similar to the duodenum, this section has shorter villi, less lymphoid tissue, and is suspended in the mesentery.

Meckel's Diverticulum

There is a duodenal loop as found in mammals. The vestige of the yolk sac (Meckel's diverticulum) is about midway in the small intestine and is used to indicate the division between jejunum and ileum. This projection is where the yolk sac was attached during the development of the embryo.

Large intestine

The large intestine is short and similar to the small intestine, running straight beneath the vertebrae and ending at the cloaca. It is sometimes called the colon or rectum. Most birds have right and left caeca at the small-large intestine junction. The bursa of Fabricius, present above the cloaca in young birds, disappears by one year of age.

Caeca: Two caeca or blind pouches are about 16-18 cm long in the adult. They extend along the line of small intestine towards liver & are closely attached to small intestine along their length by mesentery. Each caecum has three main parts: A narrow base with thick walls arising at ileo-colic-caecal junction; Middle part with thin walls and Wide blind apex with fairly thick walls.

Cloaca: The large intestine ends at the cloaca, a tubular cavity that serves as a common passage for the digestive and urogenital tracts. Structurally similar to the intestine, the muscularis mucosa disappears near the vent. The cloaca is divided into three chambers: **Coprodaeum** – a continuation of the colon-rectum. **Urodaeum** – the middle section where the ureters and genital ducts open. **Proctodaeum** – the terminal section, opening to the exterior through the vent.

Secretions and Digestion: In chickens and turkeys, the salivary glands secrete mucinous saliva (7–30 mL/day) primarily for lubrication. The esophagus and crop also produce mucus for this purpose. In gallinaceous birds, the crop aids starch digestion through bacterial action and amylase activity. More thorough mechanical and chemical digestion occurs in the stomach and intestines.

The proventriculus contains two main gland types: simple mucosal glands (secreting mucus) and compound submucosal glands (secreting mucus, HCl, and pepsinogen). Pepsinogen converts to pepsin in the acidic environment (pH 0.5–2.5), initiating protein hydrolysis. The gastric content pH is higher (4.8) due to ingesta. The gizzard (ventriculus) grinds food, mixing it with digestive fluids. Grit (sand/stones) aids grinding but is unnecessary with commercial feeds, as they are pre-ground for efficient digestion.

Intestinal, Pancreatic & Biliary Secretions

The small intestine is the main site for chemical digestion, aided by pancreatic enzymes, microbial activity, and intestinal secretions. The exocrine pancreas secretes lipase, amylase, and proteolytic enzyme precursors (trypsinogen, chymotrypsinogen, and procarboxypeptidase). Trypsinogen is activated by enterokinase, which then activates other proteolytic enzymes. Aminopeptidases and dipeptidases in enterocytes hydrolyze oligopeptides. The main carbohydrate in poultry feed, starch, exists in two forms: amylose and amylopectin. Both forms of starch are rapidly hydrolyzed by pancreatic α-amylase into maltose. The key carbohydrate enzymes are maltase and isomaltase. Fats are broken down into fatty acids and glycerol by pancreatic lipase, with bile salts aiding emulsification and enzyme activation. The exocrine pancreas also secretes bicarbonate, which neutralizes acidic gastric chyme. The intestinal pH increases from 5.6 to 7.2, with an optimal range of 6–8. Bile secretion into the duodenum further aids in chyme neutralization, and bile salts are efficiently reabsorbed and recycled.

Absorption

Most absorption of carbohydrates, amino acids, and fatty acids occurs in the duodenum and proximal jejunum. Glucose is primarily absorbed via passive transport in the duodenum and jejunum, with active transport occurring in the ileum and proximal cecum. Amino acids and peptides are absorbed through

Na+/K+-ATPase co-transport in the duodenum and jejunum. Fatty acids are absorbed in the distal jejunum and ileum, re-esterified into triglycerides, and packaged into portomicrons for transport through the hepatic portal system. Unlike mammals, birds do not have lacteals in their villi. Volatile fatty acids (VFAs), mainly acetate, are produced by microbial breakdown of uric acid in the ceca and are absorbed via passive transport.

Utilization of Yolk

Avian embryos derive energy from lipids stored in the yolk, which makes up about 50% of the yolk material. The yolk contents in chicks and poults are utilized through two main processes:

i) **Lipid transfer:** Lipids are transferred from the yolk to the bloodstream via endocytosis by endodermal cells of the yolk membrane. This process starts early in embryonation, accelerates in the final week of incubation, and continues after hatching.

ii) **Yolk secretion:** Yolk material is secreted through the yolk stalk into the small intestine in pulses. In chicks, this occurs for the first 72 hours, and in poults, for 120 hours. The yolk is distributed via peristalsis and anti-peristalsis, with lipids in the proximal intestine being hydrolyzed and utilized, while this does not occur in the ileum and cecum. By 72 hours post-hatching, lymphocytes accumulate in the yolk stalk, partially occluding it and stopping yolk passage. The yolk stalk is fully occluded by day 4th and later transforms into lymphopoietic tissue by day 14, potentially serving as a site for extramedullary haematopoiesis. The remnant of the yolk stalk is known as Meckel's Diverticulum.

Reproduction in birds

One of the most apparent differences between birds and mammals is the absence of a well-defined estrous cycle or pregnancy in birds. Being oviparous, birds must ensure that the embryo has all the necessary resources at the time the egg is laid. Consequently, their reproductive physiology and anatomy exhibit significant modifications compared to mammals.

Ovary

The ovary consists of an outer cortex containing follicles and an inner medulla composed of connective tissue, blood vessels, and nerves. In domestic birds, only the left ovary and left oviduct are functional, positioned along the dorsal wall, anterior to the left kidney and posterior to the left lung. The ovary features distinct follicles protruding from its surface, with the largest and most mature follicle designated as F1, followed by F2, F3, and so on, down to F5.

This arrangement, known as follicular hierarchy, reflects a sequential growth pattern, with each follicle differing from the next by approximately one day of development.

The ovary contains intermediate-sized follicles and numerous microscopic ones. Immature follicles consist of an oocyte surrounded by follicular cells, with a nucleus, organelles, and a surrounding membrane. As follicles mature, yolk accumulates within the ovum, enclosed by the perivitelline membrane secreted by follicular cells. Mature follicles comprise multiple layers, including the vitelline membrane, zona radiata (innermost layers), perivitelline layer, granulosa layer, theca interna, theca externa, connective tissue, and epithelium(outermost). Ovulation occurs through the avascular stigma. Unlike mammals, no corpus luteum (CL) forms after ovulation, and follicular remnants persist only briefly.

Follicle growth

The avian ovary contains follicles at various developmental stages. During the breeding season or photostimulation, yolk-filled follicles form a hierarchy, with the largest follicle ovulating next. Yolk production, regulated by estradiol, occurs in the liver and is deposited in concentric layers, creating colored rings over time. The oocyte is enclosed by the vitelline membrane, with the germinal diskcontaining nuclear materialvisible beneath it. In birds, females (ZW) determine embryo sex at ovulation via the first polar body extrusion. The granulosa cell layer, an avascular membrane, surrounds the vitelline membrane, followed by a basement membrane and the vascularized theca layer. Each follicle is suspended on a stalk.

Ovarian Hormone

Steroid hormones are produced by follicular granulosa and thecal cells. Before the first ovulation, the ovary contains only small white follicles without yolk. Thecal cells of these small follicles primarily secrete estrogen and androgens, while granulosa cells of the follicular hierarchy (F5 to F1) produce ovarian progesterone, which enters the bloodstream. Some progesterone is converted to estrogen in nearby cells, but this conversion decreases as follicles mature, with F1 follicles secreting only progesterone.Rising plasma progesterone levels trigger LH release for ovulation, while LH stimulates steroidogenesis in granulosa and thecal cells. The role of FSH in birds remains unclear, whereas prolactin (PRL) inhibits ovarian steroidogenesis. Granulosa cells of F1 follicles also produce prostaglandins (PG) involved in uterine contractions and oviposition, along with inhibin (which suppresses FSH) and activins, whose function is not yet fully understood.

Oviduct

The avian oviduct serves as a conduit from the ovary to the cloaca, with specialized regions for different functions. The oocyte spends varying durations in each section. During ovulation, the fimbriated end of the infundibulum becomes active, capturing the ovum. Fertilization occurs here, where sperm storage glands are also present, and the egg remains for about 15–30 minutes. The magnum, the longest section, deposits albumen in response to estrogen over 2–3 hours. The eggremains in the isthmus for 1–1.5 hours and this where the innerand outer shell membranes are deposited. The egg then spends approximately 20 hours in the shell gland (uterus), where the shell, along with water, salts, and pigmentation, is deposited. The uterus is separated from the vagina by uterovaginal sphincter. The vagina contains tubular glands (Sperm storage tubules) that are capable of prolonged storage of spermatozoa. It functions in transport of spermatozoa and transport of eggs. Sperm is stored for up to 32 days in chicken and 70 days in turkeys.

Egg Formation

The entire process from oocyte release to egg laying takes approximately 25–26 hours. The egg spends about 15 minutes in the infundibulum (site of fertilization and transport), 3 hours in the magnum (where albumen is secreted in response to yolk passage), 1.5 hours in the isthmus (where two shell membranes form), 20 hours in the shell gland (where the shell, fluid, and pigmentation are added), and less than a minute in the vagina before being laid.Fluid added in the shell gland during the first 5 hours doubles the albumen volume, leading to stratification. The shell, composed of 98% calcium carbonate and 2% glycoproteins, forms in the final 15 hours, featuring pores for gas exchange. A proteinaceous cuticle coats the shell, preventing bacterial entry.

Ovulatory Cycle: The cycle of follicular growth and hormonal changes culminating in ovulation is known as the ovulatory cycle, which lasts about 24–26 hours in hens. Continuous ovulatory cycles without interruption form a clutch, but anovulatory days may occur before ovulation resumes.In young, high-yielding hens, egg formation takes around 24 hours, resulting in longer clutches. As hens age, the interovulatory interval extends to 25–26 hours, leading to shorter clutches. Chicken clutches vary from 1 to 30 eggs, though 10–20% of ovulated ova are released into the abdominal cavity without resulting in eggs. The first egg of a sequence is laid in the morning, with subsequent eggs laid progressively later each day until the last egg is laid in the late afternoon. Ovulation occurs approximately 30 minutes after the previous egg is laid. Since egg formation takes over 24 hours, ovulation and oviposition gradually shift to later times on successive days.

Regulation of Ovulatory Cycle

LH is the key hormone driving the ovulatory cycle, triggering ovulation with a surge 4–8 hours prior. This LH surge occurs during the dark period. Under a 15-hour light and 9-hour dark cycle, the ovulatory cycle lasts 25–26 hours, causing the LH surge to shift progressively later in the dark phase with each successive ovulation in a clutch. If the surge coincides with the light period, LH release is inhibited. Progesterone and LH exhibit a positive feedback loop. The F1 follicle primarily secretes progesterone, which stimulates LH release, further enhancing progesterone production. Both hormones peak 4–8 hours before ovulation and decline afterward. Androgen levels also rise alongside LH and progesterone, contributing to LH stimulation. Estradiol shows a smaller increase. The role of FSH in the ovulatory cycle remains unclear, while prolactin levels may rise as LH and progesterone decline, potentially acting as a regulatory mechanism to suppress LH release.

Broodiness

Broodiness occurs when egg production declines, and the hen begins incubating eggs, accompanied by reduced food intake. It is rare in laying hens, observed in broiler strains, and common in turkeys, where it poses a commercial issue. During broodiness, gonadotropin levels drop, ovarian regression occurs, and prolactin levels rise.

Oviposition

Each ovulation is followed by oviposition, typically 24–26 hours later. Except for the last egg in a sequence, ovulation occurs soon after oviposition. Prostaglandins from both preovulatory and postovulatory follicles play a role, with the postovulatory follicle being more significant, as its removal delays oviposition for days. Additionally, arginine vasotocin from the posterior pituitary rises at oviposition.

Photoperiodism

Photoreceptors: Photoperiod is the primary environmental cue for reproductive activity in birds. While factors like nutrition, rainfall, and social interactions influence some species, most temperate-zone and domesticated birds rely on light exposure. Photoreceptors in the hypothalamus detect light, triggering the release of gonadotropin-releasing hormone (GnRH), which in turn stimulates the release of LH and FSH which causes growth of follicles and production of estrogen and androgen from the small follicles. Estrogens stimulate oviductal development and formation of medullary bone; the and rogens cause expression of secondary sex characters. Egg production follows 2 – 4 weeks after photo stimulation.

Photo stimulation: Light stimulates reproductive activity in birds by ensuring reproductive competence and regulating cycle timing. For most domesticated species, a long day exceeds 12 hours of light, which need not be continuous. Photostimulation occurs if light exposure aligns with the photosensitive phase, typically at least 12 hours after dawn, even if interrupted by darkness. This sensitivity recurs daily.

Photo refractoriness: A bird's response to light depends on its previous photoperiod exposure. Increasing daylength is stimulatory, while decreasing daylength induces regression. Prolonged long-day exposure leads to absolute photo refractoriness, where the bird no longer responds to that light duration. Exposure to a period of short daylengths is necessary to break both absolute and relative photo refractoriness and to cause the bird to regain photosensitivity.

Male Reproduction

Testes: The testes inbirds are paired and internal. They are located near the cephalic end of the kidneys and ventral to them. There is a small epididymis and a ductus deferens that conducts sperm to the cloacal opening. The left testis is generally larger than the right. Androgens stimulate comb and wattle growth in roosters. The testes contain seminiferous tubules with spermatogonia, developing germ cells, and Sertoli cells, which respond to FSH and testosterone. Interspersed Leydig cells produce androgens, primarily testosterone. As sexual maturity nears or under stimulatory photoperiods, rising LH levels promote Leydig cell differentiation, enabling androgen production. Birds have no seminal vesicles, bulbourethral and prostate glands.The vas deferens and receptaculum are the storage place for the spermatozoa. The phallus is small and does not function in intromission. The semen is transferred to the female by touching the phallus to the everted vagina. The sperms taken from the testes are functionally mature and are capable of fertilizing the egg.

Spermatogenesis

In most domestic avian species, spermiogenesis involves eight to twelve morphological changes as spermatids transform into sperm cells. Spermatogenesis occurs within the seminiferous tubules, where germ cells progress through distinct developmental stages. Maturation moves from the periphery to the lumen, forming a spermatogenic wave. Despite being internal, avian testes maintain temperature-sensitive spermatogenesis, adapted to the bird's warm body temperature (~41°C).

Endocrine regulation of the testis

In male chickens, sexual maturity is not marked by the first sperm production, as early sperm have poor fertilizing ability. By 18–19 weeks, testicular size and fertility peak. Leydig cells primarily produce testosterone and androstenedione. FSH levels rise from 11 weeks, plateauing at 19 weeks, while LH increases earlier (7–15 weeks). Males maintain higher plasma LH levels than females throughout development.

The testes in birds are regulated by a conventional negative feedback mechanism involving LH and FSH. Testis removal leads to elevated gonadotropins, and unilateral removal results in compensatory hypertrophy of the remaining testis. Full compensation in size and sperm output occurs if removal happens before 4 weeks of age, with partial compensation observed up to 8 weeks. Testis size directly correlates with sperm production, influenced by factors like photoperiod, genetics, and age. Thyroid hormones also play a role in seasonal reproduction.

Semen

Poultry semen is a highly concentrated, low-volume fluid, averaging 0.5 ml in cocks with a sperm concentration of 4 billion per ml. The sperm head is narrow and elongated with a small acrosome. Poultry sperm are immotile before ejaculation and gain motility afterward. In hens, the vagina contains sperm storage tubules that maintain viable sperm for several days, ensuring fertilization. Spermatogenesis involves mitotic division followed by meiosis, producing haploid cells. Spermatogonia develop into spermatocytes and spermatids at a core body temperature of 40–41°C. This process is regulated by gonadotropins and gonadal steroids, with testosterone from Leydig cells, under LH control, playing a key role. Immotile sperm, after spermiation, are transported via seminiferous fluid through the excurrent ducts of the testes, where they acquire fertilizing ability. Unlike mammals, poultry sperm do not require capacitation before fertilization, and polyspermy is common.

12

Important Indian and Exotic Breeds of Cattle, Buffalo, Sheep, Goat, Pig and Poultry

Anshuman Kumar, Vineeth M.R., Amitosh Kumar and Anuradha Kumari

Faculty of Veterinary and Animal Sciences, I. Ag.Sc., RGSC-Banaras Hindu University (BHU), Barkachha, Mirzapur, Uttar Pradesh-231001

India is home to a vast and diverse genetic pool of farm animals, with a remarkable population of 536.36 million livestock and 851.81 million poultry. According to the 20th livestock census (2019), out of total livestock available in the country, around 36.04% are cattle, 27.74% are goat, 20.74% are buffaloes, 13.83% are sheep and 1.69% are pigs. To date, the National Bureau of Animal Genetic Resources (ICAR-NBAGR) has registered 230 distinct breeds of livestock and poultry including 53 for cattle, 21 for buffalo, 41 for goat, 46 for sheep, 8 for horses& ponies, 9 for camel, 15 for pig, 4 for donkey, 5 for dog, 2 for yak, 20 for chicken, 4 for duck,1 for geese and 1 for synthetic cattle.The ICAR-NBAGR, established in 1984 and located at Karnal (Haryana), is the nodal agency for registration of breeds in India.

Breeds of Cattle

The cattle breeds can be classified into three categories based on their utility: milch (Sahiwal, Tharparkar, Rathi, Red Sindhi and Gir), draft (Amritmahal, Hallikar, Kangayam, Khilar, Ponwar, Kherigarh, Krishna Valley etc.) and dual purpose (Kankrej, Hariana, Gangatiri, Deoni, Dangi etc.).The 54 registered cattle breeds along with their native tract have been listed in table 1.

Table 1. India cattle breeds and their distribution (Source: ICAR-NBAGR, Karnal)

S. No.	Breed	Home tract	S. No.	Breed	Home tract
1	Amritmahal	Karnataka	28	Lakhimi	Assam
2	Bachaur	Bihar	29	Malnad Gidda	Karnataka
3	Badri	Uttarakhand	30	Malvi	Madhya Pradesh

4	Bargur	Tamil Nadu	31	Masilum	Meghalaya
5	Belahi	Haryana & Chandigarh	32	Mewati	Haryana & Rajasthan
6	Binjharpuri	Odisha	33	Motu	Odisha
7	Dagri	Gujarat	34	Nagori	Rajasthan
8	Dangi	Maharastra & Gujarat	35	Nari	Rajasthan & Gujarat
9	Deoni	Maharastra & Karnataka	36	Nimari	Madhya Pradesh
10	Gangatiri	Uttar Pradesh & Bihar	37	Ongole	Andhra Pradesh
11	Gaolao	M.P & Maharastra	38	Poda Thurpu	Telangana
12	Ghumusari	Odisha	39	Ponwar	Uttar Pradesh
13	Gir	Gujarat	40	Pulikulum	Tamil Nadu
14	Hallikar	Karnataka	41	Punganur	Andhra Pradesh
15	Hariana	Haryana	42	Purnea	Bihar
16	Himachali Pahari (Gauri)	Himachal Pradesh	43	Rathi	Rajasthan
17	Kangayam	Tamil Nadu	44	Red Kandhari	Maharastra
18	Kankrej	Gujarat & Rajasthan	45	Red Sindhi	At organized herds only
19	Kathani	Maharastra	46	Sahiwal	Punjab & Rajasthan
20	Kenkatha	Uttar Pradesh& M.P	47	Sanchori	Rajasthan
21	Khariar	Odisha	48	Shweta Kapila	Goa
22	Kherigarh	Uttar Pradesh	49	Siri	West Bengal & Sikkim
23	Khillar	Maharastra	50	Tharparkar	Gujarat & Rajasthan
24	Konkan Kapila	Maharastra & Goa	51	Thutho	Nagaland
25	Kosali	Chhatisgarh	52	Umblachery	Tamil Nadu
26	Krishna Valley	Karnataka & Maharastra	53	Vechur	Kerala
27	Ladakhi	Jammu & Kashmir	54	Frieswal (Synthetic)	U.P & Uttarakhand

Important Indigenous Breeds of Cattle

1. Gir

Breeding tract: Amreli, Bhavnagar, Junagadh district of Gujarat

Distinctive characteristics: Mostly pure red in color however some are speckled red. They have a broad convex forehead and the horns have a unique curved shape, starting at the crown base and extending downwards, backwards, twisting slightly upwards and forwards and finally taking inward sweep and terminating, resulting in a half-moon appearance. Long and pendulous ears are folded like a leaf.

2. Sahiwal

Breeding tract: Amritsar, Ferozpur district of Punjab, Ganganagar and Hanumangarh district of Rajasthan

Distinctive characteristics: They are usually reddish dun but pale red or brown is also seen. Horns are stumpy and short. Dewlapsk in is large and loose. Teats are large and pendulous naval flap is long.

3. Kankrej

Breeding tract: Ahmadabad, BanasKantha, Kheda, Mahesana, Sabar- Kantha, Kutchchh district of Gujarat and Barmer and Jodhpurdistrict of Rajasthan

Distinctive characteristics: Color varies from silver-grey to iron grey or steel grey. They have broad chest and pendulous ears. In males fore & hind quarters and hump are slightly darker than the rest of the body. Horns are strong and lyre shaped and curved outward and upward. It is the heaviest breed of cattle in India.

4. Ongole

Breeding tract: Vizianagaram, Vishakhapatnam, Guntur, Nellore, Parkasham district of Andhra Pradesh

Distinctive characteristics: They have a glossy white coat. Males have dark grey markings on head, neck and hump, black points on knees and pasterns. They have majestic gait, stumpy horns growing outward and backward and large fan shaped and fleshy dew lapserrated with smooth flowing folds instead of narrow constrictions.

5. Gangatiri

*Breeding tract:*Varanasi, Mirzapur, Ghazipur and Ballia districts of Uttar Pradesh and Bhojpur and Buxar district of Bihar

Distinctive characteristics: The animals are tall and of complete white (Dhawar) or Grey (Sokan) color. The forehead is prominent, straight and broad with shallow groove in the middle.

Important Exotic Breeds of Cattle

6. Holstein Friesian

Origin: Two Northern provinces of Netherlands; West Frieze Land and North Holland

Distinctive characteristics: Black and white in color, have a long narrow and straight head and a white switch.They are the largest and best dairy breed of cattle.Average milk production is around ~6000-7000 liters per lactation and the fat percentage is ~3.5%.

7. Jersey

Origin: Island of Jersey, a small Island off the coast of Normandy in English Channel in the United Kingdom

Distinctive characteristics: They are reddish-grey to brown in color with darker color in some areas of body. They have a double dished face and a compact, angular body and straight top line. The tail has a black switch. This is one of the smallest and oldest dairy breeds. The average milk yield is ~4000-5000 kg per lactation and fat percentage is ~5.5%.

Breeds of Buffalo

Buffaloes are multipurpose animal providing milk, meat, draught power and transport. Buffaloes contribute to more than half of total milk production in India. On the basis of habitats and genetic constitution they are classified as riverine and swamp buffalo. The riverine buffaloes have 50 chromosomes and prefer to live in clear water of river, canals and ponds. They are good milk producers. The swamp buffaloes have 48 chromosomes and are not good milk producers and reared for draft purpose. The Luit buffalo of Asaam are the only registered swamp buffalo breed in India. The 21 registered buffalo breeds along with their native tract have been listed in table 2.

Table 2. Indian buffalo breeds and their distribution (Source: ICAR-NBAGR, Karnal)

S. No.	Breed	Home tract	S. No.	Breed	Home tract
1	Banni	Gujarat	12	Manda	Odisha
2	Bargur	Tamil Nadu	13	Marathwadi	Maharastra
3	Bhadawari	Uttar Pradesh and M.P	14	Mehsana	Gujarat
4	Chhatisgarhi	Chhatisgarh	15	Murrah	Haryana
5	Chilika	Odisha	16	Nagpuri	Maharastra
6	Dharwadi	Karnataka	17	Nili Ravi	Punjab
7	Gojri	Punjab & Himachal Pradesh	18	Pandharpuri	Maharastra
8	Jaffrabadi	Gujarat	19	Purnathadi	Maharastra
9	Kalahandi	Odisha	20	Surti	Gujarat
10	Luit	Assam	21	Toda	Tamil Nadu
11	Manah	Assam			

Important Indigenous Breedsof Buffalo

1. Murrah

Breeding tract: Rohtak, Jind, Hisar, Jhajhar, Fatehabad district of Haryana

Distinctive characteristics: Jet Black color, Thin Skin, Jet black color, tightly curled horns, heavy body size, long tail with white switch reaching up to the fetlock

2. Nilli Ravi

Breeding tract: Gurdaspur, Amritsar, Ferozpur district of Punjab

Distinctive characteristic Walled eyes, coat color is mostly black with white markings on forehead, muzzle, face, tail switch and legs. Due to possession of these white markings it is also knownas "Panch Kalyani". Long tail touches almost to the ground.

3. Bhadawari

Breeding tract: Etawah and Agra district of Uttar Pradeshand Bhind and Morena district of Madhya Pradesh

Distinctive characteristics: Blackish copper to light copper color. Two white lines "Chevron" locally called as Kanthy are present on the lower side of the neck.

4. Jaffarabadi

Breeding tract: Junagarh, Bhavnagar, Jamnagar, Porbandhar, Amreli and Rajkot district ofGujarat

Distinctive characteristics: Mostly black colored but some animals are grey also. Animals are massive with prominent bulky head and heavy horns. Horns droop on each side of neck and come out upward and finally inward forming a ring like structure. This makes eyes to look small - termed as study eye.

5. Surti

Breeding tract: Kheda, Baroda, Bharuch and Surat district of Gujarat

Distinctive characteristics: Color varies from rusty brown to silver-grey. Horns are flat, sickle shaped Medium in size. Two white bands below the neck are present.

6. Mehsana

Breeding tract: Mehsana, Sabarkantha, Banaskantha district of Gujarat

Distinctive characteristics: Mostly black colored, but some animals are black brown or brown. Characters are a mix of Murrah and Surti breeds as it was produced from their cross. Horns are generally sickle shaped with the curve more upward than in theSurti breed but less curved than in the Murrah breed.

Eyes are very black and prominent, bulging from their socketswith folds of skin on upper lids.

Breeds of Goat

Goats make valuable contribution in form of precious animal protein (milk and meat) and fibre and skin and provide livelihood to marginal and poor farmers. They are widely distributed across various agro-ecological zones ranging from high altitude cold arid region to hot humid and coastal regions and contributing significantly to the rural economy. Goats are mainly used for meat production but, Beetal, Gohilwadi, Jakhrana, Jamunapari, Kahmi, Mehsana, Surti and Zalawadi are of dairy type. Changthangi and Chegu breeds produce finest quality pashmina fibres. The 41 registered goat breeds along with their native tract have been listed in table 3.

Table 3. Indian Goat breeds and their distribution (Source: ICAR-NBAGR, Karnal)

S. No.	Breed	Home tract	S. No.	Breed	Home tract
1	Anjori	Chhatisgarh	22	Kanni Adu	Tamil Nadu
2	Andamani	Andaman & Nicobar	23	Karuli	Rajasthan
3	Assam Hill	Assam and Meghalaya	24	Kodi Adu	Tamil Nadu
4	Attapady Black	Kerala	25	Konkan Kanyal	Maharastra
5	Barbari	Uttar Pradesh & Rajasthan	26	Kutchi	Gujarat
6	Beetal	Punjab	27	Malabari	Kerala
7	Berari	Maharastra	28	Marwari	Rajasthan
8	Bhakarwali	Jammu & Kashmir	29	Mehsana	Gujarat
9	Bidri	Karnataka	30	Nandidurga	Karnataka
10	Black Bengal	West Bengal	31	Osmanabadi	Maharastra
11	Bundelkhandi	Uttar Pradesh and Madhya Pradesh	32	Pantja	Uttarakhand
12	Changthangi	Laddakh	33	Rohilkhandi	Uttar Pradesh
13	Chaugarkha	Uttarakhand	34	Salem Black	Tamil Nadu
14	Chegu	Himachal Pradesh	35	Sangamneri	Maharastra
15	Gaddi	Himachal Pradesh & Jammu Kashmir	36	Sirohi	Rajasthan
16	Ganjam	Odisha	37	Sojat	Rajasthan
17	Gohilwadi	Gujarat	38	Sumi-Ne	Nagaland
18	Gujari	Rajasthan	39	Surti	Gujarat
19	Jakhrana	Rajasthan	40	Teressa	Andaman & Nicobar

20	Jamunapari	Uttar Pradesh	41	Zalawadi	Gujarat
21	Kahmi	Gujarat			

Important Indigenous Breeds of Goat

1. Barbari

Breeding tract: Bharatpur district of Rajasthan and Aligarh, Agra, Etawah district of Uttar Pradesh

Distinctive characteristics: Color is white with tan or dark red spots. Horns are twisted and directed upward and backward. Ears are small erect and tubular.

2. Beetal

Breeding tract: Amritsar and Gurdaspur district of Punjab

Distinctive characteristics: Black coat color is common, brown with white spots is also found. Horns are carried horizontally with slight twist. Nose is roman (convex) and ears are long, flat and droopy.

3. Black Bengal

Breeding tract: West Bengal

Distinctive characteristics: Short legged and compact animals. Mostly black or brown colored, grey and white are also found.Twins and triplets are common.

4. Jakhrana

Breeding tract: Alwar district of Rajasthan

Distinctive characteristics: Mainly black colored with white spots on ears and muzzle. Horns are broad and flat, going backwards. Forehead is narrow and slightly bulging. Well-developed large size udder with conical teats.

5. Jamunapari

Breeding tract: Agra, Mathura, Etawah district of Uttar Pradeshand Bhind, Morena district of Madhya Pradesh

Distinctive characteristics: Tallest goat breed of country. General White colored and sometime white with brown patches on ears, head and neck. Nose is large and roman (convex). Tufts of hairspresent on thighs. Ears are large, leafy and pendulous.

6. Sirohi

Breeding tract: Sirohi district of Rajasthan

Distinctive characteristics: Coat color is predominantly brown with light or dark brown patches. Few individuals are completely white. Flat and leafy drooping ears.

Important Exotic Breeds of Goat

1. Toggenberg

Origin: Toggenberg valley in north Switzerland

Distinctive characteristics: Light brown to dark chocolate color with two white stripes down the face from eye to the muzzle and white markings on legs and tail. The ears are erect and carried forward. Average milk production is ~5.5 kg per day.

2. Sannen

Origin: Sannen valley of Switzerland

Distinctive characteristics: Color is white or light cream. This is known as milk queen of goat. The nose is straight and ears are erect pointing upward and forward. Udder is well developed. Average milk production is 2-5 kg per day. Both sexes are normally polled but sometimes horns do appear which is saber-shaped.

3. Alpine

Origin: Alps Mountain in France

Distinctive characteristics: No distinct color has been established, it range from pure white to shades of fawn, gray, brown, black, red, bluff, piebald, or various shadings or combinations of these colors. They have curved horns, perfectly erect ears and straight nose. Average milk yield is 2-3 kg.

Breeds of Sheep

Sheep breeds are classified on the basis of agro ecological regions: a) North temperate region, b) North Western arid and semi-arid region c) Southern peninsular region, and d) Eastern region. The main product from the sheeps of North temperate region is apparel and carpet wool. The sheeps from North Western arid and semi-arid region and eastern region produce mutton and carpet wool, however the sheeps of southern peninsular regionare of mutton type. The 46 registered sheep breeds along with their native tract have been listed in table 4.

Table 4. Indian sheep breeds and their distribution (Source: ICAR-NBAGR, Karnal)

S. No.	Breed	Home tract	S. No.	Breed	Home tract
1	Balangir	Odisha	24	Kilakarsal	Tamil Nadu
2	Bellary	Karnataka	25	Macherla	Andhra Pradesh
3	Bhakarwal	J & K	26	Madras Red	Tamil Nadu
4	Bonpala	Sikkim	27	Magra	Rajasthan

5	Changthangi	Leh & Laddakh	28	Malpura	Rajasthan
6	Chevaadu	Tamil Nadu	29	Mandya	Karnataka
7	Chokala	Rajasthan	30	Marwari	Rajasthan
8	Chottanagpuri	West Bengal & Jharkhand	31	Mecheri	Tamil Nadu
9	Coimbatore	Tamil Nadu	32	Muzaffarnagri	Uttar Pradesh & Uttarakhand
10	Deccani	Maharastra & Karnataka	33	Nali	Rajasthan & Haryana
11	Gaddi	J & K, Himachal Pradesh	34	Nellore	Andhra Pradesh
12	Ganjam	Odisha	35	Nilgiri	Tamil Nadu
13	Garole	West Bengal	36	Panchali	Gujarat
14	Gurez	J & K	37	Patanwadi	Gujarat
15	Hassan	Karnataka	38	Poonchi	J & K
16	Jaisalmeri	Rajasthan	39	Pugal	Rajasthan
17	Jalauni	Madhya Pradesh & U.P	40	Ramnad White	Tamil Nadu
18	Kajli	Punjab	41	Rampur Bushair	Himachal Pradesh
19	Karnah	J & K	42	Shahabadi	Bihar
20	Katchaikatty black	Tamil Nadu	43	Sonadi	Rajasthan
21	Kendrapada	Odisha	44	Tibetan	Sikkim & Arunachal
22	Kenguri	Karnataka	45	Tiruchi Black	Tamil Nadu
23	Kheri	Rajasthan	46	Vembur	Tamil Nadu

Important Indigenous Breeds of Sheep

1. Changthangi

Breeding tract: Changthang region of Ladakh (Jammu & Kashmir)

Distinctive characteristics: Color varies from predominant white to black and brown or mixture of these colors. The legs are short, head is convex, face is long tapering and ears are pendulous. They have good fleece cover with extra ordinarily long staple.

2. Garole

Breeding tract: Sunderban region of South 24- Parganas district of West Bengal

Distinctive characteristics: Predominant color is grey and white. Animals are small sized with relatively low body weight. Body is compact and square with small head, medium ears and short and thin tail. The udder is fairly developed and multiple births are common.

3. Muzaffarnagri

Breeding tract: Muzaffarnagar, Bulandshahar, Saharanpur, Meerut, Bijnor districts of Uttar Pradesh and Dehradun district of Uttarakhand

Distinctive characteristics: Face line is slightly convex. Face and body color is white with occasional patches of brown or black. Ears are long and drooping. Tail is extremely long and reaches fetlock.

4. Rampur Bushair

Breeding tract: Shimla, Kinnaur, Bilaspur and Lahaul spiti districts of Himachal Pradesh

Distinctive characteristics: The fleece is white with varying proportion of brown, black and tan.Ears are long and drooping. Face is convex with romannose. Horns are curved backward, downward and outward. Tail is short and thin.

5. Nellore

Breeding tract: Nellore, Prakasham and Ongoledistrict of Andhra Pradesh

Distinctive characteristics: Color is red or white with or without black color on ventral side. Ears are long and drooping. Tail is short and thin. Rams are horned ewes are polled. This is the tallest sheep breed of India, resembling goats in appearance.

Important Exotic Breeds of Sheep

1. Merino

Origin: Spain

Distinctive characteristics: It has white face with white feet and pink skin. Rams have large, heavy, spirally turned horns while ewes are polled. Most of the head and legs are covered with wool. It is classified as a fine wool breed.

2. Rambouillet

Origin: Rambouillet in France

Distinctive characteristics: This breed was formerly called French Merino as it was developed from Merino sheep. They are large and rugged breed. It has white face and legs and pink skin. Ewes are polled and rams have well balanced horns curving backward and outward. They are also considered a fine wool breed.

3. Karakul

Origin: Bokhara, situated between Pakistan and Afghanistan

Distinctive characteristics: The name is derived from the area Karakul (black lake) which is not very favourable for raising livestock. Face, ear and leg of the animal is black or brown.The main use of this breed is for fur/pelt production. The pelt is taken from a prematurely born lamb or lamb killed within a short time after birth.

4. Suffolk

Origin: Suffolk, Essex and Norfolk in South Eastern England

Distinctive characteristics: Face, ears and legs is black. Head, ears, portion below knees and hocks are devoid of wool. It is classified as a mutton breed.

5. Dorset

Origin: Dorset and Somerset in Southern England

Distinctive characteristics: Face, ears and legs are white. The fleece is very white, strong, close and free from dark fiber. Dorset ewes are great milkers and mothers and twins are common.They are known for good muscle conformation and ability to breed naturally out of season. It is also a mutton breed.

Breeds of Pig

India's contribution to Asian pig production is ~1% and China alone produce more than 80% to it. The highest pig population is observed in Eastern & North-Eastern states due to their cultural preferences for pork consumption. The indigenous breeds constitutes more than three-fourth of total pig population in India is well-adapted to local conditions but constrained by low productivity and growth rates. The exotic breeds like Large white Yorkshire, Hampshire, Duroc, and Landrace etc. have been integrated in to indigenous germplasm through crossbreeding to improve the productivity per animal. The 46 registered pig breeds along with their native tract have been listed in table 5.

Table 5. Indian Pig breeds and their distribution (Source: ICAR-NBAGR, Karnal)

S. No.	Breed	Home tract	S. No.	Breed	Home tract
1	Agonda Goan	Goa	9	Manipuri Black	Manipur
2	Andamani	Andaman & Nicobar	10	Niang Megha	Meghalaya
3	Banda	Jharkhand	11	Nicobari pig	Andaman& Nicobar
4	Doom	Assam	12	Purnea	Bihar
5	Ghoongroo	West Bengal	13	Tenyi vo	Nagaland
6	Ghurrah	Uttar Pradesh	14	Wak Chambil	Meghalaya
7	Karkambi	Maharastra	15	Zovawk	Mizoram
8	Mali	Tripura			

Important Indigenous Breeds of Pig

1. Ghoongroo

Breeding tract: Darjeeling, Jalpaiguri, Cooch Behar, Dakshin Dinajpur, Uttar Dinajpur district of West Bengal

Distinctive characteristics: Color is Black. They have thick coarse and long hair coat, long tail, upwardly curved snout, large and heart shaped ears resembling those of elephant and upwardly curved (concave) snout.

2. Tenyi Vo

Breeding tract: Kohima, Phek and Dimapur districts of Nagaland

Distinctive characteristics: Mostly black in color and have a pot-bellied appearance. Females have a sagging back and pendulous belly touching the ground. They have long, strong and tapering snout, small erect ears, and bright, alert eyes. Tail is straight and ending with a white switch reaching up to hock joint, white stockings, and white markings on the forehead and ventral part of body.

3. Doom

Breeding tract: Dhubri, Bongaigaon, and Kokrajhar districts of Assam

Distinctive characteristics: These are black in color and have a short, concave snout.

Body is large and belly is flat. They have short erect ears, and a straight topline with long bristles which extends to the thoraco-lumbar area.

Important Exotic Breeds of Pig

1. Large white Yorkshire

Origin: Yorkshire in England.

Distinctive characteristics: They are white in colour with small ears tilting in front and long neck. This is a good bacon breed.

2. Landrace

Origin: Denmark

Distinctive characteristics: They are white in colour although black skin spots 'freckles' are common. They have a long narrow body, relatively short legs, large drooping ears and long snout.

Breeds of Poultry

The poultry sector in India is a significant part of the country's economy, especially in rural areas. Globally India ranks second in egg and fifth in meat production. In India, there are two types of poultry production system: rural poultry production system/backyard poultry production and commercial poultry production. The indigenous chickens are well known for their adaptability to local climatic and geographical conditions, but are generally low producers. They vary greatly in their plumage pattern, comb type, body conformation and other physiological attributes. They are, however, good brooders and efficient foragers. The 19 registered chicken breeds along with their native tract have been listed in table 6.

Table 6. Indian chicken breeds and their distribution (Source: ICAR-NBAGR, Karnal)

S. No.	Breed	Home tract	S. No.	Breed	Home tract
1	Ankaleshwar	Gujarat	11	Kadaknath	Madhya Pradesh
2	Aravali	Gujarat	12	Kalasthi	Andhra Pradesh
3	Aseel	Andhra Pradesh & Odisha	13	Kaunayen	Manipur
4	Busra	Maharashtra & Gujarat	14	Kashmir Favorolla	Jammu & Kashmir
5	Chittagong	Meghalaya and Tripura	15	Mewari	Rajasthan
6	Danki	Andhra Pradesh	16	Miri	Assam
7	Daothigir	Assam	17	Nicobari	Andaman & Nicobar
8	Ghagus	Karnataka and Andhra Pradesh	18	Punjab Brown	Haryana & Punjab
9	Hansali	Odisha	19	Tellichery	Kerala & Pandicherry
10	Haringhata Black	West Bengal	20	Uttara	Uttarakhand

Important Indigenous Breeds of Chicken

1. Kadaknath

Breeding tract: Dhar and Jhabua district of Madhya Pradesh

Synonym: Kalamasi

Distinctive characteristics: The plumage color is Jet black, ranging from silver to gold spangled to blue black without any spangling. They are characterized bypresence of black pigments on external and internal surfaces. The comb, wattle, tongue are dark purple in color. The flesh is also black.

2. Assel

Breeding tract: Khammam district of Andhra Pradesh, Koraput & Malkangiri district of Odisha and Bastar and Dantiwara district of Chattishgarh.

Distinctive characteristics: It is game bird, known for its high stamina, majestic gait and fighting qualities and are biggest in size among all native breeds.The plumage color are mostly Red (or Brown) and black. The neck is long and uniformly thick but without flesh.

3. Ankaleshwar

Breeding tract: Bharuch and Narmada districts of Gujarat

Distinctive characteristics: Plumage color ranges widely from a combination of white and light grey to brown and golden. Golden yellow plumage is predominant in cocks while black golden is more common in hens.

Important Exotic Breeds of Chicken

1. White Leghorn

Origin: Italy

Distinctive characteristics: It has small, oval and compact body, relatively long neck, prominent breast and long shanks. The single red comb stand erect with five uniform deeply serrated points. It is the world's number one egg producer. Annual hen-housed egg production per bird is 290-320 eggs.

2. Rhode Island Red

Origin: Rhode Island of New England.

Distinctive characteristics: It has long, rectangular,broad and deep chest and flat back body. The plumage is rich reddish brown in color and the color is uniform throughout their body. The comb is single and bright red in color. The annual egg production ranges from 225-260 eggs per bird.

3. Australorp

Origin: Australia

Distinctive characteristics: The Plumage is lustrous greenish black. They are small sized having deep oval body with long backand the body slopes gradually towards the tail. The comb is single red. Shank and toes are black. It is an excellent layer. The annual egg production per bird ranges from 250-280.

Photographs of some of the breeds (Courtesy: Livestock Farm Complex, FVAS; Principal Investigator P-26/124, M-21/218 and P-07/1268, BHU)

Jamunapari

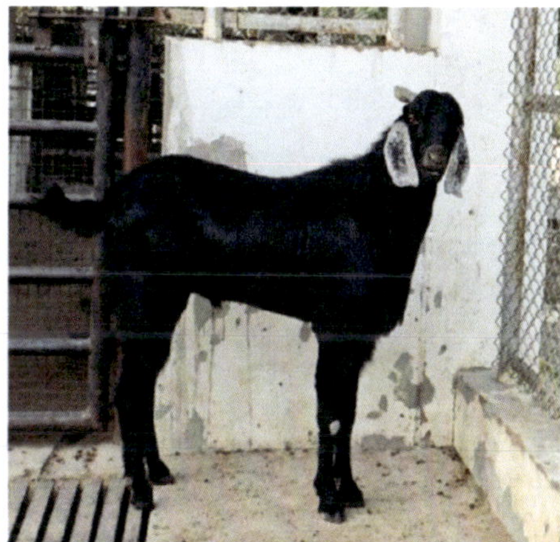

Jakhrana

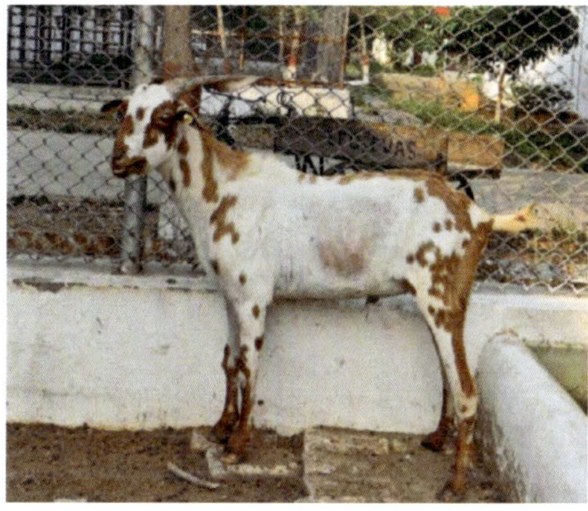

Barbari

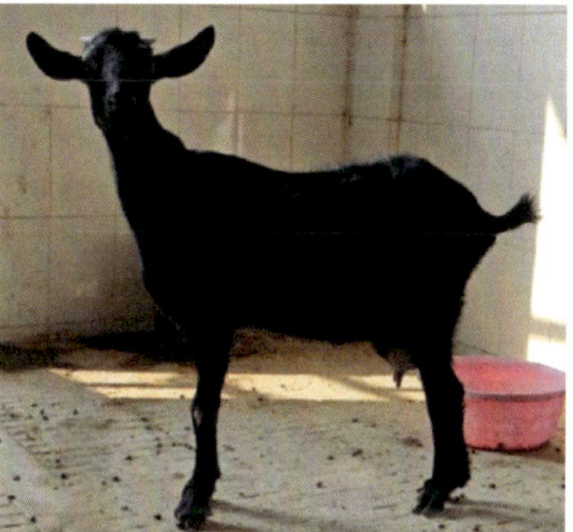

Black Bengal

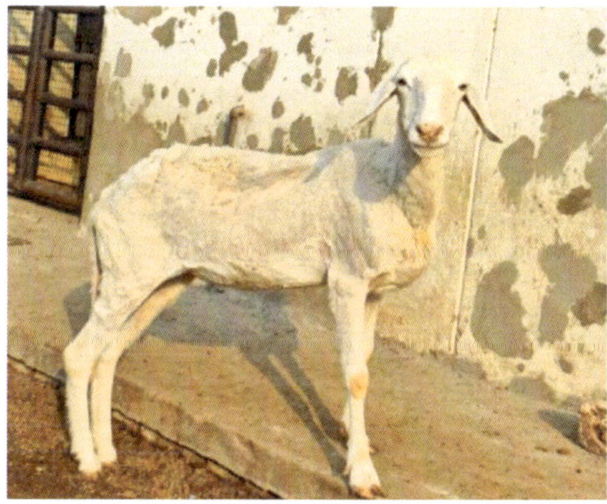

Muzaffarnagari

Kadaknath chicken

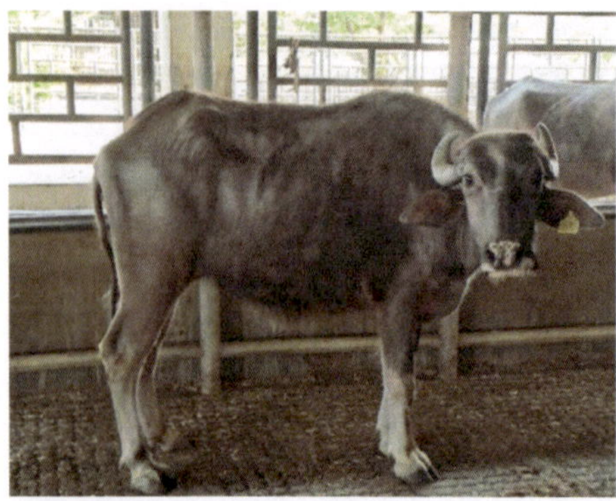
Bhadawai

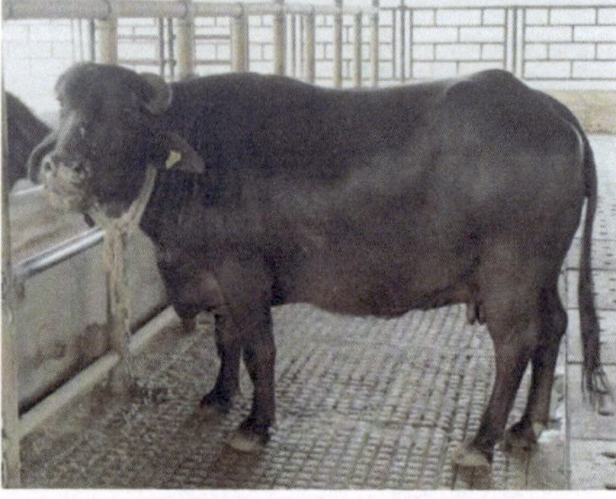
Murrah

Gangatiri

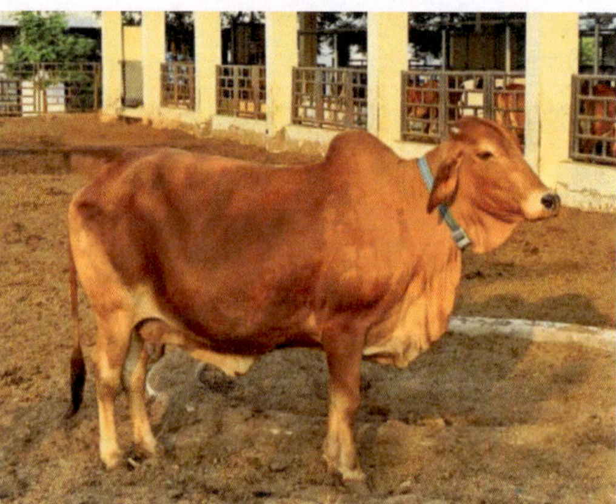
Sahiwal

13

Improvement of Farm Animals and Poultry

Anshuman Kumar, Vineeth M.R., Amitosh Kumar and Utkarsh Kumar Tripathi

Faculty of Veterinary and Animal Sciences, I. Ag.Sc., RGSC-Banaras Hindu University (BHU), Barkachha, Mirzapur, Uttar Pradesh-231001

The animal breeders are always trying to improve the performance of their animals for desirable traits or to improve the future generation of animals. **The fundamental tools available with them are selection and mating.** For this task animal breeders have two basic tools: selection and mating. In selection, we decide which individuals become parents, how many offspring they may produce, and how long they remain in the breeding population. In mating, we decide which of the males we have selected will be bred to which of the females we have selected.Selection and mating system for the genetic improvement has been discussed in details in the following sections.

Selection

Selection is choosing parents to produce the next generation. During selection process, some individuals in a population are given chance to produce offspring while others are not. So, selection is an alternate form of culling. In genetic terms, selection is differential transmission of genetic material to the next generation due to differences in the survival and fertility of individuals with different genotypes within a population.

The process of differential survival and reproduction ofindividual is called selection.

Selection is of two kinds

I. **Natural selection:** In natural selection, nature plays a role and not the man. The main driving force responsible for natural selection is "survival of fittest" so, the individuals that are better adapted to an environment are more likely to survive, reproduce and pass on the genes to next generation than individuals that are less well adapted.

Example: Animals with lethal genetic defects are naturally selected against-they don't survive to become parents and transmit these genes in their progeny.

Natural selection can be classified in three different ways

i) **Directional selection:** It is a form of natural selection in which any one of the extreme phenotype is favored over other phenotypes.

ii) **Disruptive / Bidirectional /Diversifying selection:** Natural selection selects for two or more extreme phenotypes that each have specific advantages and both/all the extreme phenotypes have higher fitness and are favored over the intermediate phenotypes.

iii) **Stabilizing selection:** Natural selection favors an average or intermediate phenotype and selects against extreme phenotypes. It reduces variation in a population.

II. **Artificial selection:** It is the process by which intentional selection is done by humans or specifically the animal breedersin order to improve the trait of interest. Animal breeder allows only those individuals with desired characteristics to produce the next generation and cull the remaining animals.

Selection criteria/Basis of selection/Aids to selection

The idea behind selection is to let the individuals with the best sets of genes reproduce so that the next generation has, on average, more desirable genes than the current generation. The animals with the best sets of genes are said to have the best breeding values. Breeding value is the genetic merit of an individual for transmitting desirable traits to its offspring. But it is not easy to identify the animals with best breeding values as it can't be measured directly. Instead, it can be only be estimated based on observed traits (phenotypic value), which is influenced by both genetic and environmental factors. Since, environmental influences varies, the phenotypic value of observed traits of an animal can't reflect its genetic potential. Moreover, not all of the genetic effects present in the parents will be transmitted to the next generation. Non-additive gene action, including dominance and epistasis, is influenced by specific gene combinations and the presence of other alleles within an individual. Consequently, these interactions are not directly inherited in a straightforward manner, as their expression can vary depending on the genetic context of the offspring, making the prediction of breeding values more complex.

$P = G + E$

$P = A + D + I + E$

Where,

P = Phenotypic value

G = Genotypic value

A = Additive Genetic value or Breeding value

D = Dominance deviation

I = Interaction deviation

E = Environmental deviation

The ability to identify the animals which possess the best genes or to estimate their breeding values can be done on the basis of their own phenotypic records or on the basis of records of their direct or collateral relatives who have their genes in common. These different sources of information which are used to estimate their breeding values (called Estimated Breeding Values (EBVs) are called aids or basis or tools or criteria of selection.

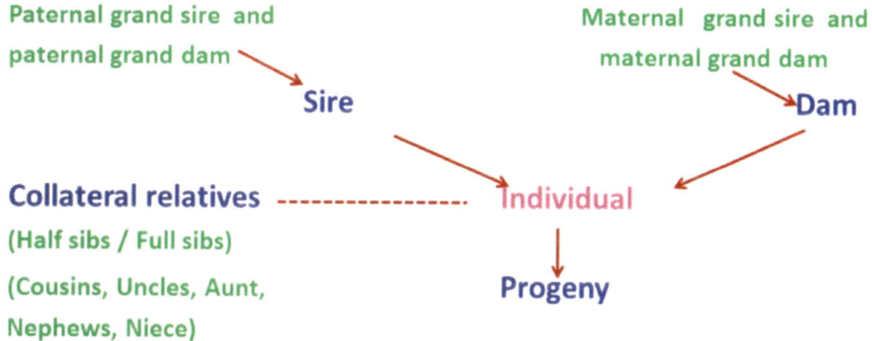

Fig. 1. Relationship of individual for the estimation of breeding value

The Selections based on these aids are known as

1. Individual selection or Mass selection
2. Pedigree selection
3. Family and sib selection
4. Progeny testing

1. Individual selection

The individual own performance/ phenotypic value for a trait can be used to estimate the breeding value of that character for that individual. This is also known as mass selection when the selected individuals are put together *en masse* for matingwithout keeping records of individual mating E.g. *Drosophila*and *Tribolium* in a bottle, poultry and lab animals etc.. The term individual selection is used when the individual mating recorded for example in case of large animals. Selection based on individual records is mainly phenotypic and phenotype is taken as an estimate of an individual's genotypic value. However, as not all of the phenotypic variation is transmitted to the next generation (only the additive genetic proportion i.e. heritability is transmitted) thus, making it difficult to accurately estimate the underlying genotypic value solely on the basis of observed phenotypic values of traits. Individual selection is effective only when the traits selected are expressed in both sexes at early stage of life and traits with high heritability such as growth rate, body conformation etc. Repeated records on traits which are repeated during life of an individual like milk yield, litter size, fleece yield etc. increases the accuracy of selection.

2. Pedigree selection

Pedigree of an animal is a list of all its ancestor in past few generations, which includes sire, dam, paternal and maternal grand sire, paternal and maternal grand dam, and earlier ancestors. When, the breeding value of an individual is estimated on the basis of performance of its ancestors who transmit their genes to the individual, it is called pedigree selection. The degree of genetic relationship between ancestors and an individual is determined using the halving process, where the proportion of ancestral genes is halved with each successive generation. Hence, it is important to consider the closest ancestors than the distant ones. An individual's own performance based on single record is more important than pedigree information and pedigree should only be preferred when the individual records are not available. However, it allows selection at younger stage and useful for sex limited traits like milk and egg production, traits expressed in later stage of life or even after death of the individual like carcass/slaughter traits.

3. Family or sib selection

Family selection is based on performance of collateral relatives. The individual does not receive any gene from collateral relatives as the direct relatives but share common genes.The procedure to estimate the breeding value of an individual on the basis of family mean is called the family selection.It is called sib selection when individual's own record is not included in estimating the family mean. Family in animal breeding consists of full sib and half sib families. Selection is based on mean phenotypic value of the family (individual's own record is included in calculating family mean). In family selection whole of the family is selected or rejected on the basis of mean phenotypic value of the family and within family deviations are given zero weightage. The efficiency of family selection depends on the fact that environmental deviations of the individuals tend to cancel each other out in the mean value of the family, hence it is useful for selecting for lowly heritable traits which are having larger portion of environmental variance. The presence of environmental effects common to all members of a family but different between families, known as common environmental effects (denoted as *c-effects*), creates resemblance within family members in addition to the resemblance due to shared genes. This contributes to variation between families and decreases the accuracy of selection. Large family size increases the accuracy of selection.

Some characters which can't be measured on the individual that are to be selected as parents (e.g. carcass/slaughter traits, sex limited traits etc.) so, the individual's records are not included in calculating the family mean and selection is only based on the records of relatives such selection criteria is called Sib-selection.

4. Within family selection

It is the selection criteria when individuals are selected on the basis of their performance expressed as deviation from family mean. The individuals which exceed their family mean by greatest value are selected. It is opposite to family selection because here family mean is given no weightage. This selection criteria is preferred when *c-effect* are present. E.g. Birth weight and pre weaning growth in species with large litter size, where large part of variation is due to mother (maternal effect). Within family selection reduces the rate of inbreeding as the selected individuals are from different family.

5. Combined selection

The selection of an individual on the basis of two or more sources of information is called combined selection. It is better to select on the basis of an index combining information from various relatives (dam, sibs, progeny etc.)

a) **Osborne index:** Osborne (1957) developed this index based on combining information of records of the individual and its family members (sire family and dam family average) by giving proper weights in poultry for improving egg production

b) **Abplanalb index:** The index of combined selection for selection of pullets and cockerels was proposed by Abplanalp (1974) for egg production.

Index for selection of pullets: own, dam family average and sire family average

Index for selection of cockerels: dam family average and sire family average

6. Progeny testing

Progeny testing is a method of estimating the individual's breeding value for any specific trait based on the average performance of its progeny. The selection criteria are the mean phenotypic value of individual's progeny is compared to mean phenotypic value of contemporaries.The principles of the progeny test come from the sampling nature of inheritance i.e. each progeny of a sire carries a sample half of his genes and if large number of progenies are studied the average is close to the sire's true genetic merit. The breeding value of the sire is twice the mean deviation of the progenies from the population mean. It gives the best and most accurate and reliable estimates of breeding value of the sire. It is useful for sex limited and carcass traits and traits with low heritability. It has major limitation that it takes long time and increases the generation interval and thereby reduces the annual genetic gain. Moreover, it is expensive due to the need to maintain a large number of breeding bulls for a longer period while awaiting test results while conducting milk recording on a large number of progeny.

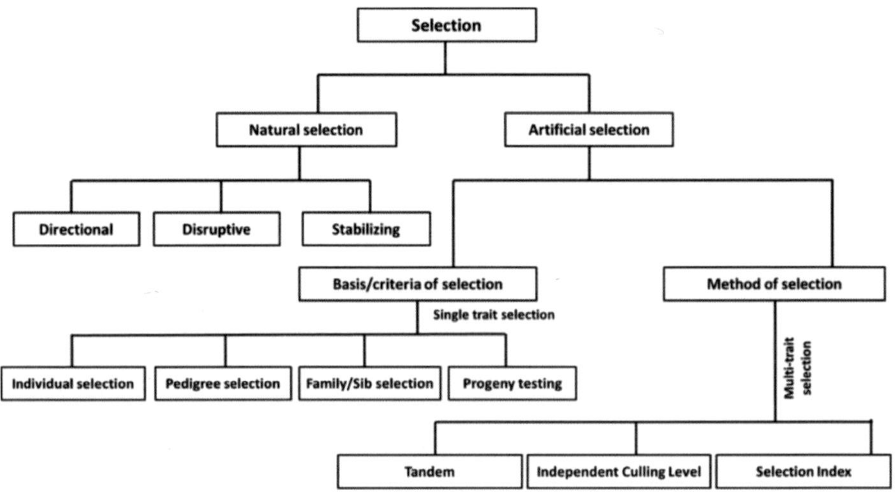

Fig. 2. Selection criteria and basis of selection

Methods of selection (multi-trait selection)

The economic value of an animal depends upon several traits (known as overall performance or net genetic merit of animal). Animal breeders are more interested in improving a number of traits like production traits, reproduction, growth, health etc. The simultaneous selection for many traits on individual's own performance can be done by adopting any of following methods of selection.

i) **Tandem method:** In this method selection is practiced for only one trait till the time improvement has been made in the trait. After improvement in this trait the selection is directed towards improvement in the second trait, then third trait and so on. The efficiency of this method of selection depends on the genetic correlation between traits. So, if there is positive correlation between traits selected then the result will be desirable and if there is negative correlation the effects will be undesirable.

ii) **Independent Culling Method:** In this method of selection few traits are consideredat a time andfor each trait a minimum standard is set so the animal must meet minimum standards for each of the trait to get selected otherwise they are not used for breeding. If an animal is excellent for one trait but miss the threshold for other trait by a slight margin, it is rejected. Hence, this method doesn't allow compensation between traits.

iii) **Selection index or total score method:** The most effective method of selection is selection index method. It gives a single numerical value for each animal, which is the total of scores given for each trait considered in the selection. Each trait is given a weightage depending on the heritability of the traits, genetic correlation between the traits and economic value of the traits. Selection index construction is not an easy task because of complex computations involved.

Mating system

Mating system can be classified in to two major groups (1) mating based on the phenotypic resemblance and (2) mating based on the genetic relationship. In mating based on the phenotypic resemblance, mates are chosen on the basis of external appearance or phenotype for a particular character concerned. If mated pairs are of the same phenotype more often than would occur by chance it is called as assortative mating or like to like mating. If mated pairs are of the same phenotype less often than would occur by chance is called as disassortative mating i.e. mating of individuals of unlike phenotype.

Mating system based on genetic relationship can be divided in to two groups: Inbreeding and Outbreeding.

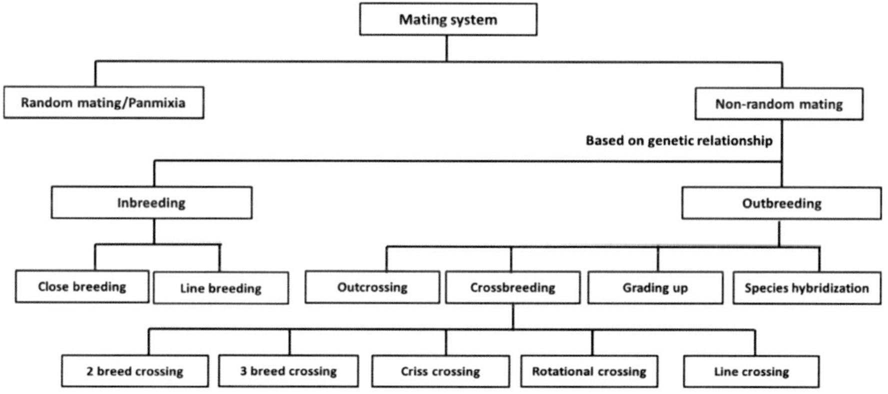

Fig. 3. Classification of mating system

Inbreeding

According to Lush, Inbreeding is defined as *mating between animals more closely related to each other than the average relationship between all individuals of the population concerned.* Inbreeding is mating between animals that are related by ancestry. Two animals are said to be genetically related when they have one or more ancestors in common in the first 4 to 6 generations of their pedigree.

Inbreeding can be classified as

i) **Close inbreeding:** Close inbreeding is a severe or strict form of inbreeding and the mating are made between close relatives. Example: Mating between full sibs or between parents and offspring i.e. sire and daughter or son and dam.

ii) **Line breeding:** Line breeding is milder form of inbreeding and the relationship is not as close as in case of close inbreeding. Line breeding is defined as *aform of inbreeding in which the relationship of an individual or individuals is kept as close as possible to an outstanding ancestor.* The ancestor is usually a male because a male can produce more number of progenies during its life time than a female.

Line breeding can be done in ways

1. Half-sib mating or cousins mating – Here, the rate of inbreeding is less than close inbreeding.
2. Mating of animals is done in such a way that their descendants are mated to outstanding animal (sire) up to 3 to 4 generations. Here, generally a sire is not mated to his own daughter and mating are made between half-sibs or grand sire and grand-daughter to avoid close inbreeding.

Inbreeding increases homozygosity. So, the lethal deleterious recessive genes get uncovered in homozygous conditions which were otherwise hidden as carrier in outbred herds (as heterozygotes). Moreover, there is reduction in the mean phenotypic value of characters associated with reproductive capacity, viability, and physiological efficiency. This phenomenon is known as inbreeding depression.

Outbreeding

The mating of unrelated individuals is called as outbreeding or it is the mating of animals which are less closely related to each other than the average of the population i.e. the mated individuals have no common ancestors in the proceeding 4 to 6 generations of their pedigree. It is complementary or opposite to the inbreeding. The outbreeding is useful in order to utilize the hybrid vigour or heterosis (superiority of outbred individuals over the average of their parents)and to produce commercial animals.

Outcrossing

It is the mating of unrelated animals within a breed i.e. mating within a purebred but parents are not related for last 4 to 6 generations. When outcrossing is done within a herd by use of selected sires it is known as *selective breeding*. This system of selection and outcrossing is very effective for highly heritable traits.

Top crossing

It is a term used for mating of purebred males with unrelated females of the same breed. It is a form of outcrossing in which different families within a purebred are combined. For example, improvement of Indian Jerseys by importing Jersey from United Kingdom.

Grading up

Mating of sires of pure breed to non-descript females. The crossbred females are back crossed to purebred sires generation after generation so that after 6-7 generations population exactly resembles the breed from which sires were used.The non-descript population is converted to “pure bred”.

Crossbreeding

The mating of animals from distinct breeds is called crossbreeding.

Crossbreeding is done for the following purposes

a) To exploit hybrid vigour or heterosis
b) To combine good qualities of two or more breeds(complementarity in crossbreeding)

Crossbreeding can be classified as two breed, three breed and four breed crosses depending on the number of populations (inbred lines, strains or breeds) crossed.

Two breed crosses

i) ***Inter se* mating:** It is the crossing of crossbred progeny having the same level of inheritance of two breeds Example: crossing of F_1 with F_1
ii) **Back crossing:** Back crossing is the mating of crossbred animals to purebred animals of either breed which was used to produce the crossbred progeny. Example: Crossing of F_1 with any of the parent
iii) **Criss crossing:** It is similar to back crossing (mating of crossbred animals to purebred animals) except that both the parental breeds are used alternately in each generation

Three breeds crosses

In this system of cross breeding three breeds are used.For example: First generation (F_1) crossbred female produced from breed A and B (AB) are crossed with males of third breed C i.e. AB female X C male

Four breed crosses or Double two breed crosses

It is crossing of crossbred females produced by crossing two breeds (A and B) with crossbred males produced by crossing another two breeds (C and D)

Table 1. Crossbred strains of Cattle developed in India

Strain	Breeds used			Location
	Native	Exotic	Exotic Inheritance	
Taylor	Local cows	Shorthorn and Jersey	50-75%	Around Patna
Jersind	Red Sindhi	Jersey	37.5-62.5%	Allahabad
Jerthar	Tharparkar	Jersey	50%	Bangalore
Sunandini	Local non-descript cattle	Brown Swiss	50%	Kerala
Karan Swiss	Sahiwal & Red Sindhi	Brown Swiss	50-62.5%	NDRI, Karnal
Karan Fries	Tharparkar	Holstein Friesian, Brown Swiss	75%	NDRI, Karnal
Frieswal	Sahiwal	Holstein Friesian	37.5-62.5%	PDC, Meerut
Phule Triveni	Gir	Holstein Friesian, Jersey	75%	MPKV, Rahuri, Maharashtra
Vrindavani	Hariana	Holstein-Friesian, Jersey, Brown Swiss	50-75%	IVRI, Izatnagar

Table 2. Crossbred strains of Sheep developed in India

Strain	Breeds used			Purpose	Location
	Native	Exotic	Exotic Inheritance		
Bharat Merino	Chokla, Nali	Rambouillet, Merino	75%	Fine wool	CSWRI, Avikanagar
Avivastra	Chokla, Nali	Rambouillet, Merino	50%	Fine wool	CSWRI, Avikanagar
Avikalin	Malpura	Rambouillet	50%	Carpet wool	CSWRI, Avikanagar,
Avimans	Malpura, Sonadi	Dorset, Suffolk	50%	Mutton	CSWRI, Avikanagar
Indian Karakul	Marwari, Malpura, Sonadi	Karakul	75%	Pelt	CSWRI, ARC Bikaner

Kashmir Merino	Gaddi, Bhakarwal Poonchi	Deline Merino, Soviet Merino, Rambouillet	50-75%	Fine wool	Jammu and Kashmir
Hisardale	Bikaneri	Australian Merino	75%	Apparel Wool	GLF,Hisar
Nilgiri Synthetic	Nilgiri	Soviet Merino	62.5-75%	Apparel Wool	TNUVAS, Sandynallah
Patanwadi Synthetic	Patanwadi	Rambouillet, Soviet Merino	50%	Carpet wool	GAU, Dantiwada

Table 3. Crossbred varieties of chicken developed in India

Variety name	Breeds involved		Purpose	Location
Vanraja	Cornish	Synthetic broiler population	Dual	ICAR,Hyderabad
Grama priya	Desi	White leghorn	Dual	ICAR,Hyderabad
Giriraj	Desi	White leghorn	Dual	UAS,Bangalore
Narmadanidhi	Kadaknath	Jabalpur colour (Colour broiler)	Dual	NDVSU,Jabalpur
Gramalakshmi	Australorp	White leghorn	Dual	KAU,Kerala
Cari-Nirbhik	Assel	CARI red	Dual	CARI, Bareilly
Cari-Shyama	Kadaknath	CARI red	Dual	CARI, Bareilly
Upcari	Frizzle	CARI red	Dual	CARI, Bareilly
Hitcari	Naked neck	CARI red	Dual	CARI, Bareilly

Table 4. Crossbred pig varieties developed in India

Variety	Breeds used		Location
	Exotic breed	Indigenous	
Rani	50% Hamshire	50% Ghungroo	NRC Pig, Assam
Mannuthy White	75% Large White Yorkshire	25% Desi	KVASU, Kerala
Jharsuk	50% Tamworth	50% Desi	BAU, Ranchi
Landley	75% Landrace	25% Desi	IVRI, Bareilly
Asha	50% Duroc 25% Hamshire	25% Ghungroo	NRC Pig, Assam
Lamsniang	75% Hampshire	25% Niangmegha	ICAR-RC for NEH, Meghalaya

14

Digestion in Ruminant Animals

Manish Kumar, Thulasiraman P, Anuradha Kumari and Utkarsh Kumar Tripathi

Faculty of Veterinary and Animal Sciences, I. Ag.Sc., RGSC-Banaras Hindu University (BHU), Barkachha, Mirzapur, Uttar Pradesh-231001

The primary function is to serve as ameans to digest and absorb the essential nutrients of the diet required tosustain the body. Digestion is the process of breakdown of complex food into simpler form by the activities of the alimentary tract and glandular secretions for absorption of nutrients and the rejection of their residues. Most foods consumed are more complex and insoluble to be absorbed into blood or lymph.

Food

Food is a complex mixture of substances like carbohydrates proteins, fats, vitamins, inorganic salts and water to meet the nutritive requirements of an animal. Herbivorous animals (Cattle, sheep, horse, goat etc.,) derive their nutritive requirements from plant sources. Carnivorous animals (dog, cat etc.,) obtain their food from animal sources. Omnivorous animals (man, pig, etc) get their foods from both animals and plants sources.

Digestive tracts

An animal's digestive tract can be thought of as a hollow sac or a tube with an entrance and exit, lying within body & specialized for initial processing of food items. Nutrients are only considered "inside" the body once absorbed. The gut tube is divided into foregut, midgut, and hindgut, performing key functions such as digestion, absorption, immune defense, and homeostasis by transferring nutrients, water, and electrolytes. It must prevent pathogen entry while tolerating commensal microbes aiding digestion. The digestive system includes the tract and accessory organs, contributing to osmoregulation, endocrine function, immunity, and detoxification. In vertebrates, it consists of the mouth, pharynx, esophagus, stomach (or rumen/proventriculus-gizzard in birds), small intestine (duodenum, jejunum, ileum), large intestine (cecum, colon, rectum), and anus/cloaca-all continuously connected.

Accessory Digestive Organs

In addition to the specialized gut tubes, there are many accessory organs that aid digestion. The accessory digestive organs of most vertebrates include the Salivary glands, Exocrine pancreas, and Biliary system, which consists of the liver and, in some species, the gallbladder.

Ruminant Stomach

Ruminant Stomach: e.g. Sheep, Cows, Buffaloes and Goats, it occupies ¾ of the abdominal cavity. **Rumen:** Composes 70 to 80% of ruminant stomach in mature bovine animals & 30% in young animals. **Reticulum:** Composes about 5% of bovine stomach, prevents indigestible objects from entering the stomach. **Omasum:** Composes 7-8% of bovine stomach, involve in absorptions of mostly water. Omasum is not well developed in sheep and goat, but it is absent in Camel and Llama. **Abomasum:**also known as "true" stomach. Composes 7-8% of stomach in mature animals and 70% in young animals.

The rumen, the largest compartment, is lined with papillae resembling shag carpet. The reticulum has honeycomb-shaped projections. Both serve as storage sites for ingesta and house fermentative bacteria that break down cellulose and hemicellulose. They are lined with stratified squamous epithelium, which absorbs VFAs, electrolytes, and minerals. The omasum contains leaf-like structures covered by stratified squamous epithelium, aiding in the absorption of VFAs and water as digesta moves toward the abomasum. The abomasum, functionally similar to a monogastric stomach, may have more folds but performs the same digestive role.

Small Intestine: (Duodenum, Jejunum, & Ileum): Long, coiled tube connecting stomach with the large intestine. Covered by villi which increases surface area to increase absorption. Food moves through by muscle contractions called peristaltic movement. Final breakdown and absorption of nutrients occurs in Small Intestine.

Large Intestine: (Cecum, Colon & Rectum) it involves in mainly absorptions of water.

Prehension

Prehension is the act of seizing and conveying food into the mouth, aided by oral structures for grasping, mastication, and swallowing. Cattle use a strong, rough tongue and lower incisors, while swine utilize a pointed lower lip.

Movement of Jaw

In herbivores, lateral movement of the lower jaw and some to and fro movement are important. In ruminants, the upper incisors are absent, but modified as dental pad and in the lower jaw they are loosely and obliquely placed in the sockets.

Regurgitation

Regurgitation involves an additional contraction of the reticulum, followed by a primary contraction. This process is triggered by a temporary negative pressure, created when the glottis closes before an inspiratory cycle. Simultaneously, the lower esophageal sphincter opens, the upper esophageal sphincter relaxes, and digesta flows into the esophagus. Reverse peristalsis propels the semiliquid digesta upward. Upon reaching the mouth, the liquid portion is immediately re-swallowed, while the solid material remains for mastication and reinsalivation, primarily by the parotid gland on the chewing side.

Rumination (chewing of cud)

Mastication is the voluntary grinding of food by molar teeth. Rumination, unique to true ruminants (deer, giraffes, bovids) and pseudoruminants (camels, llamas), involves a prolonged chewing phase followed by brief re-swallowing. The duration of rumination depends on species and diet composition, higher roughage intake increases chewing time. Cattle on high-roughage diets ruminate for about 10 hours daily, while finely ground low-roughage diets reduce this to 3 hours or less.

Rumination follows a circadian rhythm, peaking in the afternoon and late at night, often occurring during drowsiness, suckling, or milking. If solid material remains, it is chewed slowly for about 40 seconds before swallowing. Typically, chewing is confined to one side of the jaw, though camels may alternate sides with each chew. Cud chewing increases salivation three to five times, enhancing mixing and eructation. It is a key behavioral state, occupying about one-third of a ruminant's life. The presence of cud triggers slow, rhythmic chewing (50-55 chews/min), increased salivation, and heightened reticulorumen motility. Rumination is a vital indicator of health.

Fermentative Digestion in Rumen

For most ruminant animals, the primary nutrient source is grass in the form of roughage. Cellulose is a polysaccharide composed of β-1,4-linked glucose units, forming the main structural component of plant cell walls. Hemicellulose consists primarily of β-1,4-linked xylose units, contributing to plant structure. Pectin, an intracellular structural polysaccharide, is mainly composed of β-1,4-galacturonan, a polymer of galactose and uronic acid. The enzyme (cellulase, hemicellulase, pectin, lysase and fructosanases) necessary to hydrolyze β-1-4 linkages are found only in microbes and plants. It is for this region that herbivores require a symbiotic fermentative relationship with suitable microbes.

Only microbes, including bacteria, fungi, and protozoa, along with a few invertebrates, can digest cellulose and hemicellulose, while lignin remains almost entirely indigestible. In roughage, approximately 60% of the total dry matter consists of structural carbohydrates, whereas in concentrates, it is primarily non-structural carbohydrates. In legume seeds, neither dominates, as protein is the major component. In all the feed, the lipid content remains around 4%.

Fermenting mass

Fermenting mass in the ruminoreticulum consists of ingesta that is subjected to mechanical breakdown due to chewing during ingestion and during rumination and chemical breakdown during microbial fermentation. Efficient rumen function relies on several key factors: the regular intake of macerated food, the presence of appropriate microbial populations, and the continuous removal of fermentation end products such as volatile fatty acids (VFAs), bacterial byproducts, and gases. Proper mixing facilitates substrate addition, buffering, and VFA absorption while preventing the accumulation of inhibitory byproducts. Unfermented material must be propelled onward to the abomasum and intestines. Maintaining anaerobic conditions, along with stable intraruminal temperature, osmotic pressure, redox potential, and pH, is essential. Additionally, mechanisms must align fore-stomach physiological functions with fermentative requirements.

Primary bacteria are those that degrade the actual constituents of the diet and, depending on their preference for cellulose or for starch, are termed cellulolytic or amylolytic, respectively. Secondary bacteria use as their substrate the end products of the primary bacterial degradations. This group includes the lactate-utilizing propionate bacteria, which produce some of the propionate, and the hydrogen-utilizing methanogenic bacteria, which produce methane gas.

Rumen Bacteria: Bacterial number is about 10^{10} to 10^{11} per ml rumen fluid.

Table 1. Classification of ruminal bacteria.

Sl. No.	Species	Types of Bacteria
1.	*Cellulolytic Species*	*i. Fibrobacter succinogenes* *ii. Ruminococcus albus & R. flavifaciens* *iii. Butyrivibriofibrisolvens*
	Hemicellulolytic Species	*i. Butyrivibriofibrisolvens* *ii. Ruminococcus sp. &* *iii. Bacteroides ruminicola*
2.	*Pectinolytic Species*	*i. Butyrivibriofibrisolvens* *ii. Bacteroides ruminicola* *iii. Succinivibriodextrinosolvens*

3.	*Amylolytic Species*	*i. Bacteroides amylophilus* *ii. Streptococcus bovis* *iii. Succinomonasamylolytica*
4.	*Ureolytic Species*	*i. Succinivibriodextrisolvens* *ii. Selenomonas sp.* *iii. Butyrivibriosp* *iv. Bacteroides ruminicola*
5.	*Methane producingSpecies*	*i. Methanobrevibacterruminantium* *ii. Methanobacteriumformicicum* *iii. Methanomicrobium mobile*
6.	*Sugar-Utilising Species*	*i. Treponema bryanti* *ii. Lactobacillus sp.*
7.	*Acid-Utilising Species*	*i. Megaspheraelsdenii* *ii. Selenomonasruminantium*
8.	*Proteolytic Species*	*i. Bacteroides amylophilus* *ii. B. ruminicola* *iii. Butyrivibriofibrisolvens* *iv. Streptococcus bovis*
9.	*Ammonia Producing Species*	*i. Bacteroides ruminicola* *ii. Megaspheraelsdenii*
10	*Lipid-Utilising Species*	*i. Anaerovibriolipolytica* *ii. Butyrivibriofibrisolvens* *iii. Micrococcus sp.*

Rumen Protozoa

Many species of rumen protozoa actively contribute to fermentation within the rumen. These protozoa are strictly anaerobic and are found exclusively in ruminants. The majority are ciliates, which play a significant role in digesting starch, engulfing bacteria, and regulating microbial populations, while flagellates are present in very limited numbers. Rumen protozoa aid in fiber digestion, protein metabolism, and VFA production, influencing overall rumen efficiency and fermentation balance. Their populations fluctuate based on diet composition and feeding practices.

The protozoa feed on ruminal bacteria, plant starch granules, and other readily digestible materials, including perhaps the dietary polyunsaturated fatty acids (PUFAs) and linoleic and linolenic acids. Protozoa are very sensitive to abnormal intraruminal conditions, and their presence in a ruminal fluid sample is a good indicator of normality. Most of the protozoa are located in the raft of fibrous digesta in the dorsal ruminal sac. Here they (1) form a reservoir of microbial protein useful at times of intermittent food supply; (2) help to prevent an over proliferation of bacteria in times of starch loading by engulfing starch particles and thereby curbing the undesirably high rates of starch degradation by amylolytic bacteria.(3) when they pass out of the forestomach and undergo digestion in the lower gastrointestinal tract, they provide the ruminant host animal:-(a) with a higher-biological-value microbial protein than in the case of bacteria, (b) with small amounts of unfermented stored starch, and (c) possibly with some PUFAs that would otherwise have been hydrogenated by the ruminal bacteria and therefore made unavailable to the ruminant itself.

pH optima for different microbes

Rumen protozoa, primary cellulolytic bacteria, and most secondary bacteria thrive at a pH of 6.2 or higher, while primary amylolytic bacteria are more active in slightly acidic conditions, around pH 5.8. At even lower pH levels, lactic acid-producing Lactobacilli, typically insignificant, can become dominant, further acidifying the rumen and creating unfavorable conditions for the normal microbial population. Generally, microbial numbers increase with both the quantity and quality of the diet, with protozoa being especially abundant in starch-rich diets. When major dietary changes occur suddenly, it takes approximately two weeks for the microbial population to reestablish a new balance. Stratification of the ruminoreticular contents occurs and the digesta is not a homogeneous mass. The most recently ingested food is added to a "raft" of fibrous material that occupies most of the dorsal ruminal sac and floats on the underlying "soupy" fluid, in which there are suspended only very fine particles, each no longer than 2mm. Above the raft is a layer of ruminal gas, composed mainly of CH_4 and CO_2.

Ruminal digesta may be considered as being distributed into four fractions, each having a very different composition; (1) Fibrous raft particles with high densities of microbes inside, the microbes moving into fresh particles as the old ones disintegrate; (2) Liquid fraction of the raft, which shuttles mainly soluble materials in (e.g., salivary constituents) and out (e.g., fermentative end products); (3) Boundary layer lying against the luminal surface of the ruminoreticulum, across which certain blood-rumen fluid exchanges occur in both directions and in which urease-rich bacteria are located; and (4) Soupy material that occupies the reticulum and the cranial and ventral sacs of the rumen.

This material receives all ingested food and water, saliva, reswallowed cuds, and the end products from the raft. It floats light, particulate ingesta across from the cardia to join the raft. It provides both the material for the cud "bolus" and the material that flows out of the reticulum into the omasum through the reticulu-omasal opening. Clinically, it is the soupy material that is sampled, but this approach may not give an accurate indication of the microbial composition and fermentative activity of the raft.

Fermentation of Cellulose

The breakdown of β-1-linked compounds, such as cellulose, hemicelluloses, fructosans, and pectinis carried out by various species of primary cellulolytic bacteria, except for methane production, which is performed by methanogenic bacteria. Cellulose fermentation is a slow process due to the low metabolic rate of cellulolytic bacteria, which have a doubling time of approximately 18 hours, leading to gradual population changes. For protein synthesis, these bacteria do not require amino acids but instead rely on ammonia (NH_3) and small amounts of isoacids, which originate from the deamination of branched-chain amino acids found in plant proteins. Their optimal pH range is 6.2 to 6.8, aligning with the typical ruminal pH of roughage-fed animals. Methanogenic bacteria share a similar pH optimum with cellulolytic bacteria and require formate, CO_2, and reducing equivalents (2H) to produce methane, as well as a supply of amino acids to meet their protein needs. The interaction between cellulolytic and methanogenic microbes results in the production of CO_2, CH_4, and volatile fatty acids (VFAs). The VFAs derived from the fermentation of cellulose-acetate, propionate, and butyrate are generally in the ratio 75:15:10, results in respectively.

Fermentation of Starch

The degradation of the α-1-linked starches (amylose and amylopectic) and the simple sugars (e.g. sucrose, maltose) is performed by several species of primary amylolytic bacteria. Unlike the cellulolytic bacteria, the amylolytic bacteria have faster fermentation rates, have much shorter doubling times (0.25 to 4 hours), and have a lower pH optimum of 5.5 to 6.6. This matches the lower ruminal pH values of ruminants on high concentrate (starch-rich) diets and is due to higher VFA concentrations with an increase in the relative proportions of propionate, giving a typical acetate/propionate/butyrate ratio of 70:25:5, respectively. Amylolytic bacteria require not only a supply of NH_3 but also some amino acids for protein synthesis.

Secondary bacteria include methanogenic bacteria, which aid in methane formation, and propionate bacteria, which convert lactic acid into propionate. Both require amino acids, have a long doubling time (~16 hours), and thrive at a higher pH (6.2–6.8) than amylolytic bacteria. A rapid shift from roughage to concentrate increases amylolytic bacteria, accelerating VFA and lactic acid production, lowering ruminal pH. This favors amylolytic bacteria but inhibits secondary bacteria, disrupting microbial balance. Protozoa increase with starch-rich diets, regulating bacterial amylolysis, but become inactivated when pH drops below 5.5 and eventually die.

Fermentation of Dietary Protein

Proteolytic bacteria constitute 12–38% of the ruminal microbiota, degrading about half of dietary protein. Proteins are classified as rumen degradable (RDPs) or rumen undegradable (RUPs), with some bypassing ruminal degradation but hydrolyzed in the gastrointestinal tract. Bacterial proteolysis begins extracellularly, breaking proteins into peptides, which are absorbed and hydrolyzed into amino acids. Some amino acids are used by microbes, while others are deaminated into ammonia and metabolic acids, which are fermented into volatile fatty acids (VFAs), including branched-chain VFAs (isobutyrate, isovalerate) essential for cellulolytic bacteria. Ammonia also comes from nonprotein nitrogen (NPN) sources like urea, nitrites, and plant amides. Urea, entering via saliva and diffusing into the rumen, is rapidly hydrolyzed into ammonia, which supports microbial protein synthesis if adequate α-ketoglutarate, VFAs, and fermentable carbohydrates are available. In practice, feeding regimens must, first, provide sufficient crude protein (true protein plus NPN) and readily fermentable carbohydrates. These ensure that the ruminal microbes have adequate amino acids, ammonia, carbon skeletons, and available energy to meet the requirements of microbial protein synthesis for the maintenance of population numbers. Second, feeding regimens must ensure that excessive protein breakdown to VFAs and ammonia does not occur. Feeding protein in excess is a wasteful input of an expensive commodity, and it leads to the over-production of ammonia, which takes energy to convert it to urea (in the liver) and also creates risk of ammonia toxicity.

In addition to the fermentation of dietary protein, there is a continuous recycling of microbial protein from dead microbes, particularly within the fibrous raft. Amino acids produced in the forestomach are not immediately available to the ruminant. Instead, as material moves from the forestomach to the abomasum and small intestine, the ruminant acquires unfermented dietary proteins along with microbial proteins. Microbial protein has a higher biological value, meaning it contains a greater proportion of essential amino acids compared to the plant proteins originally consumed in the diet.

Fermentation of Dietary Lipids

Dietary lipids occur as structural lipids in the leaves of forage plants and as storage lipids in oil seeds. The forage plant lipids are found mainly in cell membranes and comprise 3 to 10 percent of the dry mater. less than 50 percent of the total lipids are free fatty acids (FFAs), and the majority are phospholipids, with palmitic, linoleic, and linolenic acids being the predominant fatty acids. In oil seeds, 65 to 80 percent of the lipids are triglycerides, with palmitic, oleic, and linoleic being their predominant fatty acids. Ruminal microbes rapidly hydrolyze dietary lipids and, using the unsaturated fatty acids (oleic, linoleic, and linolenic) as hydrogen acceptors, quickly convert most of them to stearic acid. Most plant unsaturated fatty acids are in the cis form. Ruminal microbes also synthesize microbial lipids from VFAs, and many of these are in the trans form.

Ruminant adipose tissue therefore contains fatty acids with both trans and cis forms. Ruminant diets generally do not contain more than 5 percent dry matter (DM) as lipid. Higher values have adverse effects on (1) food palatability: (2) cellulolytic activity: (3) food appetite and forestomach motility, (4) the physical consistency of concentrate pellets at high and low temperatures; and (5) the shelf-life of concentrates, lipids being prone to the development of rancid flavors. Protozoa have an important role in ruminal lipid metabolism. They absorb some of the PUFAs, lock them away in their own structures, and thereby protect them from hydrogenation. The protozoa that subsequently flow out of the rumen and undergo intestinal digestion release their content of PUFAs, this being probably the main source of PUFAs for the ruminant.

Volatile Fatty Acids

The primary fermentative end products of carbohydrate digestion in ruminants are acetic, propionic, and butyric acids. A starch-rich diet increases propionic acid production. Protein fermentation also yields valeric acid and branched volatile fatty acids (VFAs), which constitute less than 5% of total VFAs and primarily support microbial protein synthesis. VFAs are absorbed across the forestomach wall, with half diffusing passively in the undissociated state and the rest absorbed as anions via facilitated diffusion in exchange for bicarbonate ions. Carbonic anhydrase in the forestomach epithelium facilitates this exchange, aiding VFA absorption and buffering ruminal acidity. Additional neutralization occurs via salivary alkali. Butyric acid is largely metabolized into β-hydroxybutyrate (β-OH Bu) during absorption and in the liver, serving as an energy source and a precursor for short- and medium-chain fatty acids in milk. About 30% of propionate is converted to lactic acid in the forestomach, with both lactic acid and propionate utilized in the liver for gluconeogenesis. Acetate, the most abundant VFA, is largely unchanged during absorption and liver passage, serving as the primary metabolic substrate for energy production, fatty acid synthesis in milk, and body fat formation. Unlike non-ruminants, which rely on glucose and dietary fatty acids for fat synthesis, ruminants primarily use acetate.

Gases

Gas production in cattle peaks at up to 40L/h between 2 to 4 hours after feeding, when fermentation is most active. The primary gases are CO_2 (60%), CH4 (30–

40%), and smaller amounts of N_2, H_2S, H_2, and O_2, with nearly all eliminated via eructation. CO_2 is produced through fermentation decarboxylation and bicarbonate neutralization, while CH_4 results from CO_2 and formate reduction by methanogenic bacteria. This process helps recycle reducing equivalents but also leads to energy loss, as methane elimination accounts for about 8% of total digestible energy. H2S, derived from sulphates and sulfur-containing amino acids, is potentially toxic even in small amounts. H_2 remains in trace amounts unless abnormal fermentation occurs due to excessive concentrate intake. O_2 enters the rumen with food, water, and via blood diffusion but is rapidly consumed by facultative anaerobic bacteria, maintaining the strictly anaerobic environment essential for ruminal microbes.

15

Digestion in Non-Ruminant Animals

Manish Kumar, Thulasiraman P, Mukesh Bharti and Utkarsh Kumar Tripathi

Faculty of Veterinary and Animal Sciences, I. Ag.Sc., RGSC-Banaras Hindu University (BHU), Barkachha, Mirzapur, Uttar Pradesh-231001

The primary function is to serve as ameans to digest and absorb the essential nutrients of the diet required tosustain the body. Digestion is the process of breakdown of complex food into simpler form by the activities of the alimentary tract and glandular secretions for absorption of nutrients and the rejection of their residues. Most foods consumed are more complex and insoluble to be absorbed into blood or lymph.

Food

Food is a complex mixture of substances like carbohydrates proteins, fats, vitamins, inorganic salts and water to meet the nutritive requirements of an animal. Herbivorous animals (Cattle, sheep, horse, goat etc.,) derive their nutritive requirements from plant sources.Carnivorous animals (dog, cat etc.,) obtain their food from animal sources. Omnivorous animals (man, pig, etc) get their foods from both animals and plants sources.

Digestive tracts

An animal's digestive tract can be thought of as a hollow sac or a tube with an entrance and exit, lying within body & specialized for initial processing of food items. Nutrients are only considered "inside" the body once absorbed. The gut tube is divided into foregut, midgut, and hindgut, performing key functions such as digestion, absorption, immune defense, and homeostasis by transferring nutrients, water, and electrolytes. It must prevent pathogen entry while tolerating commensal microbes aiding digestion. The digestive system includes the tract and accessory organs, contributing to osmoregulation, endocrine function, immunity, and detoxification. In vertebrates, it consists of the mouth, pharynx, esophagus, stomach (or rumen/proventriculus-gizzard in birds), small intestine (duodenum, jejunum, ileum), large intestine (cecum, colon, rectum), and anus/cloaca—all continuously connected.

Accessory Digestive Organs: In addition to the specialized gut tubes, there are many accessory organs that aid digestion. The accessory digestive organs of most vertebrates include the Salivary glands, Exocrine pancreas, and Biliary system, which consists of the liver and, in some species, the gallbladder.

Mammalian Stomach

In addition to the circular and longitudinal smooth muscle layers found of the digestive tract, the muscularis externa of the stomach has an additional inner oblique or transverse layer. Anatomically, the mammalian stomach (ruminant abomasum) is the organ between the esophagus (or fore-stomachs in ruminants) and the duodenum. Physiologically, the stomach can divided into four distinct functional compartments.

1. **Esophageal stomach** (non-glandular) is lined by stratified squamous epithelium. No mucus, acid, or proteolytic enzymes are produced from this part.
2. **Cardia stomach** (glandular stomach) lined by simple columnar epithelial cells that produce a thick mucus and buffer that adheres to the cells to protect the epithelium from the proteolytic enzymes and acid produced in another compartment of the stomach.
3. **Proper or fundic stomach:** It has very deep invaginations in the submucosa lined by a variety of cells that produce acid, proteolytic enzymes, hormones, & mucus. All mammals have a fundic stomach and it is generally the largest compartment within the stomach.
4. **Pyloric stomach** contains moderately deep glands lined with epithelial cells that secrete mucus and buffers, but not acid or proteolytic enzymes. It houses enteroendocrine cells, including G-cells, which produce gastrin. In all mammals, the pyloric sphincter regulates stomach emptying.

Small Intestine: (Duodenum, Jejunum, & Ileum): Long, coiled tube connecting stomach with the large intestine. Covered by villi which increases surface area to increase absorption. Food moves through by muscle contractions called peristaltic movement. Final breakdown and absorption of nutrients occurs in Small Intestine.

Large Intestine: (Cecum, Colon & Rectum) it involves in mainly absorptions of water.

Functions of the digestive tract

1. **Ingestion:** This is the active process of bringing material into the oral cavity.
2. **Propulsion:** Ingested materials are moved through the digestive tract by **swallowing&peristalsis**, which involves alternating waves of

contraction & relaxation of muscles along the digestive tract wall, and is major propulsive mechanism moving food through tract.

3. **Mechanical processing:** Physically breaking down food to increase surface area for digestion, including chewing and rumination in ruminants. This begins in the oral cavity where food is **crushed** and **sheared** before being propelled along the digestive tract.
4. **Digestion:** Following reduction in size, ingested nutrients are chemically broken down into particles small enough for absorption.
5. **Secretion:** Water, mucus, acids, enzymes, buffers, and salts are released into the lumen of the digestive tract along its length.
6. **Absorption:** The digestive tract absorbs nutrients, water, electrolytes, and secreted materials. Failure in absorption leads to dehydration
7. **Excretion:** The digestive tract is a site of elimination of waste products. Such waste products can be eliminated via defecation.
8. **Immunity:** The digestive tract must provide a substantial barrier to prevent the entry of pathogens into the body.

Prehension

Prehension is the act of seizing and conveying food into the mouth, aided by oral structures for grasping, mastication, and swallowing. In bipeds, hands serve this function, while dogs and cats use forelimbs and jaw movements. Horses rely on their upper lip, tongue, and incisors, whereas sheep use a clefted upper lip for close grazing, unlike goats with an unclefted lip. Cattle use a strong, rough tongue and lower incisors, while swine utilize a pointed lower lip.

Suckling

It is effected by the creation of negative pressure in the mouth largely by the back ward movement of the base of the tongue. Milk is forced from the teat into the mouth due to pressure gradient.

Drinking: Horses, cattle, and humans create negative pressure to suck water, while cats, dogs, and wild relatives lap water by forming a ladle with their tongue. Dogs lift about 10–15 ml per lap. Birds dip their beak and lift their head, relying on gravity to move water back.

Mastication

The first step in the digestive process is mastication (chewing), the motility of the mouth that involves slicing, tearing, grinding, and mixing ingested foods by specialized mouthparts such as teeth. It is the mechanical reduction of food by the grinding action of the molar teeth. Carnivores swallow food with minimal chewing, while ruminants briefly chew before swallowing and later remasticate during rumination. Horses thoroughly chew before swallowing. Chewing is regulated by the medulla oblongata.

Movement of Jaw

In carnivores, vertical movement of the lower jaw is important for mastication. In carnivores and omnivores the upper and lower jaws are equal width and the teeth are relatively simple. In herbivores, lateral movement of the lower jaw and some to and fro movement are important. In herbivores, the upper jaw is wider than the lower jaw and hence mastication can occur on only one side a time. The molar teeth are chisels shaped with sharp edged lower teeth pointed towards inner side and that of upper teeth towards outer side. The incisor teeth are used to cut or lacerate food. In ruminants, the upper incisors are absent, but modified as dental pad and in the lower jaw they are loosely and obliquely placed in the sockets.

Deglutition or Swallowing

Swallowing, or deglutition, is a complex reflex that moves food or fluids from the mouth to the stomach while preventing entry into the respiratory tract. It begins voluntarily but quickly becomes an involuntary reflex controlled by the medulla. The process involves two stages:

1. **Oropharyngeal Stage** – The tongue pushes the bolus to the pharynx, activating pressure receptors that signal the swallowing center, which coordinates muscle movements to direct the bolus into the esophagus while closing off other openings.
2. **Esophageal Stage** – The bolus travels down the esophagus, a muscular tube extending to the stomach. In ruminants and dogs, the esophagus is entirely striated muscle, while in birds and humans, it consists of smooth muscle. Horses and cats have smooth muscle in the lower third of the esophagus.

The esophagus is regulated by the pharyngoesophageal sphincter at the top and the gastroesophageal (cardiac) sphincter at the bottom, which control food passage and prevent backflow.

Cardia

The point of opening of oesophagus into the stomach is called cardia. It is provided with a sphincter muscle known as cardiac sphincter which is capable of automatic activity under the control of CNS. It prevents back flow of food from stomach to oesophagus. The cardiac sphincter is well developed and powerful in horse.

Salivary Gland and Saliva

Saliva is produced by three major paired salivary glands: Parotid, Submaxillary (Mandibular), and Sublingual, along with numerous smaller glands in the mouth's mucous membrane. In ruminants like sheep and cattle, additional glands include inferior molar, buccal, labial, palatine, and pharyngeal glands. Salivary glands are classified by secretion composition:

1. Alkaligenic glands (Parotid, Inferior molar, Buccal, Palatine) secrete more bicarbonate (HCO3) and less mucin.
2. Mucogenic glands (Submaxillary, Sublingual, Pharyngeal) secrete more mucin with lower bicarbonate.

Saliva is a mixture of these secretions. A cow can produce 100-200 L/day, with parotid glands secreting 2mL/min at rest and 30-50 mL/min during rumination. Salivary glands are categorized as serous, mucous, or mixed based on their secretion type, with serous glands producing thin watery fluid, mucous glands producing mucin, and mixed glands providing both. The parotid gland is primarily serous in most animals, though some lack enzymes. The submaxillary (mandibular) gland is mixed in humans, dogs, ruminants, and cats, but serous in rodents. The sublingual glands are mixed in horses, cattle, pigs, dogs, and cats, while mucous in rodents. In cats and dogs, salivary secretion aids in evaporative cooling. Ruminant saliva is rich in bicarbonate (HCO3) and phosphate (PO4) buffers, offering strong neutralizing capacity. In carnivores like rats, pigs, and humans, saliva contains salivary amylase, an enzyme that splits starch and functions best at neutral pH.

Mixed saliva is a colorless, slightly opalescent liquid containing electrolytes, proteins, salivary alpha-amylase (in some animals), desquamated cells, and lymphocytes. Dog parotid saliva is thin, while submaxillary saliva is highly viscous. Saliva is produced by acinar glands along the mandible and maxilla, and is hypotonic, reducing the osmotic concentration of ingesta. Salivary secretion is controlled by the glossopharyngeal nerve (parotid glands) and facial nerve (submaxillary and sublingual glands), both carrying parasympathetic fibers. Parasympathetic tone regulates saliva production, with ruminant saliva pH rising to about 8.5 during chewing.

Functions of Saliva

Saliva aids in mastication but plays a more important role in lubricating food boluses before swallowing. The most important functions of saliva include the following:

1. **Moistening:** Saliva moistens food, aiding in swallowing and forming a bolus. It also keeps the oral cavity moist and protects the mucosa.

2. **Digestion:** In most animals (except ruminants), saliva begins starch digestion via salivary amylase, breaking down starches into maltose. Some animals also produce lingual lipase for fat digestion.
3. **Defense:** Saliva has antibacterial properties through enzymes like lysozyme, salivary agglutin, and lactoferrin, and by washing away bacteria.
4. **Taste:** Saliva dissolves molecules that activate taste buds, enabling taste sensation.
5. **Neutralization:** Saliva contains bicarbonate buffers that neutralize acids from food and bacteria, and in ruminants, it helps neutralize rumen acidity.
6. **Thermoregulation:** In animals without sweat glands, saliva aids in temperature control via evaporative cooling (e.g., panting in dogs).
7. **Poisons:** Venomous reptiles secrete toxic saliva that affects the circulatory, nervous, or muscular systems of prey.
8. **Pheromones:** Some animals release pheromones in saliva to communicate, such as male boars producing androsterones to stimulate mating behavior.
9. **Provides alkaline buffering and fluid:** Bicarbonate and phosphate in the saliva can neutralize acidic feedstuffs.
10. **Removes wastes:** Metabolic waste products such as urea and uric acid are excreted in the saliva.

Control of Salivary Secretion

Salivary secretion in vertebrates is primarily controlled by the glossopharyngeal (parotid glands) and facial (submaxillary and sublingual glands) nerves, both carrying parasympathetic fibers. In ruminants, parotid glands secrete saliva spontaneously without neural stimuli. Two types of salivary reflexes exist:

1. **The simple, or unconditioned**, salivary reflex occurs when chemoreceptors and pressure receptors within the oral cavity respond to the presence of food. On activation, these receptors initiate impulses in afferent nerve fibers that carry the information to the salivary center located in the medulla oblongata. The salivary center, in turn, sends impulses via the extrinsic autonomic nerves to the salivary glands to promote increased salivation.
2. **The acquired, or conditioned**, salivary reflex occurs without oral stimulation. A zoo mammal hearing the sound of a food cart, or a dog watching the preparation of a meal, initiates salivation through this reflex.

Unlike the autonomic nervous system elsewhere in the vertebrate body, sympathetic and parasympathetic controls from the salivary center to the salivary glands are not antagonistic. Both sympathetic and parasympathetic stimulation increase salivation, but with differing effects: parasympathetic stimulation enhances blood flow, producing abundant serous saliva, while sympathetic stimulation reduces blood flow, leading to a smaller volume of mucus saliva.

Secretory Function of Stomach

The food when swallowed is received by the stomach where it is subjected to gastric digestion. Stomach is a hollow, saclike organ made up of four layers namely Serous, Muscular, Submucosa & Mucosa form outside to inside. Physiologically, the stomach mucosa of domestic animals, is divided into four distinct functional compartments. 1. Oesophageal stomach "nonglandular", and glandular stomach includes 2. Cardia stomach, 3. Proper or Fundic stomach and 4. Pyloric stomach.

Fundic stomach secretions: The fundic stomach contains gastric pits lined with mucus-secreting cells, forming a protective gel layer that shields the stomach from acid and digestive enzymes. This mucus also contains a sodium bicarbonate buffer for additional protection. Two distinct cell types, unique to the fundic stomach, are dispersed throughout the gastric glands.

1. **Chief cell:** Secretes the proteolytic enzyme precursor pepsinogen into the lumen of the gastric gland, is rapidly conveyed into the lumen of the stomach. The enzyme is secreted in an inactive form to avoid digestion of the chief cell. Chief cells also produce a proteolytic enzyme known as rennin, important in neonates as it helps digest milk proteins & forms milk curds within the stomach.
2. **Parietal cells (or oxyntic cells):** These cells produce stomach acid (HCl), which aids in digestion and eliminates ingested bacteria. In most species, parietal cells also secrete intrinsic factor, a protein that binds dietary vitamin B12 and facilitates its absorption in the ileum through a specialized transport system.

Gastric Enzymes

Pepsin: Proteolytic enzymes, hydrolysis of proteins into polypeptides. Synthesized as inactive form Pepsinogen and is activated by HCl. Optimum pH required for pepsin activity is 1.5 to 3.0.

Rennin (Chymosin, Rennet): It is a milk coagulating enzyme, present in young animals (Calf, Lam, Piglet etc.). Secreted as Prorennin, inactive form and is activated by HCl.

Gastric Lipase: Require the optimum pH of 5.5 to 7.5. Acts on emulsified fat & hydrolyse the fat into fatty acids & glycerol. It is in lower in concentration in carnivorous animals. It is absent in birds and ruminants.

Control of Gastric Secretion

In man and horse, the secretion of gastric juice is continuous but it is intermittent in dogs and cats. The rate of secretion increases during feed intake. The secretions of gastric glands are regulated by nervous and chemical mechanisms. The gastric secretion takes place in three phases:Cephalic Phase, Gastric Phase and Intestinal Phase.

1. **Cephalic Phase:** Stimulation of sensory endings in the mouth and pharynx, or conditioned reflexes triggered by the sight, smell, or thought of food, can initiate the cephalic phase of secretion, accounting for approximately 45% of gastric secretion. This response is well-developed in dogs and also occurs in pigs.
2. **Gastric Phase:** Entry of food into stomach causes copious secretion of gastric juice. This constitutes about 45% of the gastric secretory response.Basically, two principle stimuli are responsible for gastric phase secretion, i.e. Mechanical Stimulation and Humoral / hormonal stimuli.
3. **Mechanical Stimulation:** Due to the contact of food bolus with the receptors of stomach mucous membrane and the distension of the stomach causes the release of acetylcholine. Both "G" cells and Parietal cells of the gastric glands are stimulated by acetylcholine results in increased secretion of gastrin & HCl respectively.
4. **Hormonal stimulation:** Gastrin, primarily produced by the "G" cells of the pyloric glands, is also released in small amounts from the fundus, duodenum, and small intestine. Its release is stimulated by vagal activation during food anticipation and stomach distension, leading to increased HCl secretion. Histamine in the gastric mucosa and acetylcholine from parasympathetic nerve endings also enhance HCl secretion. The vagus nerve conditions parietal cells and amplifies gastrin's effect on acid production.
5. **Intestinal phase:** Accumulation of food in the intestine excites gastric secretion and this is due to a humoral mechanism & due to entry of intestinal gastrin & Cholecystokinin from duodenum into gastric gland through blood stream. It contributes about 10% of the gastric secretion.

Cholecystokinin (CCK) & several other hormones produced in duodenum in response to entry of fats & amino acids or changes in osmolarity have a minor role in inhibiting acid production by binding to their respective receptors on parietal cells.

Secretin always works to correct low pH in the duodenum by increasing the production of alkaline secretions by the salivary glands, pancreas, & duodenal submucosal glands (Brunner's glands) that neutralize or buffer the acid.

Inhibition of Gastric Secretion

Gastric secretion can be inhibited by higher brain centers in response to unappetizing food, as well as by the autonomic nervous system during pain, anger, or emotional distress. Excess stomach acid (pH < 2.5) suppresses HCl production by inhibiting gastrin release. The entry of fat, sugar, or high-acid content into the duodenum triggers the release of gastric inhibitory polypeptide (GIP), which travels via the bloodstream to inhibit gastric secretion. Additionally, secretin and CCK, released from the duodenum in response to acidic stomach contents, further suppress gastric secretion.

Enteric Nervous System

The neural control of gut function is regulated by the autonomic nervous system (ANS) and the enteric nervous system (ENS). The ENS, functioning from the esophagus to the anus, consists of two nerve plexuses: the Submucosal Plexus (Meissner's plexus) within the submucosa and the Myenteric Plexus (Auerbach's plexus) between the circular and longitudinal muscle layers. Sensory neurons in the ENS detect changes such as stretch, pH, osmolarity, and toxins, relaying information to neurons that regulate gut responses autonomously. The parasympathetic system, primarily via the vagus nerve (cranial nerve X), dominates gastrointestinal control, while the sympathetic system, known for "fight or flight" responses, plays a secondary role.

Gastrointestinal Hormones

The gastrointestinal tract, the body's largest endocrine organ, contains enteroendocrine cells derived from the embryonic neural crest, which are widely distributed throughout the gastric, intestinal, and pancreatic tissues. The highest concentration of these cells is found in the proximal intestine, decreasing distally. The GI tract secretes hormones such as Gastrin, Cholecystokinin (CCK), Secretin, Motilin, Enteroglucagon, and Peptide YY (PYY). Gastrin, CCK, and Secretin, produced in the stomach and proximal small intestine, regulate gastric and intestinal digestion. Enteroglucagon and PYY, located in the distal small intestine, contribute to the "ileal brake" mechanism, slowing gastric emptying and intestinal transit to enhance nutrient absorption, particularly in response to undigested fats and carbohydrates.

Gastrin: Protein in the stomach stimulates release of gastrin, which has following functions:

1. It stimulates parietal and chief cells to increase secretion of HCl and pepsinogen, two substances of primary importance in initiating digestion of the protein that promoted their secretion.
2. It enhances gastric motility, stimulates ileal motility, relaxes the ileocecal sphincter, & induces mass movements in colon-functions that are all aimed at keeping contents moving through tract on arrival of a new meal.
3. It is tropic not only to stomach mucosa but also to small-intestine mucosa, helping maintain a well-developed, functionally viable digestive tract lining. Gastrin secretion is inhibited by an accumulation of acid in the stomach and by the presence in the duodenal lumen of acid and other constituents that necessitate a delay in gastric secretion.

Secretin: As the stomach empties into the duodenum, the presence of acid in the duodenum stimulates the release of secretin into the blood, which performs the following functions:

1. It inhibits gastric emptying to prevent further acid from entering the duodenum until the acid that is already present is neutralized.
2. It inhibits gastric secretion to reduce amount of acid being produced.
3. It stimulates pancreatic duct cells to produce a large volume of aqueous $NaHCO_3$ secretion, which is emptied into the duodenum to neutralize the acid.
4. It stimulates secretion by the liver of a $NaHCO_3$-rich bile, which likewise is emptied into the duodenum to assist in neutralization of the acidic chyme.

This helps prevent damage to the duodenal walls and provides a suitable environment for the optimal functioning of the pancreatic digestive enzymes, which are inhibited by acid.

5. Secretin and CCK are both trophic to the exocrine pancreas.

Cholecystokinin (CCK)

As chyme is emptied from the stomach, fat and other nutrients enter the duodenum. These nutrients, especially fat and to a lesser extent protein products, cause the release of cholecystokinin (CCK) which performs various functions:

1. It inhibits gastric motility and secretion, allowing adequate time for nutrients already in the duodenum to be digested and absorbed.

2. It stimulates the pancreatic acinar cells to increase secretion of pancreatic enzymes, which continue digestion of these nutrients in duodenum.
3. It causes contraction of the gallbladder and relaxation of the sphincter of Oddi so that bile is emptied into the duodenum to aid fat digestion and absorption.
4. CCK has also been implicated in long-term adaptive changes in the proportion of pancreatic enzymes produced in response to seasonal changes in diet.
5. CCK is an important regulator of food intake. It plays a key role in satiety or fullness.

Gastric Inhibitory Peptide (GIP)

A hormone released by the duodenum, GIP, helps promote metabolic processing of the nutrients once they are absorbed. It was believed to inhibit gastric motility and secretion, similar to secretin and CCK. This is remarkably adaptive in anticipation or "feed forward" fashion. The metabolic activities of this post absorptive phase are largely under the control of insulin.

Motilin

The hormone motilin, secreted by gland cells in the crypt of the small intestine, triggers peristaltic pumping in the mammalian intestine between meals. In support of its housecleaning function, motilin stimulates the contraction of the antrum and fundus of the stomach and accelerates gastric emptying.

Ghrelin and Obestatin: The peptide hormone ghrelin (the "hunger hormone") is secreted by epithelial cells lining the stomach fundus, primarily during fasting periods just before normal mealtimes (in rats & humans).

It stimulates release of growth hormone from the anterior pituitary, and- perhaps most importantly- greatly increases Appetite. The gene that encodes grehlin also expresses the opposing hormone obestatin (the "fullness hormone"), whose function may be to decrease appetite.

PYY3-36

Peptide PYY3-36 is secreted by cells lining the ileum and colon. It is released in proportion to the amount of food inside it. Locally, like CCK, it inhibits gastric motility and stimulates bile and pancreas secretion. It also travels to the brain (hypothalamus), where it suppresses appetite.

Nefstatin-1: It has been found in gastric mucosa and pancreas, although its function there is unclear.

Cells within the small intestine crypts

Six cell types exist within the crypts.

i) **Crypt stem cells:** The base of the crypts houses a lifelong pluripotent stem cell population that continuously divides, giving rise to various crypt cells. These include secretory cells, mucus-secreting goblet cells, enteroendocrine cells, and Paneth cells. Their apical surface features microvilli, significantly increasing surface area.

ii) **Crypt Enterocytes:** Their main role is to secrete chloride, sodium, and water into the crypt lumen, aiding absorption by villus enterocytes. As crypt enterocytes migrate up the basal lamina to the villus, they transition from a secretory to an absorptive phenotype.

iii) **Goblet cells:** These cells are derived from the crypt stem cells and migrate out of the crypts to populate the villi. They become more and more numerous from duodenum to ileum & secrete mucus.

iv) **Enteroendocrine cells:** Derived from crypt stem cells and located near the base of the crypts, these cells monitor the pH, osmolarity, and composition of the ingesta through their contact with the lumen. They contain secretory granules with hormones, some of which have paracrine effects on neighboring cells. Enteroendocrine cells produce various hormones, including CCK and secretin, along with somatomedins, vasoactive intestinal peptide, serotonin, and enteroglucagon, all playing key roles in gastrointestinal physiology.

v) **Paneth cells:** These cells, derived from crypt stem cells, remain at the base of the crypts and are long-lived. They are believed to protect the crypt stem cells by producing antibacterial substances, such as lysozyme and phospholipases, which are released into the lumen of the crypt. These substances help defend against a broad range of bacteria, fungi, and some enveloped viruses. Notably, Paneth cells are absent in dogs, cats, and pigs.

vi) **M-cells or dome cells:** Origin of this cells are not known. They can be found interspersed among the enterocytes in the crypt and even villous areas. They should be considered cells of the immune system. They capture particles (bacterial and viral antigens) and pass them on unchanged to the dendritic cells and lymphocytes within the lamina propria and within the lymphoid follicles in the mucosa and submucosa.

Cells within the villus of the small intestine: Three cell types are present in small intestine villi.

i) **Villous absorptive enterocytes:** These cells, derived from crypt secretory enterocytes, cease their secretory activity as they migrate up the villus. Once on the villus, they begin to produce enzymes in their apical membrane microvilli, known as the brush border, which are essential for the final stages of digestion. Additionally, these cells start expressing the transport proteins required for nutrient absorption.

ii) **Goblet cells:** These migrate up the crypt and are responsible for secretion of mucus. Mucus secretion can be greatly increased on stimulation by prostaglandins released by damaged mucosal cells in the area.

iii) **M-cells or dome cells:** All the cells lining mucosal surfaces of the intestinal tract (esophagus to anus) have an apical and a basolateral membrane. Adjacent cells are linked to one another on all sides by **"tight junctions"** that form a seal between cells that is relatively impermeable to bacteria, viruses, and large molecules that have been ingested.

Secretory Function of Pancreas

The pancreas is a mixed gland containing both exocrine and endocrine tissue. The endocrine portion, made up of the islets of Langerhans, secretes hormones such as glucagon (from alpha cells), insulin (from beta cells), and somatostatin (from delta cells) into the bloodstream. The exocrine pancreas consists of tubuloalveolar glands that secrete digestive enzymes into the duodenum. These enzymes, produced by acinar cells, are necessary for the digestion of starches, proteins, and triglycerides, while duct cells secrete an aqueous NaHCO3 solution.

Pancreatic enzyme production is stimulated by the hormone CCK, which is released in response to fats and amino acids in the duodenum. Ductule cells increase the alkalinity of pancreatic secretions by secreting sodium and some potassium and removing chloride. The resulting pancreatic juice is slightly alkaline with a pH of 7.8. Under the influence of secretin, produced in response to low pH in the duodenum, chloride removal is increased, raising the pH to 8.2. This fluid neutralizes the acidic chyme leaving the stomach, protecting the intestinal mucosa and optimizing enzyme activity, as most pancreatic enzymes function best at a pH between 7 and 8.

The principal types of vertebrate pancreatic enzymes are

(1) Proteolytic enzymes, which are involved in protein digestion; (2) Pancreatic amylase and chitinase (in some vertebrates), which contributes to carbohydrate digestion; and (3) Pancreatic lipase, for fat digestion.

The pancreas secretes three major proteolytic enzymes in an inactive form: trypsinogen, chymotrypsinogen, and procarboxypeptidase. Trypsinogen is activated to its active form, trypsin, in the duodenal lumen by enterokinase, an enzyme located in the luminal border of duodenal mucosal cells. To prevent premature activation within the pancreas, trypsinogen is also accompanied by trypsin inhibitor, which blocks trypsin's action. Once activated, trypsin converts chymotrypsinogen and procarboxypeptidase into their active forms, chymotrypsin and carboxypeptidase, respectively, in the duodenum.

Pancreatic Amylase

Like salivary amylase, pancreatic amylase plays an important role in carbohydrate digestion by converting polysaccharides into disaccharides. Amylase is secreted in the pancreatic juice in an active form as active amylase does not have any danger to the secretory cells.

Pancreatic Lipase

Pancreatic lipase is crucial for fat digestion in vertebrates, as it is the primary enzyme capable of breaking down fats. Unlike other enzymes, lipase is secreted in its active form. A deficiency in pancreatic enzymes leads to incomplete food digestion, particularly of fats, since the pancreas is the only significant source of lipase. This results in pancreatic exocrine insufficiency, with steatorrhea (excessive undigested fat in the feces) being the main clinical sign, with up to 60–70% of ingested fat remaining undigested. Digestion of proteins and carbohydrates is less affected, as salivary, gastric, and small-intestinal enzymes also contribute to the breakdown of these nutrients.

Pancreatic Aqueous Alkaline Secretion

Pancreatic enzymes function best in a neutral or slightly alkaline environment. This acidic chyme must be quickly neutralized in duodenal lumen, not only to allow optimal functioning of the pancreatic enzymes but also to prevent acid damage to duodenal mucosa. The alkaline ($NaHCO3$- rich) fluid secreted by the pancreas into the duodenal lumen serves the important function of neutralizing the acidic chyme as the latter is emptied into the duodenum from the stomach.

Regulation of Pancreatic secretion

The release of the two major enterogastrones, secretin and cholecystokinin (CCK), in response to chyme in the duodenum plays the central role in the control of pancreatic secretion.

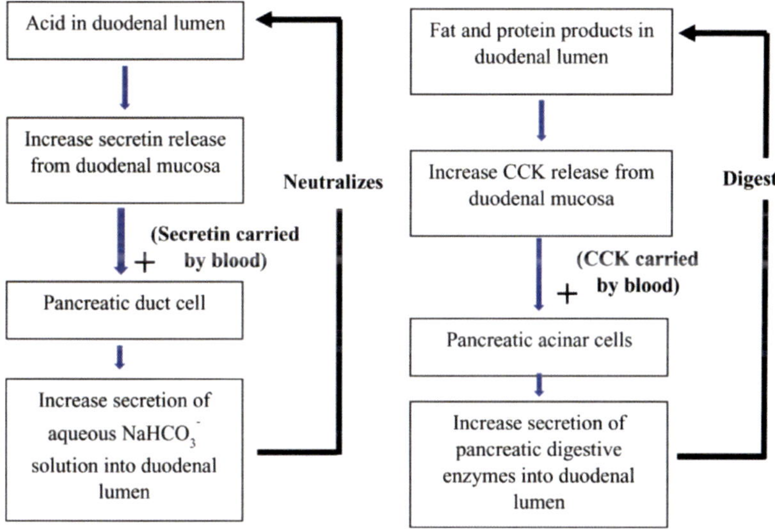

Fig. 1. Hormonal Control of Pancreatic Exocrine Secretion i.e. aqueous NaHCO3- secretion and digestive enzyme secretion:

The amount of secretin released is proportional to the amount of acid that enters the duodenum, so the amount of NaHCO3 secreted parallels the duodenal acidity.

Nervous Regulation of Pancreatic Secretion

Cephalic phase: The cephalic phase of gastric secretion refers to the increased secretion of HCl and pepsinogen that occurs in feed forward (anticipatory) fashion in response to stimuli acting in the head even before food reaches the stomach. It simultaneously transmit the impulses along the vagi to pancreas; results in pancreatic secretion rich in enzymes, but little water and electrolytes.

Gastric phase: Distension of stomach due to entry of food into the stomach causes reflex stimulation of pancreatic secretion through vagovagal reflex.

Intestinal phase: Distension of intestine following the entry of food into the small intestine, stimulates the production of pancreatic enzymes and their secretion.

This phase is controlled by both intrinsic nerves of the intestine and vagus. Acetylcholine released by the vagal action and local nerve reflexes sensitize the pancreas for the action of secretin and CCK.

Liver

The liver is considered an accessory organ to the digestive tract. It's secretions in the form of bile are vital to fat digestion. Liver also acts as a site of storage for lipids & fat-soluble vitamins A, D, & E. The liver is also home to unique macrophages known as Kupffer cells that guard the liver against bacterial and viral antigens that may enter the portal circulation.

Bile Salt

Bile salts are key components of bile, formed in hepatocytes by conjugating an amino acid with cholesterol. Taurine is commonly used for this conjugation, resulting in the bile salt taurocholic acid. These bile salts have both hydrophobic (cholesterol) and hydrophilic (amino acid) ends, enabling them to form micelles in the intestine that assist in fat digestion and absorption. Bile secretion is continuous, but in many species, bile is stored in a gallbladder for release during meals. Species such as horses, rats, deer, giraffes, camels, elephants, and pigeons lack a gallbladder, and bile flows directly into the duodenum. In most species, the bile duct joins with the pancreatic duct, and this common duct delivers both bile and pancreatic secretions into the upper duodenum. Bile production and gallbladder contraction are stimulated by the hormone CCK.

Functions of Bile Salt

Bile salts have a powerful surface tension-reducing ability, acting as strong emulsifying agents to aid in fat emulsification. This enables agitation in the intestinal tract to break fat globules into smaller sizes. Bile salts also assist in the absorption of fatty acids, monoglycerides, cholesterol, and other lipids by forming micelles, minute complexes of lipids that are highly soluble due to the electrical charges of bile salts. These micelles transport lipids to the mucosa for absorption. Without bile salts, approximately 40% of the lipids are lost in the stools.

Adaptations that Increase the Small Intestine's Surface Area

1. **Folds:** The inner surface of the small intestine is arranged in circular folds that are visible to the naked eye and that increase the surface area three-fold.
2. **Villi:** Projecting from this folded surface are microscopic, finger like projections known as villi, which increase the surface area by another 10 times.
3. **Microvilli:** Microvilli, tiny hairlike projections on the luminal surface of epithelial cells, further increase the surface area of the small intestine by about 20-fold, forming the brush border. Villi are longest in the jejunum and gradually decrease in length toward the ileum. Combined, the folds, villi, and microvilli provide the small intestine with a luminal

surface area approximately 600 times greater than if it were a tube with the same length and diameter lined by a flat surface.

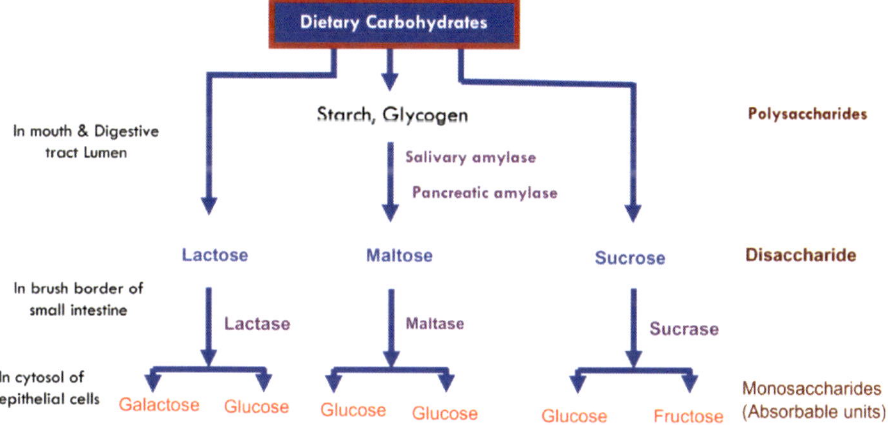

Fig. 2. Steps involved in digestion of carbohydrates:

The dietary polysaccharides starch and glycogen are converted into the disaccharide maltose through the action of salivary and pancreatic amylase. Maltose and the dietary disaccharides lactose and sucrose are converted to their respective monosaccharides by the disaccharidases (maltase, lactase, and sucrase) located in the brush borders of the small-intestine epithelial cells.

Some of these enzymes include sucrase, which converts sucrose to glucose and fructose; maltase and maltotriase, which convert maltose and maltotriose to their constitutive glucose molecules; and lactase, which converts milk lactose to glucose and galactose. Lactase is found on the villous enterocyte brush border of all mammalian neonates, but often disappears after the animal is weaned. Sucrase, is often lacking in neonates and is expressed only after the animal is several weeks old. The brush border also has its own form of α-amylase to degrade any starch that fails to be broken down by pancreatic α-amylase.

ABSORPTION OF CARBOHYDRATES

Dietary carbohydrates are primarily presented to the small intestine as disaccharides like maltose, sucrose, and lactose. Disaccharidases located in the brush borders of small intestinal cells break these down into absorbable monosaccharides: glucose, galactose, and fructose. Glucose and galactose are absorbed via secondary active transport, where symport carriers, such as the sodium-glucose cotransporter on the luminal membrane, transport both the monosaccharide and Na+ into the intestinal cell. These carriers do not directly consume energy but rely on the Na+ concentration gradient established by the energy-consuming basolateral Na+/K+ pump.

Glucose (or galactose), concentrated in the cell by symporters, exits the cell down its concentration gradient via facilitated diffusion through the GLUT-2 transporter in the basal membrane, entering the blood within the villus. A significant amount of glucose also crosses the epithelial barrier through leaky tight junctions between epithelial cells. Fructose is absorbed into the blood solely by facilitated diffusion, entering epithelial cells from the lumen via GLUT-5 and exiting into the blood via GLUT-2.

DIGESTION AND ABSORPTION OF FAT

Digestion of Dietary Fat

Dietary fats primarily exist as triglycerides, and their digestion begins in the mouth, where lingual glands release pharyngeal lipase. This enzyme breaks down triglycerides into fatty acids, monoglycerides, and diglycerides. While its impact on fat digestion is minimal in adults, it plays a significant role in the digestion of milk fat in neonates, who may not yet produce full amounts of pancreatic enzymes or bile. In the stomach, the churning action of gastric contractions causes fats to form an emulsion-a suspension of tiny fat droplets in water.

When fat enters the duodenum, it triggers the release of cholecystokinin (CCK) from enteroendocrine cells. CCK stimulates the pancreas to release digestive enzymes and signals the gallbladder to contract. The pancreatic enzymes involved in fat digestion include pancreatic lipase, colipase, pro-colipase, phospholipases, and cholesterol esterase. Bile, produced in the liver, contains bile salts, which are crucial for fat digestion.

The emulsified fat droplets entering the duodenum are too large and hydrophobic for pancreatic lipase to act upon directly. Bile salts, which are detergents formed by combining cholesterol and an amino acid, help by surrounding the fat droplets. One end of a bile salt molecule, composed of the cholesterol portion, is hydrophobic, allowing it to interact with fatty acids, while the hydrophilic amino acid end remains in the aqueous environment. This creates smaller droplets, increasing the surface area for pancreatic lipase action.

For full activation, colipase must bind to pancreatic lipase, forming a complex that allows the enzyme to break down triglycerides into monoglycerides and fatty acids at the surface of the fat droplets. Cholesterol esterase releases cholesterol from the droplet, and phospholipase breaks down phospholipids into fatty acids and monoglycerides. As the digestion process continues, bile salts encapsulate the free fatty acids, monoglycerides, cholesterol, and fat-soluble vitamins to form micelles, which facilitate their absorption in the small intestine.

Absorption of Fat

Fat must be transported from the watery chyme through the body's fluids, despite being insoluble in water. To address this challenge, fat undergoes several transformations during digestion and absorption. When stomach contents are released into the duodenum, the ingested fat is initially aggregated into large triglyceride droplets that float in the chyme. The digestion products of lipase, such as monoglycerides and free fatty acids, are also not highly water-soluble, making it difficult for them to diffuse through the aqueous chyme and reach the absorptive surface. However, bile components play a crucial role in facilitating the absorption of these fatty substances by forming micelles, which aid in their transport across the aqueous environment.

Once the micelles reach the luminal membranes of the epithelial cells, the monoglycerides and free fatty acids passively diffuse from the micelles through the lipid layer of the epithelial cell membranes into the interior of the cells. As these fat products are absorbed, the micelles can pick up more monoglycerides and free fatty acids, which are produced from the digestion of other triglycerides in the fat emulsion. Bile salts continuously perform their fat-solubilizing function along the length of the small intestine until all the fat is absorbed. Afterward, the bile salts are reabsorbed in the terminal ileum through specialized active transport mechanisms.

This process is highly efficient because relatively small amounts of bile salts can facilitate the digestion and absorption of large amounts of fat. Each bile salt performs its "ferrying" function repeatedly before being reabsorbed. Once the fat reaches the interior of the epithelial cells, the monoglycerides and free fatty acids are resynthesized into triglycerides. These triglycerides then form droplets, which are coated with a lipoprotein layer, making the fat droplets water-soluble. The largest of these lipoproteins, called chylomicrons, are secreted from the epithelial cells by exocytosis into the interstitial fluid within the villus. In mammals, the triglyceride-rich chylomicrons are released into the central lacteals and enter the lymphatic vessels, while in birds, they are absorbed into the capillaries of the villi. Fatty acids with short- or medium-chain carbon lengths enter the blood directly. While fat absorption is considered a passive process, the overall sequence of events required for fat absorption does necessitate energy.

DIGESTION AND ABSORPTION OF PROTEINS

Digestion of Proteins: Dietary proteins are large molecules, often composed of hundreds of amino acids linked by peptide bonds. For example, casein, the primary protein in milk, has a molecular weight of 23,000 and consists of around 200 amino acids. To be absorbed by the enterocytes, these proteins must be broken down to at least the dipeptide and tripeptide level. Protein digestion begins in the stomach, where the acidic environment can hydrolyze some peptide bonds. Chief cells in the gastric glands secrete pepsinogen, an inactive enzyme. When pepsinogen mixes with gastric acid, a fragment is cleaved off, converting it into the active enzyme pepsin. Pepsin primarily cleaves peptide bonds next to hydrophobic amino acids with aromatic side chains, such as phenylalanine, tryptophan, and tyrosine.

Rennin, another enzyme produced by chief cells, cleaves between phenylalanine and methionine residues in proteins and is particularly important for digesting casein in neonatal mammals. Both pepsin and rennin function best in a pH range of 2 to 3. The proteins that started as large molecules with hundreds of amino acids in the stomach are now broken down into smaller fragments (25-100 amino acids long) when they enter the duodenum. These peptides activate receptors on enteroendocrine cells lining the duodenal crypts, prompting them to secrete cholecystokinin (CCK). CCK enters the bloodstream and travels to the pancreatic acinar cells and the myoepithelial cells surrounding each acinus, triggering the release of pancreatic enzymes into the upper duodenum through the pancreatic ducts.

The pancreatic proteolytic proenzymes include trypsinogen, chymotrypsinogen, pro-elastase, and pro-carboxypeptidases A and B. CCK also reaches the villous enterocytes in the duodenum, stimulating them to secrete an enzyme called enteropeptidase (or enterokinase) into the lumen. Enteropeptidase activates trypsinogen by cleaving off a fragment to form active trypsin. Trypsin, in turn, activates the other proteolytic enzymes secreted by the pancreas by cleaving off portions of their molecules. Trypsin can also activate more trypsinogen in a positive feedback loop.

These activated enzymes trypsin, chymotrypsin, elastase, & carboxypeptidases-cleave peptide bonds between specific amino acids. By the end of the luminal phase of digestion, proteins are broken down into peptides that are typically 1–12 amino acids long. These smaller peptides and individual amino acids then move to the brush border. They are highly soluble in water and cross the unstirred water layer, entering the glycocalyx and adhering to the microvilli on the enterocytes. Several intestinal peptidases, located on the brush border, project into the glycocalyx but do not get released into the lumen. These enzymes further hydrolyze the peptides, reducing them to no more than three amino acids in length.

Absorption of Proteins

The protein presented to the small intestine for absorption is primarily in the form of amino acids & a few small peptide fragments. Amino acids are absorbed across the intestinal cells by symporters, similar to glucose and galactose absorption. The amino acid symporters are selective for different amino acids. Small peptides gain entry by means of yet another Na± dependent carrier in a process known as tertiary active transport. In this case, the symporter simultaneously transports both H+ and the peptide from the lumen into the cell, driven by H+ moving down its concentration gradient and the peptide moving against its concentration gradient.

The H+ gradient is established by an antiporter in the luminal membrane that is driven by Na+ moving into the cell down its concentration gradient and H+ moving out of the cell against its concentration gradient. The Na+ concentration gradient that drives the antiporter in turn is established by the energy-dependent Na+/K+ pump at the basolateral membrane. Thus, glucose, galactose, amino acids, and small peptides all get a "free ride" on the energy expended for Na+ transport. The small peptides are broken down into their constituent amino acids by the aminopeptidases in the brush border membrane or by intracellular peptidases. Like monosaccharides, amino acids enter the capillary network within the villus.

In addition to dietary proteins, in non-ruminants some endogenous proteins that have entered the intestinal lumen from the three following sources are digested and absorbed as well:

1. Digestive enzymes, all of which are proteins that have been secreted into the lumen.
2. Proteins within the cells that are pushed off from the villi into the lumen during the process of mucosal turnover.
3. Small amounts of plasma proteins that normally leak from the capillaries into the digestive tract lumen.

All endogenous proteins must be digested and absorbed along with the dietary proteins to prevent depletion of the body's protein stores.

16

Classification of Feedstuffs

Sandeep Kumar Chaudhary, Mahipal Choubey Abhishek Kumar Singh and Mahesh M.S.

Faculty of Veterinary and Animal Sciences, I. Ag.Sc., RGSC-Banaras Hindu University Barkachha, Mirzapur, Uttar Pradesh-231001

Introduction

Livestock production is vital to global food security and the agricultural economy, with proper nutrition being key t o animal health and productivity. Feedstuffs are different types of feed given to livestock, provide essential nutrients like carbohydrates, proteins, fats, vitamins, & minerals. Understanding different types of feedstuffs is critical for formulating balanced diets and sustainable livestock systems.

Feed accounts for 60-70% of total livestock production costs, making cost-effective and nutritionally sound feed selection essential for profitability. High-quality feed improves nutrient utilization, animal performance, and product quality while minimizing environmental impacts by reducing waste.

Feedstuffs are broadly classified into roughages and concentrates. Roughages, including forages, hay and silages, are fiber-rich and support microbial fermentation in ruminants. Concentrates, low in fiber but rich in energy or protein, include cereal grains and oilseed meals. By-products from food industries, like bran and molasses, enhance feed diversity. Mineral supplements provide vital nutrients like calcium and phosphorus, while additives boost feed efficiency and health.

Recent innovations in feed technology have introduced alternative and unconventional feed resources, promoting sustainability and reducing competition with human food supplies. By applying scientific knowledge and practical strategies, livestock producers can optimize productivity, ensure animal welfare, and enhance agricultural sustainability. This chapter provides an overview of the different classifications of feedstuffs for better understanding of their significance in livestock nutrition.

Feedstuffs: Feed and fodder covering all edible material in livestock ration and having nutritive value are feedstuffs. These include naturally occurring plants and animal products and by products of milling and other processing and artificially prepared stuffs.

Roughages or forages: These are bulky/low density feeds, containing more than 18% of crude fibre or 35% cell wall and less than 60% TDN. These are low in energy and high in fibre, e.g., hay, straw, hull (covering of seed), husk, stovers, whole plant form (except roots), pasture and green crops, tree leaves *etc.*

Hay: Forage product obtained by drying the tender stemmed leafy plant material in sunlight or in shade to have not more than 12-14% moisture are known as hay.

Straw: These are the byproduct of cereals and pulses (wheat, rice, millet) or leguminous crops leftover after harvesting, threshing and removal of grains or pulses. They supplies mainly bulk to the animals.

Fodder: Arial part of nearly mature plants with ears, husk or head e.g., corn/ sorghum/legume plant in fresh or cured form.

Stover: Arial part of nearly mature plants without ears, husk or head.

Bagasse: Fibrous material of sugarcane leftover after extraction of all juices.

Hull: Outer covering of beans, peas and cottonseed.

Husk: Dry outer covering of grains, grams, rice husk, gram husk.

Shell: Hard outer covering of nuts e.g. ground nut shells.

Corn cobs: Material obtained after removal of all corn grains.

Concentrates: They are high density feeds or feed mixtures, which are highly digestible and contain more than 60% TDN and less than 18% crude fibre.

Classification of animal feedstuff

Feedstuff are broadly classified into roughages, concentrates and additives.

Roughages

Feedstuffs with high fiber and low net energy are designated as roughages. These materials are bulky, containing more than 18% of crude fiber and less than 60% of total digestible nutrients (on DM basis). Roughages are further categorized into two types: dry roughages (>80% DM) and succulen troughages (<80% DM). Dry roughages include: hays (leguminous and non-leguminous), straw (wheat straw, paddy straw, barley straw etc.) and stovers (bajra, jowar, maize etc.). Succulent roughages primarily comprise green fodders (both leguminous and non-leguminous), silages, haylage, root crops, pasture grasses, and tree leaves.

Table 1. Common succulent green fodders for livestock

Common succulent green fodders by season		
Season	**Leguminous**	**Non-leguminous**
Kharif	Cowpea and Guar	Jowar, Maize and Sudan grass
Rabi	Lucerne, Berseem and Pea	Oat and Barley

Concentrates

Feed and mixture of feed containing less than 18% of crude fiber and more than 60% of total digestible nutrients are classified as concentrates. Being rich in either energy or protein, they are further classified into energy rich concentrates (when crude protein is less than 18-20%) and protein rich concentrates (when crude protein is more than 18-20%).

Table 2. Classification of common concentrate ingredient of livestock ration

<table>
<tr><th colspan="4">Common concentrate feedstuff</th></tr>
<tr><th colspan="2">Energy rich concentrates</th><th colspan="2">Protein rich concentrates</th></tr>
<tr><td>Grains and seeds</td><td>Maize, barley, sorghum etc.</td><td>Animal origin protein supplement (Majority over 47% CP, 1% Ca, 1.5% P & under 2.5 % CF)</td><td>Meat meal, blood meal, fish meal, feather meal etc.</td></tr>
<tr><td>Mill byproduct</td><td>Brans and Chunnies</td><td>Plant origin protein supplement (Majority under 47% CP, 1% Ca, 1.5% P & over 2.5 % CF)</td><td>Oilseed cakes & meals like groundnut cake, sesame meal, mustard cake, soybean meal, linseed oil meal etc.</td></tr>
<tr><td>Roots</td><td>Tapioca, tubers, turnip and potatoes</td><td colspan="2" rowspan="2"></td></tr>
<tr><td>Molasses</td><td></td></tr>
</table>

Additives

The third group additives mainly includes mineral and vitamin supplements and feed additives like antibiotics, hormones etc. which when added to the total feed mixture to improve the performance of the supplemented livestock.

Table 3. Different livestock feedstuff

Animal Feedstuff	Types	Examples
1. Roughages (>18% CF & <60% TDN)	**A) Dry** (>80% DM)	i) Hays (18-34% CF, 40-60% TDN) a) Leguminous hay: hay of lucerne, berseem etc. (>10.5% CP, >0.9% Ca) b) Non-leguminous hay: hay of oat, sorghum, doob grass etc. (6-10.5% CP &<0.9% Ca) ii) Straw: low quality roughage; straw of wheat, rice, oat, jowar, barley etc. (>28% CF, <52% TDN, <6% CP) iii) Stovers: Very low quality roughage, stover of bajra, jowar, maize etc. (>34% CF, <40% TDN, <6% CP)
	B) Succulent (<80% DM)	i) Green fodder: a) Leguminous: berseem, lucerne, cowpea, cluster bean and green pea (>10.5% CP) b) Non-leguminous: jowar, maize, bajra, oat and grasses of sudan, napier, guinea etc. ii) Haylage iii) Root crops (9-30% DM, <3% CF) iv) Pasture v) Tree leaves: Subabul, oak, peepal, moringa etc. vi) Silage
2. Concentrates (<18% CF & >60% TDN)	**A) Energy rich** (<18% CP)	i) Grains and seeds: maize, barley, sorghum etc. ii) Mill byproduct: brans of wheat, rice and chunies of pulses iii) Roots: turnip, potatoes, tapioca etc. iv) Molasses
	B) Protein rich (>18% CP)	i) Animal origin protein supplement (Majority over 47% CP, 1% Ca, 1.5% P & under 2.5 % CF) a) Animal byproducts: meals of blood, meat cum bone and dried whole & skimmed milk. b) Marine byproducts: meals of fish and shrimp c) Avian byproducts: feather meal, poultry manure/litter ii) Plant origin protein supplement (<47% CP, 1% Ca, 1.5% P &>2.5% CF; defatted: <17% EE & oil seeds: >17% EE) a) Oilseed cakes & meals: groundnut cake, sesame meal, mustard oil cake, soybean meal, cotton seed cake, linseed oil meal etc. b) Brewer's grains and yeast (>17% EE)
3. Mineral supplement: bone meal, dicalcium phosphate, mono calcium phosphate, phosphoric acid, chalk, oyster shell, shell grit, marble chips etc.		
4. Vitamin supplement: tocopherol acetate etc.		
3. Additives: probiotic, prebiotic, antioxidants, enzymes, hormones, flavoring and coloring agents, antibiotics, sweeteners etc.		

Table 4. Classification of roughness based on their nutritive values (on DMB)

Character	03 types		
	Non-Maintenance type	**Maintenance type**	**Productive type**
DCP %	Below 3%	3-5%	Above 5%
Examples	Straw and stovers	Non-leguminous cereal fodders, grasses and their hay	Leguminous cereal fodders and their hay

Table 5. International feed classes

Code	Class description
1	Dry forage and Roughages
2	Pasture, range plants and forage fed green
3	Silages
4	Energy feeds
5	Protein supplement
6	Mineral supplement
7	Vitamin supplement
8	Additives

17

Proximate Principals in the Feedstuff

Sandeep Kumar Chaudhary, Mahipal Choubey Abhishek Kumar Singh and Mahesh M.S.

Faculty of Veterinary and Animal Sciences, I. Ag.Sc., RGSC-Banaras Hindu University Barkachha, Mirzapur, Uttar Pradesh-231001

Introduction

Proximate analysis, also known as the Weende's System of Analysis, is a standard method for analyzing the nutritional components of animal feed or feed ingredients, developed in 1865 by Wilhelm Henneberg and Friedrich Stohmann at Weende Experiment Station in Germany. This system is used for approximating the nutritional value of a feed for feeding purpose, without actually using it in an animal feeding trial.

The system involves the separation of feed components into distinct fractions (*viz.,* water, ash, crude fat, crude protein and crude fiber), each representing different nutrient groups, based on their feeding value. These fractions are used to estimate the energy, protein, and other nutritional content of the feed, allowing for informed decisions regarding its suitability for specific livestock or poultry. It primarily classifies carbohydrates into two broad categories: crude fiber and nitrogen-free extract (NFE). Notably, NFE is not directly analyzed but calculated by difference. Proximate analysis enables valid comparisons of feeds based on specific nutrient content, allowing for an assessment of how one feed compares to another in terms of nutritional value, overall nutrient composition of a feedstuff, thereby, helping to ensure the optimal health and productivity of animals.

The major fractions typically included in proximate analysis are

1. ***Moisture:*** This represents the water content of the feed, which can influence the overall nutritional value and shelf-life of the feedstuff. The moisture content in feedstuff is determined by drying a known weight of the sample at a constant temperature (usually 100±2°C) in a hot air oven until a constant weight is achieved. The loss in weight represents the moisture content. This method is based on the principle that water

evaporates at elevated temperatures, while other feed components remain stable under these conditions.

$$\%\,\text{Dry Matter (DM)} = \frac{\text{Weight of sample after drying}}{\text{Weight of sample before drying}} \times 100$$

% Moisture = 100 – % DM

Significance of moisture estimation

- ***Determines keeping quality of hay:*** Moisture content is a critical factor in determining the preservation and shelf life of hay, as higher moisture levels can promote mold growth and spoilage.
- ***Affects silage making:*** Moisture estimation is essential in silage production, as it helps promote potential losses during fermentation, ensuring the proper preservation of nutrients.
- ***Classification of feeds:*** Accurate moisture content helps classify feeds as succulent or non-succulent, which influences their digestibility and nutrient value.
- ***Essential for feed storage:*** Moisture content plays a significant role in the proper storage of feed. Excess moisture can lead to spoilage and degradation, while low moisture may affect feed palatability and quality.
- ***Important for feed purchasing:*** When purchasing feed, moisture estimation helps determine the true cost per unit weight of feed, as high moisture content can reduce the dry matter available for animal consumption.
- ***Impact on proximate composition and nutritional value:*** The moisture content directly influences the overall proximate composition of a feed, affecting its energy, protein, and other nutrient estimates. As a result, moisture is vital in determining the feed's nutritional value.

Limitations

- ***High drying temperatures can cause artifacts:*** Drying feed above 60°C can lead to the formation of artifacts, which can interfere with the analysis of lignin, fiber, and acid detergent fiber (ADF), ultimately affecting the accuracy of these measurements.
- ***Potential underestimation of dry matter:*** Moisture estimation methods, especially those involving high drying temperatures, may result in the underestimation of the true dry matter content of feed.
- ***Loss of volatile fatty acids (VFAs) in silages:*** During the drying process, there is a potential for the loss of volatile fatty acids from silages, which are essential for evaluating their nutritional value and fermentation quality.

2. ***Total ash:*** This represents the total mineral content of the feed, including macro and micro minerals like calcium, phosphorus, and trace elements, which are essential for various physiological functions.The total ash content in feedstuff is determined by incinerating a known weight of the sample in a muffle furnace at a high temperature (typically 550-600°C) until all organic matter is completely burned off (preferably 2-3 hrs.), leaving only inorganic mineral residues. The residue left after complete combustion represents the total ash content. This method is based on the principle that organic components volatilize at high temperatures, while inorganic minerals remain as ash. The weight of ash thus obtained is expressed in terms of percentage. The organic matter (OM) content can be determined by subtracting the total ash percentage from 100.

$$\%\,\text{Total ash (DM)} = \frac{\text{Weight of sample after ashing}}{\text{Weight of sample after ashing}} \times 100$$

Limitations

- ***Non-representative of total inorganic matter:*** Ash content may not fully represent the total inorganic matter in a sample. This is because some organic carbon can be bound as carbonate and may not be captured in the ash determination.
- ***Loss of volatile minerals:*** During the ash determination process, volatile minerals like sulfur (S), selenium (Se), iodine (I), fluorine (F), and even sodium (Na) and chlorine (Cl) can be lost due to their volatilization at high temperatures.
- ***Time-consuming process:*** The process of determining total ash can take a considerable amount of time (approximately 24 hours), which can be inefficient for rapid analysis.
- ***High operational costs:*** Muffle furnaces, which are commonly used for ash determination, can be costly to run due to the high energy consumption required to maintain the necessary temperatures for long periods.

3. ***Crude Protein (CP):*** Protein is essential for growth, maintenance, and production of animals. It is often estimated by measuring nitrogen content (since proteins are rich in nitrogen) and multiplying by a factor of 6.25. The nitrogen content in feedstuff is determined using the Kjeldahl method, which is based on the principle that organic nitrogen present in the sample is converted into ammonium sulfate by digestion with concentrated sulfuric acid (H_2SO_4) in the presence of a catalyst. During this process, the nitrogen from proteins is released in the form

of ammonia. The ammonium sulfate formed is then neutralized with sodium hydroxide (NaOH), which liberates ammonia (NH_3). The released ammonia is distilled into a boric acid solution and quantified by titration with a standard acid solution (such as hydrochloric acid or sulfuric acid). The total nitrogen content thus determined is used to calculate the crude protein content by multiplying it with a conversion factor (typically 6.25 for most feedstuffs), based on the assumption that proteins contain approximately 16% nitrogen:

$$\text{Crude Protein (\%)} = \text{Total Nitrogen (\%)} \times 6.25$$

Since this method measures total nitrogen rather than protein directly and assumes that all nitrogen present in the sample comes from proteins, which may slightly overestimate the actual protein content in certain feeds containing non-protein nitrogen (NPN) compounds.

Assumptions while estimating CP

- All N in the food is present as true protein
- All food proteins contain 16% N.

4. ***Ether Extract (EE):*** This measures the fat content of the feed, providing a concentrated energy source. It includes oils, fats, and waxes. The ether extract, also known as crude fat, is determined by extracting the feed sample with a non-polar organic solvent (petroleum ether, n-hexane etc.) in a Soxhlet apparatus. The fat-soluble components, including lipids, oils, and other ether-soluble substances, are dissolved in the solvent, leaving behind the insoluble residue. After extraction, the solvent is evaporated, and the remaining residue is weighed to determine the ether extract content. The weight of the extracted fat is expressed as a percentage of the original sample weight:

$$\%\,\text{Ether Extract} = \frac{\text{Weight of extracted fat}}{\text{Weight of Sample}} \times 100$$

Limitations

- ***Assumption that all ether-soluble substances are fat:*** The ether extract method assumes that any substance soluble in ether is fat. However, it overlooks the fact that other non-fat substances, such as plant pigments, waxes, and certain resins, can also dissolve in ether. These substances are chemically distinct from fats and have different nutritional roles. For instance, plant pigments like carotenoids or chlorophylls are not fats, and while they might be important for animal health (due to their antioxidant properties), they don't provide the same energy or nutritional value that fats do. The inclusion of these non-fat components in the ether extract can lead to an overestimation of the actual fat content of a feed sample, potentially distorting the true nutritional value.

- ***Plant pigments and waxes soluble in ether but nutritionally different from fat:***

5. ***Crude Fiber (CF):*** This fraction includes the indigestible plant materials, primarily cellulose, hemicellulose and lignin. It is an important indicator of the feed's fiber content, which is necessary for maintaining gut health in herbivores. It is determined by successive acid (0.255N H_2SO_4) and alkali digestion (0.313N NaOH), which simulates the digestive processes in animals. The method involves boiling a known weight of the sample with dilute sulfuric acid (H_2SO_4) followed by dilute sodium hydroxide (NaOH) to remove soluble carbohydrates, proteins, and fats, leaving behind the insoluble fiber fraction, which consists mainly of cellulose, lignin, and some hemicellulose. The remaining residue is then filtered, dried, weighed, and ashed in a muffle furnace. The loss in weight after ashing represents the crude fiber content, which is expressed as a percentage of the original sample weight:

$$\%\,\text{Crude Fiber} = \frac{\text{Weight of dried residue before ashing - Weight of ash}}{\text{Weight of Sample}} \times 100$$

Limitations

During acid and base boiling in crude fiber analysis, compounds like hemicelluloses, pectin, and sometimes lignin are solubilized. These are not true fiber components but can dissolve into the solution, leading to their misclassification. As a result, some of these solubilized compounds are incorrectly counted as part of the nitrogen-free extract (NFE), causing an overestimation of the NFE.

6. ***Nitrogen-Free Extract (NFE):*** This is the portion of the feed that is primarily composed of carbohydrates such as starch, sugars, and other easily digestible materials. It is calculated by difference after determining all the other components. This includes all the nutrients which are not assessed by the prior methods of proximate analysis, and consists mainly of digestible carbohydrates, vitamins and other non-nitrogen soluble organic compounds. Since the result is obtained by subtracting the percentages calculated for each nutrient from 100, any errors in evaluation will be reflected in the final calculation for NFE as well.

% NFE = 100 – (% Moisture + % Crude protein + % Crude fat + % Crude fiber + % Ash)

Limitations

- Not determined analytically but is calculated by difference; therefore, it accumulates all of the errors that exists in other proximate components.

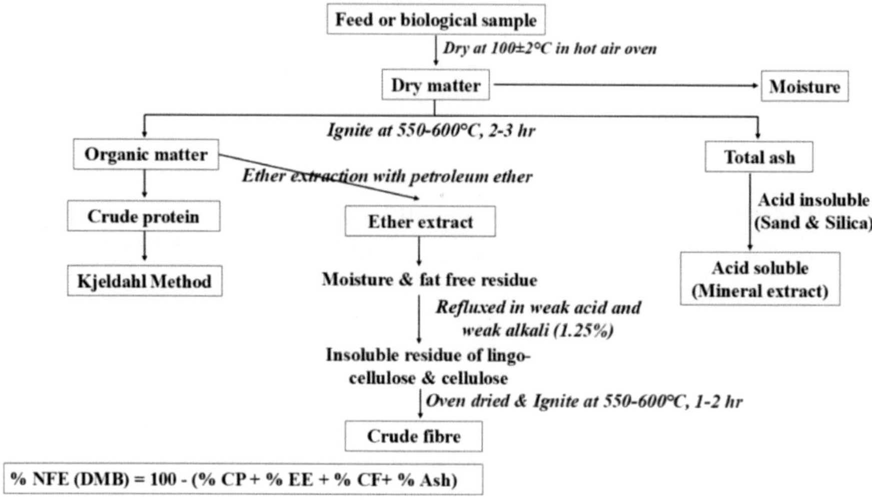

Fig. 1. Schematic diagram representing estimation of proximate principles in feed and biological samples

Analysis of fibre fractions

The Weende System of partition of forage carbohydrates into NFE and CF is neither chemically nor nutritionally accurate for ruminants. To address this limitation, P.J. Van Soest and colleagues at the USDA Station in Beltsville, Maryland, developed a rapid method to classify feed carbohydrates based on their nutritional availability to ruminants. The method separates feeds into two fractions: Neutral Detergent Soluble (NDS), which includes highly digestible components like sugars, starches, and soluble proteins, and Neutral Detergent Fibre (NDF), which contains variably digestible plant cell wall constituents such as cellulose, hemicellulose, and lignin. The process involves boiling a sample in a neutral detergent solution. Further refinement led to Acid Detergent Fibre (ADF) analysis, which separates NDF into soluble hemicellulose and insoluble cellulose, lignin, and bound nitrogen. Lignin, a key factor affecting forage digestibility, can be measured by acid or oxidation treatments. This method improves accuracy in predicting feed digestibility compared to crude fiber analysis. The method is being schematically represented here as:

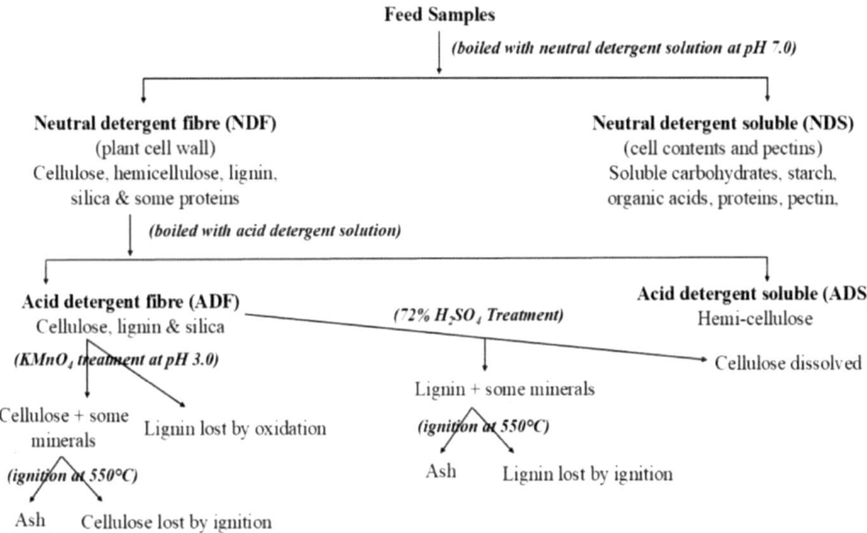

Fig. 2. Schematic diagram representing estimation of Analysis of fibre fractions of nutrient samples

Importance of proximate analysis

Proximate analysis is a fundamental method for evaluating the composition of animal feeds, fodders, and biological samples. It helps in understanding the nutritional value of feedstuffs and plays a critical role in livestock nutrition, feed formulation, and research. Its importance can be described as:

- ***Describes feeds and fodders:*** Classifies feeds based on nutritional composition, aiding in selection and usage.
- ***Basis for feed purchase & ration formulation:*** Ensures informed feed selection and balanced diet formulation for cost-effective nutrition.
- ***Starting point for detailed analysis:*** Guides further specialized studies like fiber fractionation and amino acid profiling.
- ***Feed comparison:*** Allows standardized evaluation of different feedstuffs for optimal livestock nutrition.
- ***Used in biological analysis:*** Helps assess digestion, metabolism, and health through faeces, urine, and tissue analysis.

18

Nutrients and Their Functions

Mahipal Choubey, Sandeep Kr. Chaudhary and Abhishek Kr. Singh

Faculty of Veterinary and Animal Sciences, I. Ag.Sc., RGSC-Banaras Hindu University (BHU), Barkachha, Mirzapur, Uttar Pradesh-231001

All living entities including plants, animals and microbes require nutrients as major input for their survival. These nutrients are building blocks of life essentially required for maintenance, production and reproduction. In animal husbandry practices, the nutrition of livestock and poultry is utmost important and plays pivotal role in successful management of farm. The animal nutrition deals with providing the essential nutrients, which have highly specialised metabolic roles ranging from cellular repair to energy production and immune system functions.

Nutrients: Nutrients may be regarded as chemical substances present in feed/ fodder and are essentially required by the animals for their survival to carry out the life processes. In general, nutrients are not synthesised in the body and are obtained from the environment, primarily through food and water. There are 6 basic nutrient required by animals body viz. carbohydrates, proteins, fats, vitamins, minerals and water.

Table 1. General function of nutrients

Nutrient	Generalized Functions
Carbohydrates	Provide energy @ 4 kcal/g (Starch/Sugar and fibre in ruminants) Bulkiness and propulsion of digesta (fibre/cellulose)
Proteins	Building and repair of tissues. Synthesize enzymes, hormones, & antibodies.
Fats	Provide essential fatty acids and concentrated energy (9 kcal/g) Absorb fat-soluble vitamins (A, D, E, and K). Protect and insulate visceral organs.
Vitamins	Regulate various biological processes. Support immunity and energy metabolism. Act as antioxidants.
Minerals	Support structural components (e.g., bones, teeth). Regulate fluid balance and nerve function. Facilitate oxygen transport.
Water	Regulate body temperature. Transport nutrients and waste. Solvent for biochemical reactions.

These nutrients can be classified in various ways based on certain criteria as shown below:

1. Classification Based on amount required in ration

Macronutrients: Required in large amounts to provide energy and support bodily functions. E.g. Carbohydrates, Proteins, Fats, Water

Micronutrients: Required in smaller amounts to facilitate physiological processes. E.g. Vitamins and Minerals

2. Classification Based on Energy Provision

Energy-Providing Nutrients: used for energy production.

E.g. Carbohydrates (4 kcal/g),

Proteins (4 kcal/g),

Fats (9 kcal/g)

Non-Energy Providing Nutrients: Do not provide energy, carrtyout essential physiological functions.

E.g. Water (regulates temperature and transport)

Vitamins (cofactors in metabolic reactions)

Minerals (structural roles and enzyme activation)

3. Classification Based on Organic and Inorganic nature

Organic Nutrients: Contain carbon and are derived from living organisms.

E.g. Carbohydrates, Proteins, Fats, Vitamins

Inorganic Nutrients: Do not contain carbon and may be derived from non-living sources.

E.g. Minerals (e.g., Calcium, Iron, Sodium) and Water

Functions of nutrients

1. Carbohydrates

These are organic compounds composed of carbon (C), hydrogen (H), and oxygen (O) in a ratio of 1:2:1. They are the primary source of energy in animal diets.

Classification

Table 2. Classification of carbohydrate

1	Monosaccharides: (Single sugar unit)	Glucose (blood sugar, dextrose) Galactose (component of lactose) Fructose (fruit sugar)
2	Disaccharides: (Two sugar unit)	Lactose (milk sugar): galactose and glucose, Maltose (Malt sugar): two glucose molecules, Sucrose (cane sugar): glucose and fructose

3	Polysaccharides (Many sugar unit)	Starch (form of energy storage in plants) Cellulose (form of structural carbohydrate in plants) Glycogen (limited form of storage carbohydrate in animals)

Functions of Carbohydrates

a) **Energy Source:** Carbohydrates provide the primary energy required for vital processes. In ruminants (Cow, buffalo, sheep goat), carbohydrates (e.g., cellulose) are fermented in the rumen, producing volatile fatty acids (VFAs) like acetate, propionate, and butyrate, which serve as key energy sources.

In non-ruminates (poultry and pig) Starches and disaccharides are digested enzymatically to form glucose, which serve as energy source.

b) **Bulkiness and satiety feeling:** Fibrous carbohydrate like cellulose (very low density) absorbs water to provide feel of bulkiness and satiety in animals. It also helps in forward movement of digesta.

c) **Protein-Sparing Effect:** By supplying sufficient energy, carbohydrates allow dietary proteins to be used for growth and production rather than energy metabolism.

d) **Milk Production:** In lactating animals, carbohydrates are essential for providing energy to synthesize milk and milk components, such as lactose.

e) **Growth and Weight Gain:** Adequate carbohydrate intake ensures optimal growth rates in meat-producing animals by meeting energy demands and enhancing feed efficiency.

f) **Fibre and Digestive Health:**

a) In ruminants and herbivores, fibrous carbohydrates maintain gut health by promoting proper rumen function and preventing rumen disorders.

b) In monogastric animals, dietary fibre supports intestinal motility and gut microbiome health.

Sources

a) **Forages:** Grass, hay, silage (rich in cellulose and hemicellulose).

b) **Cereal Grains:** Corn, wheat, barley, oats (rich in starch).

c) **By-Products:** Molasses, beet pulp, and brans.

2. Protein

Proteins are complex organic compounds composed of amino acids linked by peptide bonds. They contain carbon, hydrogen, oxygen, nitrogen, and sometimes sulphur or phosphorus. Proteins are essential macromolecules required for the structure, function, and regulation of cells, tissues, and organs. They are vital in animal production for growth, reproduction, lactation, and overall productivity.

Animals need protein to meet their amino acid requirements. Protein are polymer of amino acids. During digestion feed proteins are digested or broken down, to amino acids and digested. The animal body then uses these amino acids to make their own body proteins. There are about 20 amino acid required by animal's body. Based on requirement, amino acids may be classified as essential and non-essential amino acids

a) **Essential (indispensible) amino acids:** Cannot be synthesized by the body and must be obtained through the diet

   Eg. Histidine, isoleucine, leucine, lysine, methionine, phenylalanine, threonine, tryptophan, and valine

b) **Non-essential (dispensible) amino acids:** Can be synthesized by the body, and need not be taken in through diet.

   Eg. Alanine, arginine, asparagine, aspartic acid, cysteine, glutamic acid, glutamine, glycine, proline, serine, and tyrosine.

Functions of Proteins in Animal Production

1. **Tissue Growth and Repair:** Proteins are the primary building blocks of muscles, skin, organs, and connective tissues.In growing animals, proteins are essential for skeletal and muscular development.
2. **Enzyme and Hormone Production:** Proteins act as precursors for enzymes and hormones that regulate metabolic and physiological processes, such as digestion, reproduction, and immune response.
3. **Reproductive Efficiency:** Adequate protein levels in the diet support ovarian function, pregnancy maintenance, and the development of healthy offspring.
4. **Lactation and Milk Production:** In lactating animals, proteins contribute to milk synthesis, including the formation of casein, whey, and other milk proteins essential for offspring nutrition.
5. **Immune System Support:** Proteins are crucial for synthesizing antibodies and other immune molecules that help animals to resist infections and recover from diseases.
6. **Production of Specialized Proteins:** Wool, hair, feathers, and other specialized structures in animals are made of keratin, a structural protein.

7. **Energy Source (Secondary Role):** When carbohydrates and fats are insufficient, proteins can be metabolized to provide energy (4 kcal/g), though this is inefficient and diverts them from their primary functions.
8. **Feed Efficiency and Growth Rates:** High-quality proteins improve feed conversion ratios, leading to faster weight gain and increased production efficiency in meat-producing animals.
9. **Support for Microbial Growth in Ruminants:** In ruminants, dietary proteins provide nitrogen and amino acids that support microbial protein synthesis in the rumen, a critical source of nutrition for the host animal.

Sources of Proteins in Animal Feed

1. **Plant-Based Proteins:**
   a) Soybean meal, canola meal, sunflower meal, cottonseed meal.
   b) Cereal grains (e.g., corn, wheat, barley) and their by-products.
2. **Animal-Based Proteins:**
   a) Fish meal, meat and bone meal, blood meal.
   b) Milk products (e.g., skim milk, whey powder).
3. **Non-Protein Nitrogen (NPN) Sources (for Ruminants):** Urea and ammonia are utilized by rumen microbes for their own protein, which will act as protein source for ruminants..

3. Fats

Fats, also known as lipids, are ester of fatty acids and occur as main components of vegetable oils and of fatty tissue in animals. They may act as energy source and/or reserve of excess energy. They play a crucial role in animal production, providing energy, supporting bodily functions, and influencing productivity.

Classification of Fats

1. Based on Saturation

a) Saturated Fats (No double or triple bond):

- Solid at room temperature; found in animal fats (e.g., tallow, lard).
- Examples: Palmitic acid, Stearic acid.

b) Unsaturated Fats (double or triple bond):

- Liquid at room temperature; found in plant oils (e.g., soybean oil, linseed oil).
- Examples: Oleic acid, Linoleic acid, Linolenic acid.

2. Based on Essential Fatty Acids

a) **Essential Fatty Acids:**

- Cannot be synthesized by the animal and must be supplied through the diet.
- Examples: Linoleic acid, Linolenic acid, Arachidonic acid

b) **Non-Essential FatyAcids:**

- Can be synthesized by the animal and hence need not to be supplied through the diet.
- Oleic acid, stearic acid, palmitic acid

3. Complex Lipids

- Include phospholipids, glycolipids, and lipoproteins; involved in cellular functions

Functions of Fats in Animal Production

1. **Source of Energy:** Fats are a dense energy source, providing approximately 2.25 times (9 kcal/g) more energy than carbohydrates or proteins.
2. **Carrier of Fat-Soluble Vitamins:** Essential for the absorption and transportation of vitamins A, D, E, and K
3. **Supply of Essential Fatty Acids (EFAs): EFAs are essentially required for c**ell membrane integrity, hormone production (e.g., prostaglandins), and immune function.
4. **Improves Feed Palatability and Texture:** Increases feed intake by enhancing flavor and reducing dustiness in feeds.
5. **Insulation and Protection:** Fats stored in adipose tissue help to regulate body temperature and act as cushion vital internal organs.
6. **Milk and Meat Quality:** Enhances milk fat percentage and energy-corrected milk yield in dairy production, while improves marbling and carcass quality, along with flavor and tenderness in meat production
7. **Enhances Reproductive Performance:** Supports hormonal activity, aiding in fertility and reproduction

Fat Inclusion in Diets of Farm Animals

1. **Ruminants:**

   a) High fat levels (>7% of diet) can impair rumen microbial activity and fiber digestion.

b) Bypass fats (rumen protected or inert fats) are used to prevent negative energy balance in high yielding dairy animals

2. **Non-Ruminants and poultry:** May tolerant to dietary fats but require proper balance of other nutrients for optimal performance. Fats are crucial for high-energy diets to support egg production and growth in poultry.

Sources of fat in animal diet

1. **Plant-Based Fat Sources:**
   a) Vegetable Oils: soybean oil, canola oil, sunflower oil, corn oil, palm oil (rich in unsaturated fats and linoleic acid).
   b) Oil cakes and meals: soybeans, cottonseeds, sunflower seeds, flaxseeds (contain both oil and protein).
   c) Rice Bran & Wheat Germ – Contain natural oils with beneficial fatty acids.
2. **Animal-Based Fat Sources:**
   a) Tallow (Beef Fat) & Lard (Pork Fat) – Provide saturated fats and energy.
   b) Fish Oil – A rich source of omega-3 fatty acids (EPA & DHA), beneficial for growth and immunity.
   c) Milk & Dairy Byproducts (Butterfat, Cheese whey) – Provide fats along with protein and lactose.
   d) Rendered Animal Fats – Byproducts from meat processing industries used in livestock and pet foods.

4. Minerals

Minerals are essential nutrients required for various physiological functions in animal production. They are classified into macrominerals (required in larger amounts) and microminerals (trace minerals) (needed in smaller quantities).

General functions of minerals in animal production

1. **Growth & Development:** Essential for bone formation, enzyme activation, and metabolic functions.
2. **Milk Production:** Calcium, phosphorus, and magnesium are critical for high-yielding dairy animals.
3. **Reproduction & Fertility:** Zinc, manganese, and selenium play key roles in fertility and pregnancy maintenance.
4. **Immune Function:** Copper, zinc, and selenium enhance disease resistance.

5. **Meat Quality:** Zinc and iron contribute to muscle development and meat quality.
6. **Egg Production:** Calcium is crucial for eggshell formation in poultry.

Table 3. Function and sources of important macro minerals

Macrominerals (Required in larger amounts)		
Mineral	**Functions**	**Sources**
Calcium (Ca)	Bone and teeth formation, muscle function, blood clotting, enzyme activation	Limestone, dicalcium phosphate, bone meal, legumes
Phosphorus (P)	Bone development, energy metabolism (ATP), DNA synthesis	Bone meal, grains, fish meal, dicalcium phosphate
Magnesium (Mg)	Enzyme activation, muscle and nerve function, prevention of grass tetany	Magnesium oxide, dolomite, legumes
Sodium (Na)	Maintains osmotic balance, nerve transmission, muscle function	Common salt (NaCl), plant and animal products
Potassium (K)	Muscle contractions, nerve signaling, acid-base balance	Alfalfa, molasses, grains, soybean meal
Chlorine (Cl)	Acid-base balance, formation of gastric acid (HCl)	Common salt (NaCl), fish meal
Sulfur (S)	Component of amino acids (methionine, cysteine), essential for wool growth and microbial protein synthesis in ruminants	Sulfur-containing amino acids, gypsum, molasses

Table 4. Function and sources of important trace minerals

Some micro-minerals (Trace Minerals) (Required in smaller amounts)		
Mineral	**Functions**	**Sources**
Iron (Fe)	Hemoglobin formation (oxygen transport), enzyme function	Green leafy forages, blood meal, fish meal
Copper (Cu)	Enzyme activation, hemoglobin synthesis, wool and hair pigmentation	Copper sulfate, molasses, legumes
Zinc (Zn)	Enzyme function, immune system support, wound healing, hoof and skin health	Zinc oxide, grains, legumes
Manganese (Mn)	Bone formation, enzyme function, reproduction	Manganese sulfate, grains, oilseeds
Iodine (I)	Thyroid hormone production (regulates metabolism and growth)	Iodized salt, seaweed, fish meal
Selenium (Se)	Antioxidant (prevents muscular dystrophy), immune function	Sodium selenite, grains, forages (amount depends on soil)
Cobalt (Co)	Vitamin B12 synthesis (important for ruminants)	Cobalt chloride, forages, mineral supplements
Fluorine (F)	Bone and teeth development (in small amounts, excess is toxic)	Water, rock phosphate

Feed Ingredients and supplements rich in minerals

a) **Forages & Pastures:** Good source of calcium, magnesium, and potassium.

b) **Cereal Grains (Corn, Wheat, Barley, Oats, Sorghum):** Contain phosphorus but in the form of phytate, which is poorly available to non-ruminants.

c) **Legumes (Alfalfa, Clover, Soybean Meal):** High in calcium and trace minerals.

d) **Animal By products (Bone Meal, Meat Meal, Fish Meal, Blood Meal, Eggshells):** Rich in calcium, phosphorus, iron, and trace minerals.

e) **Seaweed & Marine Sources:** Good sources of iodine, calcium, and trace elements.

f) **Trace Mineral Premixes:** Customized blends for specific species (cattle, poultry, pigs, etc.).

g) **Salt Licks & Mineral Blocks:** Often fortified with trace minerals and vitamins for free-choice consumption.

h) **Chelated Minerals:** Organic forms (e.g., zinc methionine, copper proteinate) for better absorption and bioavailability.

5. Vitamins

Vitamins are essential organic compounds required in small amounts for various physiological functions in animals. They may be classified as fat-soluble and water-soluble vitamins, each playing its own a crucial role in growth, reproduction, immunity, metabolism, and overall health.

General functions Vitamins in Animal Production

1. **Growth & Development:** B-complex vitamins and Vitamin A support healthy growth in young animals.
2. **Reproduction & Fertility:** Vitamins A, D, E, and B12 improve fertility, egg production, and fetal development.
3. **Milk & Meat Production:** Vitamin D and B vitamins are essential for metabolism and productivity.
4. **Disease Resistance & Immunity:** Vitamins A, C, and E boost immunity and help fight infections.
5. **Bone & Muscle Health:** Vitamin D supports strong bones, and Vitamin E prevents muscle degeneration.
6. **Nervous System Health:** B-complex vitamins are crucial for nerve function and preventing neurological disorders.

Table 5. Function and sources of important fat soluble vitamins

Fat-Soluble Vitamins (Stored in body fat, require dietary fat for absorption)		
Vitamin (Chemical name)	**Functions**	**Natural sources**
Vitamin A (Retinol)	Vision, reproduction, immune function, skin health, growth	Green leafy forages, carrots, fish liver oil, egg yolk, dairy products
Vitamin D (Cholecalciferol, D3; Ergocalciferol, D2)	Calcium and phosphorus metabolism, bone formation, eggshell production	Sunlight (D_3 synthesis), fish liver oil, fortified feeds
Vitamin E (Tocopherol)	Antioxidant, immune function, reproductive health, muscle integrity (prevents white muscle disease)	Green forages, vegetable oils, wheat germ, alfalfa
Vitamin K (Phylloquinone, K1; Menadione, K3)	Blood clotting, bone metabolism	Green leafy vegetables, fish meal, liver, synthesized by rumen microbes

Table 6. Function and sources of important water soluble vitamins

Water-Soluble Vitamins (Not stored in the body, require daily intake)		
Vitamin (Chemical name)	**Functions**	**Natural sources**
Vitamin B1 (Thiamine)	Energy metabolism, nerve function, prevents polioencephalomalacia in ruminants	Whole grains, yeast, green forages
Vitamin B2 (Riboflavin)	Growth, skin health, energy metabolism	Dairy products, eggs, leafy vegetables, yeast
Vitamin B3 (Niacin)	Energy production, skin health, prevents pellagra	Meat, fish, yeast, whole grains
Vitamin B5 (Pantothenic Acid)	Fat and carbohydrate metabolism, hormone production	Meat, grains, legumes, eggs
Vitamin B6 (Pyridoxine)	Protein metabolism, nervous system function	Meat, fish, grains, bananas
Vitamin B7 (Biotin)	Hoof and skin health, enzyme activation	Egg yolk, liver, yeast, legumes
Vitamin B9 (Folic Acid)	DNA synthesis, cell growth, prevents anemia	Green leafy vegetables, legumes, yeast
Vitamin B12 (Cobalamin)	Red blood cell formation, nerve function, metabolism	Animal products, *synthesized by rumen microbes*
Vitamin C (Ascorbic Acid)	Antioxidant, immune support, collagen synthesis	Fruits, vegetables, *synthesized in most livestock except guinea pigs and primates*

Vitamin Supplementation

a) **Ruminants:** Usually need supplementation of vitamins A, D, and E, as rumen microbes synthesize most B-complex vitamins and Vitamin K.

b) **Poultry & Pigs:** Require supplementation of all vitamins due to limited microbial synthesis.

c) **Horses:** Need Vitamin A, D, and E, while gut microbes provide some B vitamins.

Since vitamin content in natural feeds varies alot, commercial **vitamin premixes** must be supplemented to ensure balanced nutrition:

a) **Vitamin Premixes:** Contain fat- and water-soluble vitamins for poultry, swine, ruminants, and aquaculture.

b) **Fortified Feeds:** Livestock feed manufacturers add vitamins to formulated rations.

c) **Injectable Vitamins:** Used in cases of deficiency or stress (e.g., Vitamin B12, A, D, E injections).

d) **Mineral-Vitamin Blocks:** Often include Vitamin A, D, and E for free-choice consumption in grazing animals.

6. Water

Water is the most essential nutrient in animal production, playing a critical role in digestion, metabolism, temperature regulation, and overall health. It is often overlooked compared to other nutrients, but without adequate water, animals cannot survive or perform efficiently.

Table 7. Functions of Water in Animal Production

Function		Importance
a)	Digestion & Nutrient Absorption	Helps to dissolve and transport nutrients, enzymes, and hormones for proper digestion.
b)	Metabolism & Biochemical Reactions	Involved in energy production, protein synthesis, and enzymatic activities.
c)	Temperature Regulation	Maintains body temperature through sweating, respiration, and evaporation.
d)	Waste Excretion	Helps to eliminate waste products via urine, feces, and respiration.
e)	Circulation & Transport	Essential for blood circulation, oxygen transport, and the movement of nutrients.
f)	Milk Production	A major component of milk (about 87% in dairy cows), influencing lactation performance.
g)	Growth & Reproduction	Supports fetal development, egg formation, and overall reproductive health.

h)	Lubrication & Cushioning	Helps in joint movement and protects organs (e.g., cerebrospinal fluid in the brain).
i)	Feed Intake & Digestion	Softens feed, aids in rumen fermentation in ruminants, and prevents digestive disorders.

Water Requirements for Different Animals

Water requirements vary among species and are governed by factors like size, production stage, environmental temperature, and diet. Water requirement of animals increases during hot weather, lactation, and high-protein or high-salt diets.

Table 8. Water requirement for the livestock (litres per day)

Animal	Water Requirement (Liters per Day)
Cattle (Dairy)	30–60 L (more in lactation)
Cattle (Beef)	20–40 L
Sheep/Goats	3–8 L
Horses	30–50 L
Pigs	5–12 L
Poultry (Layers)	0.2–0.5 L per bird
Poultry (Broilers)	0.2–0.7 L per bird

Sources of Water for Animals

Animals obtain water from three different sources:

1. **Drinking Water**
   a) Freshwater from rivers, lakes, ponds, wells, or pipelines.
   b) Must be clean and free from contaminants like bacteria, toxins, and heavy metals.
2. **Feed Water (Moisture Content in Feeds)**
   a) Fresh forage & silage: 60–80% water content.
   b) Grains & dry feeds: 8–15% water content.
3. **Metabolic Water**
   a) Produced during digestion and metabolism of fats, carbohydrates, and proteins.
   b) Fats, carbohydrate and protein produces 1.07, 0.56 and 0.42 g metabolic water per g of food oxidised.
   c) Metabolic water comprises 5 to 10% of total intake of water.

19

Feed Ingredients for Ration of Livestock and Poultry

Mahipal Choubey, Abhishek Kr. Singh and Sandeep Kr. Chaudhary

Faculty of Veterinary and Animal Sciences, I. Ag.Sc., RGSC-Banaras Hindu University (BHU), Barkachha, Mirzapur, Uttar Pradesh-231001

Feed ingredients are a term used to designate all the individual components of a ration that are utilized to prepare it. It may range from concentrates and their byproducts to different forms of roughages. These feed ingredients are rich in one or more nutrients. They are generally classified according to the amount of specific nutrients they furnish in the ration.. The various feeds and fodders can be divided into following major categories.

Feed ingredients and its classification

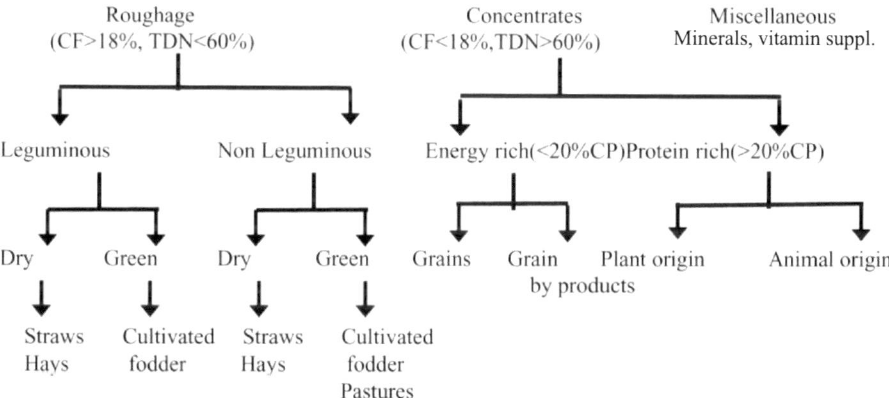

Fig. 1. Classification of feed ingredients for ration of livestock and poultry

ROUGHAGES

Roughages are bulky feeds having large amount (>18%) of crude fibre (less digestible material) and low total digestible nutrients (TDN <60%) content on dry matter (DM) basis.

1. **Succulent Roughages:** Succulent feeds usually contains moisture from 60-90%, For the sake of convenience succulent feeds may again be classified into 'pastures, cultivated fodder crops, tree leaves, silage and root crops.

a) **Pastures:** The word pastures refer to land on which different types of edible grasses and other plants grow or are grown for grazing livestock. Permanent pastures are those covered with perennial or self-seeding species of plants. Temporary pastures are those planted with quick growing crops like Sudan grasses and millet to provide supplemental grazing during lean season.

b) **Cultivated Fodder Crops:**

a) ***Leguminous***: Leguminous fodder such as Cowpea, Berseem, Lucerne etc. belong to "Leguminosae" family and very palatable to the animals. Legumes contain 2-3% DCP and 12% TDN on fresh basis. If, legumes are fed liberally to cattle and buffaloes, then there is no need to incorporate the another protein supplements in diet up to 6-7 litres of milk per day Sudden changes in feeding from non-leguminous to leguminous and early morning feeding of leguminous fodder should be avoided to reduce the possibility of bloat due to accumulation of gases in rumen.

b) ***Non leguminous***: Contains (lower percentage of nitrogen) 0.5 to 1.0% DCP and 11-15% TDN, these include many cereal crops, cultivated grasses, indigenous grasses and introduced grasses. e.g. Maize, Jowar, Bajra Oat etc.

c) **Tree Leaves/Top feeds:** Tree leaves are also fairly good source of nutrients and may serve as potential feed resource wherever feeds and fodders are in acute shortage e.g. Babul, Lucaena, Mulbery, Banayan

d) **Root Crops:** Root crops sometimes may be used as alternate feedstuffs during the natural calamity and scarcity period. They contain low crude fibre (5-11 %). e.g. Tapioca, Turnips, Sugar beet, Carrot

e) **Silage:** Silage is an anaerobically fermented feed prepared from green fodder whenever the supply of green fodder is in plenty. It possesses most the characters of green fodder and very high nutritive value.

2. **Dry Roughages:** Contains only 10-15% moisture. Dry roughages are dried plant materials, which are preserved for usage in summer and in adverse climatic conditions.

a) **Hay:** Leguminous crops harvested at preflowering or half blooming stage is air dried to reduce the moisture content $< 15\%$ and preserved in the form of hay. The moisture content of hay reduced to desirable level to check the enzymatic fermentation and fungal infestation during storage. Lucerne is best crop for hay making rather than berseem and cow-pea due to their hollow and thick stems, whereas under non-leguminous category oat is best crop for hay preparation. Due to high lignification, different nature of fibre and less palatability of non leguminous fodder than the leguminous fodder, hay is generally not prepared from non leguminous fodder.

b) **Straws/Stover:** The crop residues left after harvesting the main product of crop (i.e. grains) are known as straws/stovers. Straws are deficient in protein, mineral, vitamins and energy. Due to its high crude fibre, low TDN and high degree of lignification, strategic nutrients supplementation is pre requisite with straw feeding.

c) **Gotars:** It is the mixture of premature pods, pod covers, stems and leaves of legume crops after the harvesting seeds.

d) **Hull:** outer covering of beans, pea, cotton seed.

e) **Husk:** Dry outer covering of grains, grams, rice husk, gram husk.

f) **Shells:** Hard outer covering of nuts eg. GN shell.

CONCENTRATES

Generally contains less than 18% crude fibre and more than 60% TDN. These are less bulky and more digestible than roughages.

1. **Energy rich concentrates:** The crude protein is generally less than 20% in energy rich concentrates. These are also called as basal feeds.

a) **Cereal Grains:** Cereal grains are rich in soluble carbohydrates e.g. Maize, Barley, Rice, Oats etc.

b) **Millets:** e.g. Ragi (Nagli), Jowar and Bajra.

c) **Mill by products:** The by-products of milling cereal grains i.e. Bran (rice, wheat), flour, hulls, polishing and embryo of seeds are used as a part of concentrate mixture.

   a) **Bran:** The coarse pericarp (Outer covering) or seed coat of grain removed during processing eg. Wheat bran

   b) **Rice polish:** The final processing powdered obtained during polishing of rice seeds after initial removal of bran. This is rich in fat or oil.

d) **Molasses:** By product of sugar factories used in feeds as binding agent for pelleting and as readily available source of soluble carbohydrates.

e) **Roots and tubers:** e.g. Roots - turnip, sugar been, carrot. Tubers - potato, sweet potato.

2. **Proteins rich concentrates:** The crude protein is more than 18% in protein rich concentrates.

a) **Plant origin:** Oil seed cakes, pulses, pulse chuni, pulse churi, Brewers yeast and grains etc.

   a) **Chuni:** It is a byproduct of pulse meal industries consist primarily of brocken pieces of legume seeds

   b) **Oil cakes:** Residues which remains after extracting oil from oil seeds by mechanical or chemical process.

b) **Animal origin:** The by-products of slaughter houses i.e. meat, blood and bones are dried and made in to powders to be mixed in concentrate feeds. e.g. Meat meal, meat cum bone meal etc. The marine by-products are also used as animal protein source for feeding of livestock e.g. fish meal. Poultry-byproduct meal and feather meal are avian origin animal protein sources which can be used for livestock feeding.

FEED SUPPLEMENTS

1. **Mineral Supplements:** Those supplements which are given to the animals for providing the major or minor minerals in desired quantity. Many mineral mixtures are marketed under the different trade names. Generally, salt, calcium carbonate, zinc sulphate and copper sulphate supplements are provided in livestock and poultry ration toimprove production and reproduction.
2. **Vitamin Supplements:** Various Vitamin supplements for poultry, pigs and cattle are marketed in India under different names. For poultry *Vit* A, B and D synthetic vitamin supplements are marketed.

FEED ADDITIVES

Feed additives are a group of feed ingredients that can cause a desired animal response in a non-nutrient role such as pH shift, growth or metabolic modifier. These are the products used for the purpose of improving the quality of feed and the quality of food from animal origin, or to improve the animals' performance and health, e.g. providing enhanced digestibility of the feed materials.

a) **Hormones:** Some of the hormones have growth promoting properties like oestrogens, androgens, progestogens, thyroxine and pituitary growth hormones.

b) **Probiotics:** Live microbial feed supplement. Most commonly used products are based on *Aspergillus oryzae, Saccharomyces cerevisiae, Lactobacillus spp.*

c) **Antibiotics:** defined as the subgroup of anti-infectives that are derived from bacterial sources and are used to treat bacterial infections. At the lower intake antibiotics are known to stimulate the growth of animals when added to their feed and drinking water. In India penicillin, terramycin, tetracycline, flavomycin etc. are being used as a feed supplements in poultry, pigs and pre-ruminant calves.

d) **Prebiotics:** Prebiotics are non-digestible feed ingredients that beneficially affect the host by selectively stimulating the growth and or activity of one or limited number of bacteria in the colon that can improve the host health. Galacto oligosaccharides, fructo oligosaccharides and lactose derivatives have been used in poultry and other non-ruminants as prebiotic.

e) **Enzymes:** A protein (or protein-based molecule) that speeds up a chemical reaction in a living organism. An enzyme acts as catalyst for specific chemical reactions, converting a specific set of reactants (called substrates) into specific products. They helps to improve nutrient digestibility. eg. cellulases, phytases, xylanases, proteases, amylases etc.

f. **Dietary Buffers:** Dietary buffers are compounds, which were added in the ruminant diets to stabilize rumen pH, improved digestion of fiber, increased acetate:propionate ratio in the rumen (which improves milk fat percentage) eg. sodium bicarbonate, limestone, sodium bentonite, and magnesium oxide.

g. **Ionophores:** These are small molecular weight molecules that bind ions of various minerals and modulate their movement across cell membrane. Ionophores are antibiotics produced by a variety of *actinomycetes*, most often *Streptomyces* spp. Ionophores alters the flux of ions across biological membranes. Ionophores include monensin, lasalocid, salinomycin, and narasin. At present, monensin (*Rumensin™*) and lasalocid (*Bovatec™*) are the only ionophores approved to be fed to cattle.

h. **Arsenicals:** organic arsenicals had growth promoting properties similar to those of antibiotics when added to the diet of chicks. In addition to their use as growth stimulants, these have been used at low levels to help protect feeds from microbial destruction and to prevent and control poultry diseases. eg. Arsanilic acid, sodium arsanilate are compounds

i) **Tranquilizers**: A number of tranquilizing drugs have been usually used to combat stress due to heat or other environmental factors. Certain tranquilizers such as natural alkaloid of Rauwolfia, reserpine, hydroxyzine, chloropromazine have been shown to improve daily live weight gain in livestock.

j) **Anthelmentics, coccidiostat and antifungal:** Anthelmentics are the deworming drugs and used periodically in the feed or water to prevent parasitic infestation, specially of round warm. Out of many commercial products, 2,2 dichlorovinyl dimethyl phosphate, has both anthelmintics and separate growth stimulatory effect in cattle. Coccidiostat are routinely used in the diets of the poultry to prevent the most devastating type of disease the coccidiosis. Antifungals are natural or synthetic substances which inhibit the growth of fungi.

k) **Pigmenters and flavoring agents**: Pigmenters are usually carotenoid sources added to feed to improve pigmentation of broilers and egg yolk. Flavouring substances are also used as feed additives to improve the palatability of certain feed stuffs.

l) **Antioxidants:** Any substance that reduces oxidative damage (damage due to oxygen) such as that caused by free radicals. antioxidants include a number of enzymes and other substances such as vitamin C, vitamin E and beta carotene (which is converted to vitamin A) that are capable of counteracting the damaging effects of oxidation. Antioxidants are added for the stabilization of fats and fat soluble vitamins and also to prevent the destruction of vitamin by oxidation. Commonly used synthetic antioxidants are butylated hydroxyanisole (BHA), butylated hydroxytoluene (BHT) and ethoxyquin, which can be added @ 0.01% of fatty matter.

m) **Mineral Complexes and Chelates:** Some trace minerals are bound or complexed with an organic substance such as an amino acid, protein, or carbohydrate are called as chelates. Chelating or complexing these trace minerals will increase absorption of the mineral in the small intestine. Increased absorption of these trace minerals may improve immune response. Trace minerals that are most commonly complexed or chelated are zinc, manganese, copper, and cobalt.

n) **Mycotoxin Binders:** These are the compounds which bind with the naturally occurring fungal toxins and prevents mycotoxicosis. Several compounds, such as hydrated sodium calcium aluminosilicates (HSCAS) or mannan oligosaccharids (MOS), help alleviate the negative affects associated with aflatoxins in feed.

o) **Surfactants:** They are chemical compound which are neutral or alkaline salts or combination of both. They act like antibiotics by selecting inhibitory effects on intestinal microorganism. eg. alkyl benzene sulphonate.

Table 1. Various feeds and fodders and their botanical anme

A.	**GRASSES**	
	Sewan	*Lasirus sindicus*
	Bhurat	*Cenchrus biflorus*
	Dhaman	*Cenchrus setigerus*
	Anjan	*Cenchrus ciliaris*
	Elephant grass	*Pennisetum purpureum*
	Dub	*Cynodon dactylon*
	Para grass	*Brachiaria mutica*
	Guinea grass	*Panicum maximum*
	Napier grass	*Pennisetum purpureum*
B.	**TOP FEEDS**	
	Mango	*Mangifera indica*
	Beri (pala)	*Zizyphus nummularia*
	Neem	*Azadirachta indica*
	Ardu	*Ailanthus excelsa*
	Subabool	*Leucaena leucocephala*
	Sares	*Albizia lebbek*
	Khejri (loong)	*Prosopis cineraria*
C.	**SHRUBS**	
	Phog	*Colligonum polygonoides*
	Ker	*Caparis aphylla*
	Kheemp	*Leptadenia pyrotechnica*
	Cinia	*Crotolaria burhia*
D.	**CULTIVATED FODDER**	
	Legumes	
	Lucerne	*Medicago sativa*
	Berseem	*Trifolium alexandrinum*
	Cowpea	*Vigna sinensis*
	Sun hemp	*Crotalaria juncea*
	Non Legumes	
	Oat	*Avena sativa*
	Jowar	*Sorghum vulgare*
	Bajra	*Pennisetum typhoides*
	Maize	*Zea mays*
	Barley	*Hordeum vulgare*
E.	**CROP RESIDUES**	
	Wheat bhusa (Tudi)	*Triticum aestivum*
	Paddy straw	*Oryza sativa*

	Bajra straw (Kadbi)	*Pennisetum typhoides*
	Moth straw	*Phaseolus aconitifolius*
	Groundnut straw	*Arachis hypogaea*
	Guar straw (phalgati)	*Cyamopsis tetragonoloba*
	Moong straw	*Phaseolus aureus*
	Soyabean straw	*Glycine max*
	Sugarcane bagasse	*Saccharum officinarum*
	Sugar beet pulp	*Beta vulgaris*
F.	**CONCENTRATED FEEDS**	
	Wheat bran	*Triticum aestivum*
	De oiled rice bran	*Oryza sativa*
	Guar churi/guar korma	*Cyamopsis tetragonoloba*
	Moth churi	*Phaseolus aconitifolius*
	Groundnut cake	*Arachis hypogaea*
	Cotton seed cake	*Gossipium hirsutum*
	Tumba cake	*Citrullus colocynthis*
	Mustard cake	*Brassica campestris*
	Soybean meal	*Glycine max*

Table 2. General (average) chemical composition (%) of common feedstuffs

Feedstuff	DM	CP	EE	CF	NFE	Total Ash	Ca	P
Roughages								
Green fodder (as fed basis)								
Non-legume	25	2.2	0.8	8.0	12.20	1.80	0.11	0.05
Leguminous	20	4.0	0.8	5.0	7.70	2.50	0.40	0.08
Hay (on DM basis)								
Indigenous grasses	88	5.0	1.0	35.0	47.0	12.0	0.30	0.20
Improved pasture grass	88	7.0	1.5	30.0	51.50	10.0	0.35	0.25
Legumes	88	15.0	2.0	30.0	41.0	12.0	1.50	0.25
Straws								
Cereals	90	3.8	1.0	35.0	48.70	11.5	0.34	0.17
Legumes	90	6.0	1.0	39.0	43.50	10.5	1.0	0.12
Concentrates (on DM basis)								
Cereal grains	90	10.20	2.7	4.0	79.80	3.3	0.04	0.35
Legume seeds	90	23.30	2.3	6.8	63.60	4.0	0.16	0.41
Brans	90	12.50	3.5	13.0	64.00	7.0	0.10	1.0
Oil seed cakes	90	48.50	9.5	9.5	27.70	8.3	0.50	0.10

Table 3. Chemical composition of commonly used concentrates

Ingredients	DM%	ME (Kcal/ kg)	CP (%)	CF (%)	EE (%)	NFE (%)	Ca (%)	P (%)
Cereal grains & byproducts								
Maize	89.50	3309	9.20	2.40	3.90	82.80	0.25	0.40
Jowar	87.30	2645	10.30	3.60	4.60	78.10	0.18	0.32
Bajra	89.60	2642	12.70	2.20	4.90	78.20	0.13	0.72
Wheat	89.80	3045	10.30	2.10	2.60	82.30	0.18	0.43
Barely	96.40	2618	12.00	6.50	2.8	74.50	0.29	0.63
Rice bran (deoiled)	92.30	2235	14.10	13.80	1.70	53.40	0.37	1.80
Rice polish	91.80	2937	12.70	11.20	13.90	48.60	0.27	1.37
Wheat bran	88.90	1069	14.70	11.30	3.80	62.30	0.19	1.12
Legumes byproducts								
Soybean meal	89.90	2694	41.70	6.30	21.20	26.00	0.36	0.90
Mustard cake/ Rapeseed cake	91.30	2373	35.10	8.20	14.10	33.40	0.89	1.78
Ground nut cake	91.50	2596	47.0	6.10	5.10	33.10	0.51	0.82
Cottonseed cake	92.30	1556	27.50	18.40	9.40	38.40	0.28	1.28
Sunflower meal	89.10	2230	37.20	11.60	10.90	32.60	0.43	1.14
Animal byproducts								
Fish meal	93.80	1834	43.10	3.60	4.30	11.50	7.16	1.67
Animal fat	97.90	7700	-	-	97.90	5.6	27.00	12.11
Meat and bone meal	94.00	2111	53.80	2.30	-	-	11.25	5.39
Bone meal	95.50	1044	14.60	2.50	3.10	5.60	27.00	12.11
Poultry byproduct meal	93.00	-	56.40	0.9	17.80	7.60	3.95	1.73

Table 4. BIS specification for the poultry

Nutrient level in poultry ration							
Nutrienr,%	**Broiler**			**Layer**			
	Pre-starter	**Starter**	**Finisher**	**Chick**	**Grover**	**Layer-I**	**Layer-II**
Moisture	11 .0	11 .0	11 .0	11 .0	11 .0	11 .0	11 .0
Energy, Kcal/ kg	3000	3100	3200	2800	2500	2600	2400
Protein, %	23.0	22.0	20.0	20.0	16.0	18.0	16.0
Ether extract, %	3.0	3.5	4.0	2.0	2.0	2.0	2.0

C. Fibre, %	5.0	5.0	5.0	7.0	9.0	9.0	10.0
AIA, %	2.5	2.5	2.5	4.0	4.0	4.0	4.5
Salt, %	0.5	0.5	0.5	0.5	0.5	0.5	0.5
Calcium, %	1.0	1.0	1.0	1.0	1.0	3.0	3.5
Phosphorous, %	0.7	0.7	0.7	0.7	0.65	0.65	0.65
Source: BIS (2007) ; AIA: Acid Insoluble Ash							

Table 5. Different stages poultry ration composition with different feed ingredients

Examples of poultry ration					
Ingredients, %	**Broiler**		**Layer**		
	Starter	**Finisher**	**Chick**	**Grover**	**Layer**
Maize	58.60	64.20	58.00	47.00	59.00
Soybean meal	36.40	30.20	22.00	15.00	22.50
DORB	--	--	17.00	35.00	9.00
Fat	2.00	2.20	--	--	--
Lime Stone	1.00	1.00	1.00	1.10	8.00
DCP	1.70	1.00	1.50	1.30	1.20
Salt	0.35	0.35	0.35	0.35	0.30
Lysine	0.20	0.20	--	--	--
Methionine	0.14	0.12	0.06	0.06	0.06
Vitamin+ Trace mineral	0.25	0.25	0.25	0.25	0.25
Total	100	100	100	100	100
Source: ICAR-NIANP					

Unconventional feeds

Feeds which are traditionally not feed to the animals by livestock owners but recently based on research they have been recommended for using in livestock feeds at certain level.

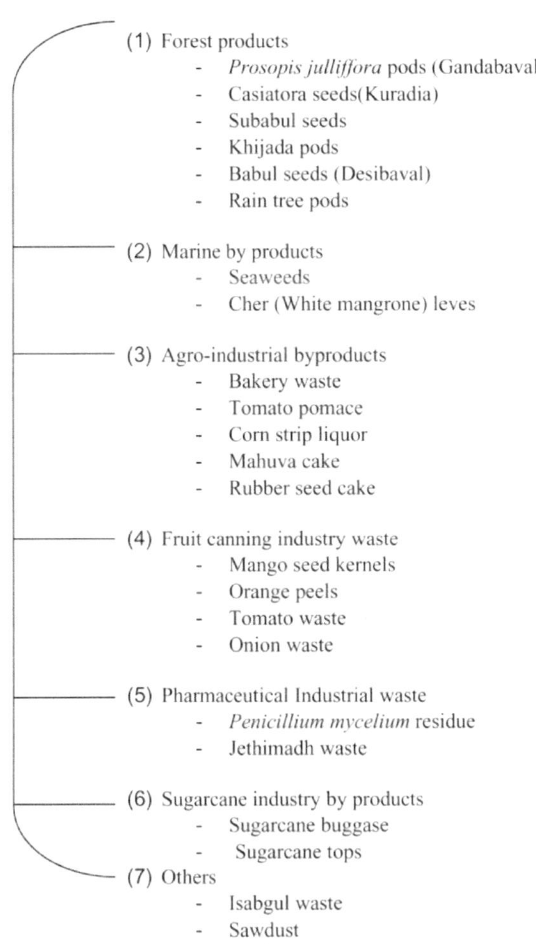

Ration Formulation

It is the process of deciding the level of different feed ingredients to generate a feed-mixture formula to achieve a target nutrient composition for specific physiological function of animal. There are so many methods to formulate feed for dairy cattle. The thumb rule of ration formulation is a quick and informal method for this purpose.

Step 1: calculation of dry matter (DM) requirement

Cattle: 2-2.5% of body weight (2-2.5 kg/100 kg BW)

Buffalo and crossbred cattle: 2.5-3.0 % of body weight (2.5-3.0 kg/100 kg BW)

Step 2: Partitioning of DM into different components of ration

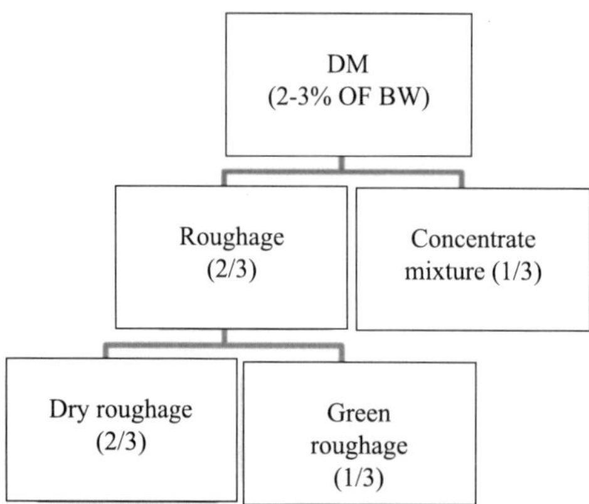

Fig. 2. Ration formulation thumb rule

Table 6. Ration composition with the available ingredients on the basis of BIS specification of Type & I Type II ration

Ingredients	**Type I**	**Type II**
Maize	20	10
Rice polish	15	15
Rapeseed (extracted)	15	18
Groundnut (extracted)	19	12
Rice bran (extracted)	15	30
Molasses	10	10
Calcite powder	1.5	1.5
Common salt	1.5	1.5
Min. mix. (type Ii)	2	2
Urea	1	--
Total	**100**	**100**

Table 7. Basal requirement of concentrate for bovine

Concentrate requirement(s)			
Status	**Basal requirement**	**Cattle**	**Buffalo**
Dry state	Maintenance as	1.25-1.5 kg	1.75-2.0 kg
Lactation	Maintenance requirement +	0.4 kg per litre of milk	0.5 kg per litre of milk
Pregnant	Maintenance + Lactation Allowance +	1 to 1.5 kg from third trimester	1.5-2.0 kg from third trimester

The above tables give an example of typical concentrate mixture and its requirement by dairy animals. The feeding of Type-I (22% CP) and Type – II (20% CP) concentrate mixture is advocated with non-leguminous and leguminous fodder, respectively. The proportion of concentrate and roughages must always be moderated and adjusted to meet the nutritional requirement of animal based on physiological status.

Feeding of Calves

Important points for consideration

a) Practice colostrum feeding at the earliest
b) Feeding whole milk/ milk replacer to calves
c) Introduction of calf starter/good quality hay from 2nd week onwards

Table 8. Calf feeding schedule (birth to 26th of week of age)

Period	Colostrum/ Whole Milk (kg/day)	Calf starter (kg/day)	Good quality hay* (kg/day)	Green fodder* (kg/day)
0-2 days	1.5-2.0 (colostrum)	--	--	--
3-4 days	1.5-2.0 (milk)	--	--	--
4-14 days	1.0-1.5 (milk)	0.1	0.1	--
3rd week	0.5-1.0 (milk)	0.2	0.15	0.75
4th week	*Milk (0.5 kg) or milk replacer (0.25 kg) can be fed, if available*	0.25	0.2	1.25
5th week		0.4	0.3	2
6th week		0.5	0.4	2.5
7th week		0.6	0.6	3
8th week		0.7	0.8	3.5
9th week		0.8	0.9	4
10th -11th wk		1	0.9	5
12th week		1.2	1	5
13th -16th wk		1.5	1.2	6
17th -20th wk	--	1.75	1.5	7.5
21st -26th wk	--	2	2	8
Note: *Requirement of hay and green fodder may vary from breed to breed & body weight of calf. Colostrum feeding is very essential during early life of calf. (**Source: NDDB**)				

Table 9. Examples of calf starter (23-26% CP, 75% TDN)

Ingredients of calf starter	
Ingredients	**Parts (%)**
Maize	42
GNC	35

Wheat or rice bran	10
Fish meal	10
Min. mix.	1.98
Salt	1
Vitamins (A,D, E)	0.02
Total	**100**

Table 10. Examples of milk replacer (economical alternate to milk)

Ingredients of milk replacer	
Ingredients	**Parts (%)**
Rice polish fine	14
Crushed maize grain	20
Maize gluten	16
Groundnut extraction	15
Soyabean extraction	12
Skimmed milk powder	10
Bypass fat	4
Molasses	6
Mineral mixture	2
Common salt	0.8
Antibiotics	0.1
Vitamins	0.02
Preservatives	0.08
Total	**100**

Mazie grain Wheat grain

Soyabean

Deoilde rice bran*

Groundnut meal*

Soybean meal

Mustard cake

Cotton seed meal*

Meal Guar

Oat grain

Lucerne hay

Bamboo leaves*

Subabul pods and leaves

Pipal leaves

Albizialebek leaves

Yellow Gulmohar leaves

Napier grass

Sheesham leaves

Fig. 3. Different type of conventional and nonconventional feed resources for livestock and poultry

* NDDB compendium on "nutritive value of commonly available feed and fodder in India"

20

Feed Supplements and Feed Additives

Abhishek Kumar Singh, Sandeep Kumar Chaudhary and Mahipal Choubey

Faculty of Veterinary and Animal Sciences, I. Ag.Sc., RGSC-Banaras Hindu University (BHU), Barkachha, Mirzapur, Uttar Pradesh-231001

Feed Supplement

A supplement is a feed ingredient used to enhance the nutritional value of basal feeds. Supplements can be included in large quantities, such as protein supplements, or in very small amounts, such as trace minerals.

When formulating a ration, the primary focus is on meeting dry matter, protein, and energy requirements. Once these are addressed, micronutrients such as essential amino acids, minerals, and vitamins are incorporated to correct any deficiencies.

Feed Additive

A feed additive is a substance added to basic feed, typically in small quantities, to enhance its nutritional profile, improve feed utilization, or support animal health. Unlike direct nutrient sources, additives may include growth stimulants, medicines, or other compounds that influence animal performance.

Generally, "feed additives" refer to non-nutritive substances that modify feed efficiency, digestion, or overall productivity. These additives and implants are categorized based on their mode of action.

1. Additives that enhance feed intake

a) **Antioxidants:**

Antioxidants

Antioxidants are compounds that prevent the oxidative rancidity of polyunsaturated fats. Once rancidity occurs, it can lead to the destruction of vitamins A, D, E, and several B-complex vitamins. Additionally, the breakdown products of rancidity may react with lysine, reducing the protein value of the ration. Ethoxyquin and butylated hydroxytoluene (BHT) are commonly used antioxidants in animal feed.

Types of Antioxidants

I. Primary Antioxidants (Interrupt and terminate free radical propagation)

- **Natural Antioxidants:** Tocopherols, ascorbates, carotenoids, flavonoids, herbal extracts, thiocyanates
- **Synthetic Antioxidants:**
  - Ethoxyquin
  - BHT (Butylated hydroxytoluene)
  - BHA (Butylated hydroxyanisole)
  - TBHQ (Tertiary butylhydroquinone)

Specific Applications

- BHA, BHT → Used in animal fats
- Ethoxyquin → Used in both animal and plant lipids
- Regulatory Limitations: The FDA has reduced the permissible level of ethoxyquin from 150 ppm to 75 ppm.
- Propyl Gallate → Forms color complexes with copper (Cu) and iron (Fe).
- TBHQ → Lacks global approval for use in animal feed.

II. Secondary Antioxidants (Prevent free radical formation)

- Metal chelators (e.g., citrate, phosphates)
- Reducing agents (e.g., ascorbates, sulfites)

  Increased polyunsaturated fatty acids (PUFAs) and omega fatty acids in feed elevate susceptibility to oxidative damage, necessitating the use of antioxidants.

b) **Flavouring Agents:** Flavouring agents are feed additives designed to enhance palatability and encourage feed intake. They are particularly useful in maintaining consumption levels under the following conditions:

- When highly unpalatable medications are mixed into the feed
- During disease outbreaks
- When animals are under stress
- When incorporating less palatable feedstuffs into the diet

Species-Specific Preferences

- **Ruminants (cattle, goats, sheep):** Prefer sweet compounds and respond positively to salts of volatile fatty acids.

- **Horses:** Often reject musty feed, even when mold levels are too low for human detection.

2. Additives That Enhance the Color or Quality of Marketed Products

In poultry farming, feed additives are sometimes used to improve the visual appeal and quality of the final product. One common example is the enhancement of the yellow color in broiler skin and egg yolks, which is achieved by incorporating xanthophylls (natural pigments) into poultry feed. Additionally, certain arsenic-based compounds, such as arsanilic acid, sodium arsanilate, and roxarsone, have historically been used to improve feed efficiency, enhance growth, and promote uniform pigmentation in poultry products. However, due to concerns about their environmental and health effects, the use of these additives has been restricted or discontinued in many countries.

3. Additives That Facilitate Digestion and Absorption

a) ***Grit:*** Unlike mammals, poultry lack teeth, meaning they cannot chew their food. Instead, digestion begins in the gizzard, a thick-walled muscular organ where feed is mechanically ground. The finer the feed is ground, the greater the surface area available for digestion and nutrient absorption. To aid in grinding, poultry are sometimes provided with grit, which consists of small, hard particles that help break down coarse or fibrous feed in the gizzard. This is particularly beneficial when birds are fed whole grains or high-fiber diets. However, when finely ground or mash feeds are used, the role of grit becomes less significant.

Common types of grit include

- Oyster shells
- Coquina shells (a type of marine limestone)
- Limestone granules

In addition to aiding digestion, oyster shells and limestone also serve as sources of calcium, which is essential for eggshell formation in laying hens.

b) ***Buffers and Neutralizers:*** During periods of high production, such as peak lactation in dairy cows, ruminants are often fed high-energy concentrate diets to meet their increased nutrient demands. However, excessive concentrate feeding can lower the rumen pH, leading to an imbalance in the microbial population.

Normally, the rumen contains a diverse community of microbes, but when the pH drops too low, acidophilic (acid-loving) bacteria proliferate, disrupting normal digestion. This can result in ruminal acidosis, a condition that negatively affects feed efficiency, nutrient absorption, and overall animal health.

To counteract this, feed buffers and neutralizers are added to maintain a stable rumen environment. These compounds help neutralize excess acidity and restore microbial balance. Some commonly used buffers and neutralizers include:

- Sodium bicarbonate (baking soda - $NaHCO_3$)
- Sodium carbonate (soda ash - Na_2CO_3)
- Magnesium oxide (MgO)
- Calcium carbonate (limestone - $CaCO_3$)
- Sodium hydroxide (NaOH)
- Phosphate salts
- Ammonium chloride (NH_4Cl)
- Sodium sulfate (Na_2SO_4)

Benefits of Buffer Supplementation in Ruminants

Studies have shown that adding sodium bicarbonate to ruminant diets can:

- Increase average daily weight gain by approximately 10%
- Improve feed efficiency by 5–10%
- Boost milk production by around 0.5 liters per animal per day

Thus, the strategic use of buffers and neutralizers helps optimize digestion, prevent metabolic disorders, and improve overall productivity in high-producing ruminants.

c) ***Chelates:*** The term "Chelate" is derived from the Greek word Chele, meaning "claw", which describes how polyvalent cations (minerals with multiple charges) are held by organic binding agents. These organic substances, before binding to a metal, are called ligands.

Formula: Ligand + Mineral = Chelated Element

Chelates are cyclic compounds that play a crucial role in the absorption of minerals. When a mineral is in its chelated form, it can either:

- Be released as an ionic form at the intestinal wall for absorption, or
- Be absorbed directly as an intact chelate, depending on the nature of the compound.

Chelates may be derived from naturally occurring substances like

Chlorophyll (magnesium chelate), Cytochromes (iron chelate), Hemoglobin (iron chelate)

Vitamin B12 (cobalt chelate), Certain amino acids (such as cysteine and histidine)

Chelates may also be synthetically created, such as ethylene-diamine-tetraacetic acid (EDTA), which is used to enhance mineral availability.

d) ***Probiotics:*** A probiotic is a live microbial feed supplement that provides health benefits to the host by improving intestinal microbial balance.

Most probiotic preparations contain beneficial bacteria such as

Lactobacillus species; *Streptococcus* species; Some also contain yeast cultures

Functions and Benefits of Probiotics

- Direct Antagonistic Effect on Harmful Microorganisms
- Probiotics produce antibacterial compounds that suppress harmful pathogens.
- They compete with harmful bacteria for nutrients, limiting their growth.
- Modify Microbial Metabolism in the Digestive Tract
- Probiotics alter fermentation patterns, leading to better digestion and absorption.
- Probiotics stimulate immune responses, enhancing disease resistance.
- Neutralize Enterotoxins

They help in breaking down toxins produced by pathogenic bacteria in the gut.

Overall Impact of Probiotics on Livestock

- Increased growth rate
- Improved feed efficiency
- Enhanced gut health and nutrient absorption
- Reduced digestive disorders

4. Additives that promotes growth and production

I. **Antibiotics:** These substances are produced by living organisms, such as molds, bacteria, or green plants, and possess bacteriostatic or bactericidal properties even at low concentrations. Initially developed for medical and veterinary applications to control specific pathogens, certain antibiotics were later found to enhance the growth rates of young pigs and chicks when included in their diets at low levels. This discovery led to extensive research on various antibiotics, revealing that penicillin, oxytetracycline (Terramycin), chlortetracycline, bacitracin, streptomycin, tyrothricin, gramicidin, neomycin, erythromycin, and flavomycin exhibit growth-promoting effects. The most significant weight gain occurs during periods of rapid growth, with the effect diminishing over time. Notably, the differences between treated and control animals are more pronounced when the diet is slightly deficient or marginal in protein, B vitamins, or specific minerals.

Mode of Action of Antibiotics

- Antibiotics "spare" protein, amino acids and vitamin on diets containing 1 to 3 % less protein, but balance experiments have often failed to show increased nitrogen retention. Growth stimulation has been greatest when the antibiotic penicillin supplement has been added to a ration containing no protein supplements of animal origin or to a ration low in vitamin B_{12}.
- Intestinal wall of animals fed antibiotics is thinner than that of untreated animals which might explain the enhanced absorption of calcium shown for chicks.
- Reduce or eliminate the activity of pathogens causing "subclinical infection."
- Reduce the growth of micro-organisms that compete with the host for supplies of nutrients.
- Antibiotics alter intestinal bacteria so that less urease is produced and thus less ammonia is formed.
- Stimulate the growth of micro-organisms that synthesise known or unidentified nutrients.

Following points should be kept in mind while using antibiotics for animal feeding

1. Antibiotics should be used only for (a) growing and fattening pigs for slaughter as pork or bacon; (b) growing chicks and turkey poults for killing as table poultry.
2. Antibiotics should not be used in the feed of ruminant animals (cattle, sheep and goats), breeding pigs and breeding and laying poultry stock.
3. While adding antibiotics at the recommended level, care should be taken that they are thoroughly and evenly mixed with the feed.
4. For best results, antibiotics should be used with properly balanced feeds. Also, the feeds containing antibiotics should be fed only to the type of stock for which they are intended.
5. Antibiotics are not a substitute for good management and healthy living conditions, or for properly balanced rations.

II. **Probiotic and Prebiotic:** The animal gut hosts a diverse microbial community, including both beneficial and harmful microorganisms. A balanced gut microbiome plays a crucial role in maintaining health, with beneficial bacteria contributing to overall well-being. These microbes, often referred to as probiotics, or the intake of nutrients that promote their growth (prebiotics), support digestive and immune functions.

Probiotics

Probiotics are live microbial supplements that help improve intestinal microbial balance and benefit the host. Prebiotics, on the other hand, are non-digestible food components that selectively stimulate the growth or activity of beneficial gut bacteria, such as Lactobacilli and Bifidobacterium. They enhance gut health by promoting the proliferation of health-supporting microbes.

Common probiotic bacteria include Lactobacilli, Bifidobacterium, Enterococcus faecium, and spore-forming Bacillus spp., while certain yeasts, such as Saccharomyces, are also utilized.

Probiotics benefit the host by

- Producing antibacterial compounds that suppress harmful micro-organisms.
- Modifying microbial metabolism in the gastrointestinal tract.
- Enhancing immune function.
- Neutralizing enterotoxins produced by pathogens.

Prebiotics

Prebiotics support beneficial gut microbes by selectively promoting the growth of favourable bacteria, including administered probiotics and indigenous microbiota. Unlike general dietary fibers that nourish a broad range of microorganisms, prebiotics specifically enhance beneficial strains such as *Lactobacillus* and *Bifidobacterium*, improving growth and feed efficiency.

The main prebiotic categories include

1. **Fructans:** Fructooligosaccharides (FOS) and inulin.
2. **Galactans:** Galactooligosaccharides (GOS).

III. **Synbiotics:** Synbiotics combine probiotics (beneficial live micro-organisms) with prebiotics (non-digestible food components that nourish beneficial microbes) to enhance gut health and overall physiological functions. By improving probiotic survival and activity, synbiotics optimize digestion, immune function, and microbial balance while reducing harmful bacteria.

They are categorized into

- **Complementary Synbiotics**: Contain independent probiotics and prebiotics that provide additive benefits.
- **Synergistic Synbiotics**: The prebiotic is specifically selected to enhance the function of the probiotic strain.

Common synbiotic formulations include combinations like *Lactobacillus* with (FOS) or *Bifidobacterium* with inulin, supporting microbial balance and intestinal health.

IV. **Postbiotics:** Postbiotics are bioactive compounds produced by probiotics during fermentation. Unlike live probiotics, postbiotics provide similar health benefits but with greater stability and safety, making them ideal for individuals with sensitivities to live microorganisms.

Key postbiotic components include

- **Short-chain fatty acids**: Improve gut barrier integrity and reduce inflammation.
- **Enzymes and peptides**: Aid digestion and exhibit antimicrobial properties.
- **Exopolysaccharides**: Support immune modulation and gut health.
- **Cell wall components**: Stimulate immune responses and protect against pathogens.

5. Additives that alter metabolism

a) Hormones

These are chemicals released by a specific area of the body (ductless glands) and are transported to another region within the animal where they elicit a physiological response.

Extensive use is being made of synthetic and purified estrogens, androgens, progestogens, growth hormones and thyroxine or thyroprotein (iodinated casein) to stimulate the growth and fattening of meat producing animals. There is concern, however, about possible harmful effects of any residues of these materials in the meat or milk for the consumers.

The whole question whether hormones should be used as growth promoters is still debatable but it seems logical that with any feeding system the economic advantages, however great should never take precedence over any potential risk to human health. These substances may induce cancer in human beings if taken over a prolonged period through products of the treated animals. The use of such substances in poultry rearing has been prohibited by law in U.S.A.

b) Implants

Implants are hormone or hormone like products that are designed to release slowly, but constantly, the active chemicals for absorption into the bloodstream. These are implanted subcutaneously in the ear. (e.g.) diethylstilbesterol (DES).

6. Additives that affect the health status of livestock

Antibloat compounds: Surfactants such as poloxalene is used as a preventive for pasture bloat, several other products have been shown to be highly effective to prevent bloat are also available in the market.

Antifungal additives: Mould inhibitors are added to feed liable to be contaminated with various types of fungi such as Aspergillus flavus, Penicillium cyclopium etc. Before adding commercial inhibitors, all feedstuff should be dried below 12 cent moisture. Propionic, acetic acid and sodium propionate are added in high moisture grain to inhibit mould growth. Antifungals such as Nystatin and copper sulphate preparations are also in use to concentrate feeds to prevent moulds.

Anticoccidials: Various brands of anticoccidials are now available in the country to prevent the growth of coccidia which are protozoa and live inside the cells of the intestinal lining of livestock.

Anthelmintics: Under some practical feeding conditions anthelmintics have also been used. The compounds act by reducing parasitic infections.

Anticaking agents

Anticaking agents are anhydrous substance that can pick up moisture without themselves becoming wet. They are added to dry mixes to prevent the particles clumping together and so keep the product free flowing. They are either anhydrous salts or substance that hold water by surface adhesion yet themselves remain free flowing:

- Salt or long chain fatty acids., Calcium phosphate, Potassium and sodium ferryocyanide Magnesium oxide, Sodium aluminium silicate, Sodium calcium aluminium silicate, Calcium aluminium silicate

Humectants

These are substance which are required to keep the product moist, as for example, bread and cakes. Anticaking agents immobilise moisture that was picked up. Humectants are not or much use in poultry feed.

Firming and crisping agents

These are substance that preserve the texture or vegetable tissues and by maintaining the water pressure inside them, keep them turgid. It prevents a loss of water from the tissues.

Sequestrants

Certain metals – copper, iron can act as pro-oxidant catalytic and there fore need to the immobilised. Sequestrants are compounds added to do this.

These compounds should have affinity to metal ions and should prevent the metal in becoming engaged in oxidative action. Most effective sequestrants Ethylene diamine tetra acetic acid.

Calcium salt of EDTA works satisfactorily as a sequestrants.

Sweeteners

It is common constitution of food but yet used as additives. Eg. Sugar

Some are poorly digestible, may cause digestive upsets.

Saccharin – extensively used during World War I. It is a compound without any calorific value.

Additives such as humectants, firming and crisping agents, sweeteners, emulsifiers, stabilisers, acid, buffers are not commonly used in poultry feeds.

Enzymes

Enzymes are protein which have the property of catalysing specific biochemical reactions. They are found in all plants and animals and are responsible for growth and the maintenance of health.

Microorganism also produce enzymes and in recent years it has been possible to produce enzymes using microorganism on an industrial scale, extract and use these enzymes in a wide range of processes for the production of feed and natural products.

Poultry feeds are largely composed or plant and vegetables materials and there are enzymes developed to degrade, modify or extract the plant polymers found in some of the cereals and their by-products. The enzymes can be used to improve the feeding of poultry in the following way:

- By improving the efficiency of the utilisation of the feed.
- By upgrading cereals by-products or feed components that are poorly digested
- By providing additional digestive enzymes to help poultry to withstand stress conditions eg. Hot climates.

The dual role of enzymes has been demonstrated in trials with barley-based feed supplemented with beta-glucanase, where the apparent increase in available energy was far in excess of that available in the beta-glucan of the barley. In this case not only was the problem of sticky dropping completely eliminated but the chicken's rate of growth was equivalent to that observed normally with feeds containing a higher energy density (eg. Wheat based).

Choice of enzyme

Because of feed is normally composed of a single raw material of constant quality, it is important that the correct choice of enzyme product be made. Even in the case of a relatively well-defined problem such as that in barley, the use of multi enzyme activity products in an advantage.

- The enzymes should fulfil the following criteria for practical application:
- The enzymes must be active at the pH of the animal's digestive system and capable of surviving transit through the stomach.
- They must be in a physical form in which they can be safely and easily mixed into all forms of animal feed.
- The products should be or a high standardised activity that will remain stable both before and after incorporation into the feed or pre-mix.
- The enzymes must be capable of surviving normal pelleting conditions.

7. Phytogenic Feed Additive

Phytogenic feed additives (PFAs), also known as phytobiotics or botanicals, are plant-derived bioactive compounds used to enhance animal nutrition, health, and performance. They consist of essential oils, herbs, spices, plant extracts, and secondary metabolites such as flavonoids, tannins, and saponins.

Types of Phytogenic Feed Additives and Their Functions

A. **Essential Oils** (Oregano, Thyme, Cinnamon, Clove, Peppermint)
   - Improve digestion by stimulating enzyme secretion.
   - Exhibit antimicrobial effects against pathogens like *E. coli* and *Salmonella.*
   - Reduce gut inflammation and oxidative stress.

B. **Herbs & Spices** (Turmeric, Garlic, Ginger, Fenugreek, Aloe Vera)
   - Possess anti-inflammatory, immunomodulatory, and appetite-stimulating properties.
   - Support gut health by modulating microbial populations.

C. **Tannins** (Chestnut, Quebracho, Tea Extracts)
   - Bind to proteins and improve protein utilization in ruminants.
   - Have antimicrobial and anti-parasitic effects, beneficial for gut health.

D. **Saponins** (Yucca, Alfalfa, Soapwort)
   - Reduce ammonia emissions by modifying gut microbial activity.
   - Improve nutrient absorption and reduce methane production in ruminants.

E. Flavonoids & Polyphenols (Citrus Extracts, Green Tea, Grape Seed)
- Act as strong antioxidants, protecting cells from oxidative damage.
- Enhance immune response and improve stress tolerance.

Benefits of Phytogenic Feed Additives

- **Enhanced Digestibility & Nutrient Utilization** – Stimulate digestive enzyme activity, improving feed conversion efficiency.
- **Modulation of Gut Microbiota** – Promote beneficial bacteria while inhibiting harmful pathogens, reducing the need for antibiotics.
- **Immune System Boost** – Strengthen the immune response, reducing disease susceptibility.
- **Growth Promotion & Performance** – Increase weight gain and productivity in poultry, swine, and ruminants.
- **Environmental Benefits** – Reduce methane emissions in ruminants and lower ammonia levels in manure, contributing to sustainable livestock production.

Applications in Different Livestock Species

- **Poultry** – Improve gut integrity, enhance feed efficiency, and prevent bacterial infections like *Clostridium perfringens*.
- **Ruminants (Cattle, Sheep, Goats)** – Support rumen fermentation, improve fiber digestion, and reduce methane emissions.
- **Swine** – Aid in gut health post-weaning, enhance digestion, and reduce stress-related disorders.
- **Aquaculture (Fish & Shrimp)** – Boost immunity, improve resistance to bacterial infections, and enhance feed utilization.

Table 1. Some of the common feed additives used in Animal Nutrition

Category	**Purpose**	**Examples**
Probiotics	Improve gut health, enhance digestion	*Lactobacillus spp., Bifidobacterium spp., Saccharomyces cerevisiae*
Prebiotics	Promote beneficial gut bacteria	Mannan-oligosaccharides (MOS), Fructo-oligosaccharides (FOS), Inulin
Synbiotics	Combination of probiotics and prebiotics	Probiotic + FOS/MOS combinations
Enzymes	Enhance nutrient digestibility	Phytase, Xylanase, Protease, Cellulase
Antibiotics (as growth promoters)	Improve growth performance, control infections	Oxytetracycline, Chlortetracycline, Bacitracin

Organic Acids	Lower gut pH, control pathogens	Citric acid, Formic acid, Lactic acid
Toxin Binders	Bind and neutralize mycotoxins	Clay-based binders, Activated charcoal, Yeast cell wall components
Ionophores	Improve rumen efficiency and reduce methane production	Monensin, Lasalocid, Salinomycin
Amino Acids	Improve protein synthesis, balance diets	Lysine, Methionine, Threonine, Tryptophan
Vitamins & Minerals	Support metabolic functions, immunity, and growth	Vitamin A, D, E, Zinc, Selenium, Copper
Phytogenics (Plant Extracts)	Enhance digestion, act as natural antibiotics	Essential oils (Thyme, Oregano), Saponins, Tannins
Yeast & Fermentation Products	Improve rumen function and fiber digestion	*Saccharomyces cerevisiae*, Aspergillus fermentation extract
Beta-Agonists	Improve lean muscle growth	Ractopamine, Zilpaterol

Table 2. Essential feed supplements used in livestock nutrition

Category	Purpose	Examples
Protein Supplements	Enhance muscle growth, improve productivity	Soybean meal, Mustard cake, Groundnut cake, Fish meal
Energy Stupplements	Provide additional energy for maintenance and production	Cereal grains (Maize, Barley, Wheat), Molasses, Vegetable oils
Mineral Supplements	Support bone health, enzyme function, and overall metabolism	Dicalcium phosphate (DCP), Limestone, Salt (NaCl), Chelated minerals
Vitamin Supplements	Improve immunity, reproduction, and metabolic functions	Vitamin A, D, E, K, B-complex (Riboflavin, Niacin)
Amino Acid Supplements	Balance protein intake, enhance growth, and improve feed efficiency	Lysine, Methionine, Threonine, Tryptophan
Fatty Acid Supplements	Improve milk yield, reproductive efficiency, and energy supply	Omega-3 (Fish oil, Flaxseed), Omega-6 (Sunflower oil)
Electrolyte Supplements	Maintain hydration and acid-base balance	Sodium bicarbonate, Potassium chloride, Electrolyte mixtures
Bypass Nutrient Supplements	Improve nutrient utilization in ruminants by protecting them from rumen degradation	Bypass fat, Bypass protein, Protected amino acids
Nucleotides & Immunostimulants	Enhance immunity, gut health, and cell regeneration	Yeast extract, Purified nucleotides, Beta-glucans

21

Feeding of Livestock and Poultry

Abhishek Kumar Singh[1], Sandeep Kumar Chaudhary[1] Mahipal Choubey[1] and Punita Kumari[2]

[1]*Faculty of Veterinary and Animal Sciences, I. Ag.Sc., RGSC-Banaras Hindu University (BHU), Barkachha, Mirzapur, Uttar Pradesh-231001*

College of Veterinary Sciences and Animal Husbandry, Birsa Agicultural University, Kanke, Ranchi, Jharkhand

Introduction

The feeding of livestock and poultry is a critical component of animal husbandry that directly influences agricultural productivity, animal health, and food security. As the global demand for animal protein continues to rise, optimizing feeding practices becomes increasingly vital. This involves understanding various feeding methods, nutritional requirements, and the economic implications of feed formulations. A high-yielding dairy cow requires a well-balanced diet that meets the nutritional demands for sustained milk production. Key nutrients include carbohydrates, proteins (amino acids), fats (fatty acids), minerals, vitamins, and water-all of which are essential for mammary gland function and milk synthesis. The foundation for a productive dairy cow begins early in life, therefore ensuring optimal growth and development from birth through maturity is crucial, starting with proper calf and heifer nutrition.

Colostrum: The First Feed

Colostrum is the initial secretion produced by the mammary gland after birth, playing a critical role in early calf nutrition, particularly for pre-ruminant animals like neonatal calves. Unlike some species, calves receive minimal antibody transfer through the placenta due to its cotyledonary structure, making postnatal passive immunity essential. Colostrum is rich in immunoglobulins (Ig), growth factors, and essential nutrients, helping calves build immunity against pathogens and ensuring early health and survival.

Milk and Milk Replacer Feeding

From the second day of life, calves are typically fed either whole milk or milk replacer. Traditionally, a feeding rate of **4 liters per day** (split into morning and evening feedings) has been standard practice for both liquid milk and reconstituted milk replacer (**10%–15% solids**). However, recent advancements in calf nutrition have led to higher feeding rates, with **Holstein calves receiving up to 6 liters per day**. Additionally, automated feeding systems now allow for **ad libitum suckling**, where calves may consume **10 liters or more per day**, promoting improved growth and early development.

Calf Starter Grain and Rumen Development

The transition from a pre-ruminant to a functional ruminant requires the introduction of **highly digestible solid feed**, such as **calf starter grain**, which plays a pivotal role in rumen development. When consumed, starter grain undergoes microbial fermentation, producing **volatile fatty acids (VFAs)**. Among these, **butyrate** is particularly important for stimulating **rumen papillae growth** and epithelial development. Since grains ferment more rapidly than forages, feeding hay is generally unnecessary at this stage.Typical calf starters consist of **whole, crimped, or steam-flaked grains** combined with pelleted components to ensure adequate nutrient intake. In some cases, **fully pelleted starters** are used, requiring supplemental **high-quality hay** to maintain fiber intake and promote digestive health.By implementing a structured feeding program from birth, dairy calves can achieve optimal growth and transition smoothly into high-producing cows, ensuring both productivity and longevity in the herd.

Table 1. Feeding schedule of calves

Age of Calf	Whole Milk	Calf Starter	Good Quality Hay
1-3	Colostrum @ 1/10 of body wt in 3 feeds	-	-
4-7	Whole milk @ 1/10 of body wt in 3 feeds	-	-
8-14	Whole milk @ 1/10 of body wt	-	-
15-21	Whole milk @ 1/10 of body wt	A little	A little
22-35	Whole milk @ 1/15 of body wt	100g	Ad lib
Up to 2 months	Whole milk @ 1/20 of body wt	250g	Ad lib
2-3 months	Milk is gradually reduced and tapered	500g	Ad lib

Table 2. Feeding the post-weaned heifer (3-6 months)

Category	Concentrate (Kg)	Roughage (Kg)
Indigenous cattle/ buffaloes	1-2 Kg	Green oat/maize-10 Kg or Berseem 1.5-2.5 Kg + Dry fodder -2 kg or Green fodder – 3 kg + Straw – 2 kg
Crossbred	1.5 – 2.0 Kg	Green oat/maize or alike fodders 5 -10 kg upto 4 months and 10- 15 kg from 4-6 months

Table 3. Feeding of cow and buffaloes 6-12 months)

Category	Concentrate (Kg)	Roughage (Kg)
Indigenous cattle/buffaloes	1- 2 Kg	Green oat/maize-15 - 20 Kg or 15 – 20 Kg of Berseem + 5 Kg Dry fodder or Green Oat 15 kg + Straw 2 – 3 kg
Exotic	2.0 – 2.5 Kg	Green oat/maize or alike fodders -15 -20 kg

Feeding of Livestock

Dry Matter-Based Feed Requirements for Maintenance

All animal feeds contain a certain proportion of **water**, and the portion remaining after removing all moisture is termed **dry matter (DM)**. The **DM intake** represents the actual amount of feed an animal consumes daily, excluding water content. Ensuring adequate **DM intake** is essential for maintaining overall health, body function, and productivity.

The daily DM requirements of livestock vary by breed and physiological status:

- **Indigenous cattle** require **2.0–2.5% of body weight** as DM.
- **Crossbred cows and buffaloes** have higher DM intake, typically **2.5–3.0% of body weight** per day.

Proportion of Concentrate and Fodder in the Diet

A widely accepted feeding practice among farmers is to balance **one-third of the total DM requirement** with **concentrates** and the remaining **two-thirds with fodder** (green and dry). When **leguminous fodder** (rich in protein) is available, the amount of **green fodder** required can be significantly reduced while still meeting the animal's nutrient needs.

Determining Total Dry Matter and Concentrate Requirements

The concentrate requirements for various physiological stages are as follows:

1. Maintenance Requirements

- Indigenous (Desi) cows require **1 kg of concentrate per day**.
- Crossbred cows and buffaloes require **1.5 kg of concentrate per day**.
- A **maintenance ration** is the minimum feed allowance required to support essential body functions without weight gain or loss.

2. Lactation Requirements

- Lactating cows require **1.0 kg of additional concentrate** for every **3.0 kg of milk produced**.
- Buffaloes require **1.0 kg of concentrate per 2.5 kg of milk production**.

3. Pregnancy Requirements

- Pregnant cows and buffaloes should receive **1.5 kg of extra concentrate per day** during the **advanced stages of pregnancy** to meet the increased nutrient demand for fetal growth.

4. Breeding Bulls

- Active breeding bulls should receive **an additional 1.0 kg of concentrate per day** to maintain **optimal health and reproductive performance**.

5. Mineral and Salt Supplementation

- **Mineral mixture and common salt** should be supplemented at **25–50 grams per day** to fulfill essential mineral requirements.

By following these feeding guidelines, livestock can achieve optimal health, productivity, and reproductive efficiency. Proper nutrition management ensures **better milk yield, growth, and reproductive performance**, ultimately improving farm profitability.

Table 4. Concentrate Feed Formula for Lactating Cattle and Buffaloes

(Designed to support milk production and body maintenance)

Ingredient	**Proportion (%)**
Maize/Bajra/Barley	33
Deoiled Rice Bran/Wheat Bran	33
Groundnut Cake/Mustard Cake	15
Soybean Meal	11.5
Molasses	5
Common Salt	0.5
Mineral Mixture	2
Total	**100**

Poultry

Poultry convert feed into food products quickly, efficiently, and with relatively low environmental impact relative to other livestock. Their high rate of productivity results in relatively high nutrient needs. Poultry require at least 38 nutrients in their diets in appropriate concentrations and balance.Most diets used to feed poultry are nutritionally "complete" and commercially mixed, ie, prepared by feed manufacturing companies, most of which employ trained nutritionists. The formulation and mixing of poultry feeds requires knowledge and experience in purchasing ingredients, experimental testing of formulas, laboratory control of ingredient quality, and computer applications. Improper mixing can result in vitamin and mineral deficiencies, lack of protection against disease, or drug toxicity.

The physical form of the feed influences the expected results. Most feeds for starting and growing birds are produced as pellets or crumbles. In the pelleting process, the mash is treated with steam and then passed through a suitably sized die under pressure. The pellets are then cooled quickly and dried by means of a forced air draft. The conditions under which pelleting occurs (eg, use of an expander rather than an extruder, exposure to high temperature, use of soft pellets) have an important effect on the nutritional quality of the pellets or of the crumbles produced by crushing the pellets.

Feeding of Broiler birds

Broiler birds require a nutritionally balanced diet to ensure optimal growth, feed efficiency, and meat quality. Their diet changes with age to meet specific protein and energy needs. Below is a structured overview of broiler feeding, including phases, nutrient requirements, feed formulation, and supplementation strategies:

Table 5. Broiler Feeding Phases and requirements

Feeding Phase	Age (Days)	Remarks
Pre-Starter	0-7	High protein for early growth
Starter	8-21	Balanced amino acids
Finisher	22-Market	Muscle development and weight gain

Table 6. Nutrient Requirements for Broilers

Nutrient	Pre-Starter	Starter	Finisher
Crude Protein (%)	23	22	20
Ether Extract	3.0	3.5	4.0
ME (kcal/kg)	3000	3100	3200
Calcium (%)	1.0	1.0	1.0
Available Phosphorus (%)	0.45-0.50	0.40-0.45	0.45-0.50
Lysine (%)	1.3	1.2	1.0
Methionine (%)	0.5-0.6	0.45-0.5	0.45
Salt	0.5	0.5	0.5

Table 7. Feed Ingredients for Broiler Diets

Ingredient	Pre-Starter (%)	Starter (%)	Grower (%)	Finisher (%)
Maize	50-55	55-60	60-65	65-70
Soybean Meal	35-40	30-35	25-30	20-25
Fish Meal	5-8	3-5	2-3	0-2
Vegetable Oil	2-3	2-3	2-3	3-4
Dicalcium Phosphate	1.5-2	1.5-2	1.0-1.5	1.0-1.5
Limestone	0.8-1.0	0.8-1.0	0.8	0.8
Vitamin & Mineral Premix	0.5-1.0	0.5-1.0	0.5	0.5

Provide up to 2 weeks 5 cm and from 3 weeks to finish 10 cm linear feeder space per bird. Raise the level of the feeder as the birds grow. Do not fill the feeder more than half. If tube feeders are used, provide 3 nos. of 12 kg capacity feeders per 100 chicks.

Water Provision

- For 0 to 2-week-old chicks, provide two waterers with a capacity of 2 liters each per 100 chicks.
- From 3 weeks until market age, increase the capacity to two waterers of 5 liters each per 100 birds.
- Ensure a continuous supply of clean, fresh water at all times to support optimal hydration and growth.

Layer Feeding

Layer feeding is divided into distinct phases based on the bird's age and physiological needs. These phases ensure optimal growth, reproductive performance, and egg production efficiency. The feeding strategy is adjusted according to nutrient requirements, particularly energy, protein, calcium, and phosphorus levels.

Table 8. Nutrient Requirements Table for Each Phase

Nutrient	Starter (0-8 weeks)	Grower (8-20 weeks)	Layer 1 (21-40 weeks)	Layer 2 (40-weeks onwards)
Metabolizable Energy (kcal/kg)	2800-2900	2700-2800	2700-2800	2700-2750
Crude Protein (%)	20	16	18	16
Crude Fiber (%)	7.0	9.0	9.0	10.0
Ether Extract (%)	2.0	2.0	2.0	2.0
Calcium (%)	0.9-1.0	0.9-1.0	3.5-4.0	4.0-4.5
Available Phosphorus (%)	0.45-0.50	0.40	0.40	0.35-0.40

Lysine (%)	1.1	0.7	0.75	0.65
Methionine (%)	0.5	0.35	0.35	0.35
Salt	0.5	0.5	0.5	0.5

Feeding and Watering Management in Poultry

1. Feeding of Chicks (0–8 Weeks)

- Chicks should be provided with chick feed from day-old to the end of the 8th week. This feed contains approximately 22% crude protein and 2700 kcal/kg metabolizable energy to support optimal growth.
- During the first few weeks, feed should be placed in feeders at least four times a day. Feeder trays must be cleaned thoroughly before refilling to prevent contamination.
- Initially, feeders should be kept full to encourage feed intake. Gradually, the feed level should be adjusted to half-full to minimize wastage.
- Each chick should be provided with a 5 cm (2 inches) feeder space and 2.5 cm (1 inch) water space to ensure adequate access to feed and water.
- Feeders should be positioned at an appropriate height to facilitate comfortable feeding and prevent spillage. They must be evenly distributed across the housing area so that chicks can access feed from any location.
- Feeders should not be placed directly under the heat source during brooding. After brooder removal, feeders should be arranged parallel to natural light rays to prevent shadow formation.

Before refilling feeders, lift one end (preferably the end far from light) to ensure leftover feed collects near the light source, making it easily accessible to chicks.

An inadequate number of feeders can lead to crowding and feed wastage. It is essential to maintain feeder height appropriately throughout the rearing period.

2. Watering of Chicks

- Chicks' mortality can be minimized by providing boiled and cooled water throughout the brooding period or at least during the first four weeks.
- Waterers must never be allowed to dry out, as even a few hours without water can slow growth rates.
- Drinking surface area is more critical than the volume of water provided. Several small waterers are preferable to a few large ones to ensure easy access.
- As chicks grow, waterers should be replaced with larger ones.

- Waterers should be positioned such that no bird has to walk more than 3 meters (10 feet) to reach water. The maximum distance between two waterers should not exceed 2.5 meters (8 feet).
- To prevent wet spots and disease outbreaks, waterers should be relocated daily.
- Only fresh water should be used, and any leftover water should be discarded.
- Waterers must be washed thoroughly with detergent every morning before refilling. Additional refilling should be done in the afternoon.

3. Feeding and Watering of Growers (9–20 Weeks)

- During this phase, birds should be provided with grower feed, which contains approximately 16% crude protein to support controlled growth and reproductive development.
- The recommended floor space per bird ranges from 0.111 to 0.139 square meters (1.2 to 1.5 square feet).
- Longitudinal or large hanging feeders should be used for efficient feed distribution.
- Each bird should be allocated 6.5 cm (2.6 inches) of feeder space to ensure adequate access.
- Water should be supplied through automatic or pan-and-grill type waterers during the initial weeks for better hygiene and convenience.

4. Feeding and Watering of Layers (21+ Weeks)

- Pullets should not be switched to layer feed until they achieve 5% egg production to ensure proper nutrient balance.
- Layers should be fed twice a day, ensuring consistent nutrient intake for optimal egg production.
- The daily feed requirement per bird varies based on breed:
- White egg layers: 110–115 grams per day.
- Brown egg layers: 120–125 grams per day.
- Fresh drinking water should be provided every morning to maintain hydration and performance.
- Layers consume 3–4 times more water than feed on a daily basis, emphasizing the need for a continuous and adequate water supply.

Restricted Feeding in Layers

Restricted feeding in layers is a **nutritional strategy** where feed intake is controlled to optimize growth, prevent obesity, and enhance egg production. It is primarily applied during the **grower phase (9–20 weeks)** to regulate body weight and delay early sexual maturity.

1. Methods of Restricted Feeding in Layers

A. Quantitative Feed Restriction (Limiting Feed Quantity)

- Birds receive a **restricted amount of feed** daily.
- Usually, **10-20% less** than ad libitum feeding.
- Prevents **excess fat deposition** and improves **feed efficiency**.
- Commonly used in **cage and deep litter systems**.

Types

1. **Skip-a-Day Feeding:** Birds are fed on alternate days.
2. **Daily Quantitative Restriction:** Birds receive a fixed but reduced daily ration.

B. Qualitative Feed Restriction (Reducing Nutrient Density)

- Birds can eat **as much as they want**, but nutrient content is adjusted.
- Lower energy (less maize, more fiber) prevents excess weight gain.
- Involves **low-energy, high-fiber diets** (e.g., rice bran, wheat bran).

Table 9. Ingredients composition of feed of layer birds

Ingredient (%)	Starter (0-8 weeks)	Grower (8-20 weeks)	Layer 1 (21-40 weeks)	Layer 2 (40-weeks onwards)
Maize	50.00	45.00	50.00	55.00
Sorghum/Bajra	5.00	10.00	8.00	6.00
Deoiled Rice Bran	5.00	10.00	8.00	10.00
Soybean Meal	25.00	18.00	15.00	12.00
Groundnut Cake	6.00	7.00	5.00	4.00
Rapeseed Meal	2.00	3.00	3.00	2.00
Fish Meal	3.00	2.00	2.00	2.00
Limestone Powder	1.00	2.00	8.00	9.00
DCP	1.50	1.25	1.00	0.75
Salt	0.30	0.30	0.30	0.30
Mineral Mix	0.50	0.50	0.50	0.50
Vitamin Premix	0.10	0.10	0.10	0.10
DL-Methionine	0.15	0.10	0.10	0.10
Lysine	0.10	0.10	0.10	0.10

22

Introduction of Livestock and Poultry Diseases

Sakshi[1], Dayanidhi Jena[2] and Saurabh Zingare[2]

[1]Division of Medicine, ICAR-Indian Veterinary Research Institute, Izzatnagar, Bareilly

[2]Faculty of Veterinary and Animal Sciences, I. Ag.Sc., RGSC-Banaras Hindu University (BHU), Barkachha, Mirzapur, Uttar Pradesh-231001

Livestock refers to the animals like cattle, buffalo, sheep, goat etc. which are reared in farms for economic purposes whereas chicken and other species of birds like turkey, goose, duck, quail are collectively known as poultry. According to World Health Organization "Health is a state of complete physical, mental and social well-being and not merely the absence of disease or infirmity." Any deviation from the health is known as disease. For greater economic gain animals and birds need to be disease free. In a broader way these diseases can be categorized into infectious and non -infectious based on the causative factor (s). Infectious diseases are the pathological conditions caused by different pathogens like bacteria, fungus, virus, parasites etc. which can be further subdivided into contagious as well as non-contagious diseases. Contagious diseases are those which can spread from animal to animal with direct contact whereas same is not true for the non-contagious diseases. Non -infectious disease are not caused by any pathogen but they are the result of any nutrient deficiency, metabolic disturbances, endocrine abnormalities etc. (Fig. 1).

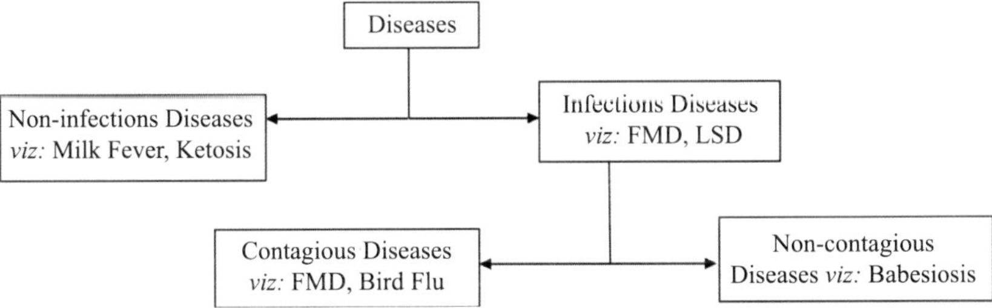

Fig. 1. Broad classification of livestock and poultry diseases.

Common diseases of livestock

There are various diseases which have found their way to inflict harmful effects upon the normal physiological health condition of the animals. Some of them doesn't show much symptoms and exists in subclinical form causing severe economic loss whereas some are very dreadful having high mortality rate. In the below section, the common occurring yet significant diseases are discussed:

1. Foot and mouth disease/ aphthous fever / aftosa

This is an acute, febrile, contagious viral disease of cloven-footed animals like cattle, buffalo, sheep, goat, pigs etc. and characterized by blister formation on mouth, nose, teat and foot. It is caused by *Apthovirus* of family *Picornaviridae.* This disease is having high morbidity rate where as low mortality rate. Total 7 serotypes of above virus can cause this disease but in India serotype "O" is responsible for > 70% cases. Inhalation and aerosol route are the most common route of disease spread along with ingestion, inoculation, abraded skin and insemination.

Clinical signs and symptoms: The common clinical findings of FMD are high temperature, fluid field blisters/vesicles on tongue, gum, lips teats and on interdigital spaces. With time vesicles will erode and develop secondary bacterial infection. There will be drooling of saliva. The morbidity rate reaches 100% whereas mortality rate remains quite low but it is high in young calves due to myocarditis.

Diagnosis: FMD can be diagnosed by clinical symptoms and some advanced tests like PCR and ELISA. It should also be differentiated from vesicular stomatitis, vesicular exanthema and swine vesicular disease.

Treatment: Till now, there is no definite treatment for this condition. However, symptomatic treatment should be done with antipyretics, antibiotics, topical antiseptics. Dextrose and Ringer's lactate should be given if animal is unable to eat due to mouth lesions.

Prevention and prophylaxis: For prevention, regular vaccination should be done. During the outbreaks, movement of human and animals should be restricted.

2. Lumpy skin disease

Lumpy skin disease (LSD) is a highly contagious viral disease of cattle. It is characterized by fever, nodules on the skin, mucous membrane and other parts of the body (fig. 1). The severity of disease depends upon immunity and age of the host strain of virus and load of virus. The etiological agent is *Lumpy skin disease virus* of genus *Capripoxvirus.* The transmission of disease from infected animal to healthy animal is by contaminated feed, utensils, water, different vectors like ticks, mosquitoes and biting flies.

Clinical signs and symptoms: The symptoms of virus include high fever, inappetence, enlargement of superficial lymph nodes, muco-purulent ocular and nasal discharge, reduced milk production in lactating animals, a firm, rounded 2-4 cm diameter nodules on the skin. These nodules can be seen on entire body.

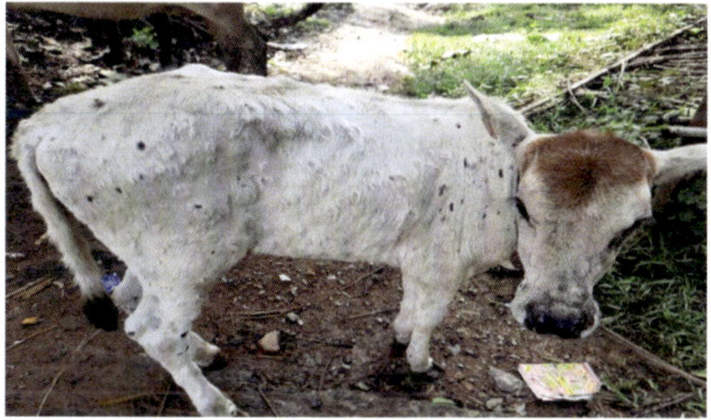

Fig. 2. A calf affected with lumpy skin disease.

Diagnosis: Diagnosis of the disease can be done based upon clinical signs and symptoms. For confirmatory diagnosis different laboratory test can be done like virus culture, PCR, ELISA and virus neutralization test. Disease should be differentiated from other conditions showing similar symptoms like Pseudo-LSD, FMD, blue tongue, Malignant catarrhal fever etc.

Treatment: There is no definite treatment for this. Symptomatic treatment should be done with antibiotics, antipyretics, analgesics. If animal is off-fed, parenteral fluid therapy with Ringer's lactate, normal saline and dextrose, can be done. In case of open nodules, topical antibacterial ointment as well as fly repellant sprays should be applied. Adequate nutrient supply should be ensuring to enhance immunity of the animal.

Prevention and prophylaxis: Control of LSD include regular vaccination, inhibiting the movement of animals and human to disease outbreak area. Care should be given not to purchase animals from affected area and affected animals must be isolated from the healthy population. Regular cleaning of the equipment with antiseptics should be done.

3. Hemorrhagic septicemia (HS)

It is also known as Pasteurellosis. It is an acute fatal disease characterized by fever, oedema of the throat region, dyspnoea and sudden death. The etiological agent is *Pasteurella multocida* type A or several serotypes of *Mannheimiahae molytica*. Both of these bacteria are part of normal microflora of animals but they cause disease when the host immunity is impaired by some stress factors like high humidity, rainfall, travel, parasitism etc.

Clinical signs and symptoms: Affected animals will show respiratory distress and septicaemia symptoms. There will be acute high fever, cough, mucopurulent nasal and ocular discharge, open mouth breathing, dilated nostrils, oedema of throat area, dyspnoea due to fibrinous bronchopneumonia and ultimately death of affected animal with 12-24 hours if not treated.

Diagnosis: Diagnosis of HS is done based on history of stress and symptoms like respiratory distress. Isolation and identification of bacteria can be done in laboratory. Complete blood count reveals neutrophilia and leucocytosis. For further confirmation of serotypes some advance tests like rapid slide agglutination test, indirect hemagglutination test, somatic antigen agglutination tests, AGID and counter immune-electrophoresis can be done.

Treatment: Treatment can be done with different antibiotics like penicillin, amoxicillin, cephalothin, ceftiofur etc. Along with antibiotics, management of shock and respiratory distress should also be done.

Prevention and prophylaxis: For prevention regular vaccination can be done and animals should be protected from any kind of stressors.

4. Brucellosis

Brucellosis is an acute or chronic, zoonotic and contagious disease of animals caused by bacterium *Brucella abortus*. Characteristic findings of brucellosis are placentitis, birth of stillborn or weak calves, retained placenta and abortion at third trimester of pregnancy. Infection spreads by contaminated food and water, abraded skin or conjunctiva, grazing on infected pasture, artificial insemination with infected bulls, and milk from infected dams.

Clinical signs and symptoms: This disease is the most common causes of abortion in farm animals that too in third trimester along with retention of placenta, persistent uterine infection and muco-purulent discharges from vagina.It is also a zoonotic disease. Animals show decreased conception rates, reduced milk production, hygroma, and limb bursitis. Males can get epididymitis and orchitis.

Diagnosis: To confirm infection, isolate and identify the bacterium from aborted fetal tissues and excretions. Serological assays such as serum agglutination, Rose Bengal Plate Test, Milk Ring Test, CFT, and ELISA can help detect sero-positive animals.

Treatment: Treatment of this bacteria is quite difficult as it is intracellular pathogen. One should go for culling of diseased animal to prevent further spread of pathogen.

Prevention and prophylaxis: For prevention, there should be vaccination of all the female calves aged between 4-6 months with live attenuated *Brucella abortus* strain 19 vaccine. All newly bought animals must undergo a 30-day quarantine period. Regular monitoring of herds with a history of abortion is recommended.

5. Mastitis

Mastitis is the inflammation of the mammary glands or udder in dairy animals (fig 2). It causes hardening of udder glandular tissue, physical and chemical abnormalities in milk quality, reduced milk production, and increased somatic cell count in milk. Mastitis can be caused by bacteria, virus, mycoplasma, or fungus, with bacteria being the most usually isolated agent. Based on severity, mastitis can be categorized as subclinical and clinical mastitis. Clinical mastitis shows visual changes in milk quality but in subclinical mastitis milk apparently will be normal but there will be reduction in milk production along with different presence of high number of somatic cell count. Transmission of pathogens can be through teat canal and blood circulation. The contagious pathogens can spread during milking through the milkers' hands or the liners of the milking machine.

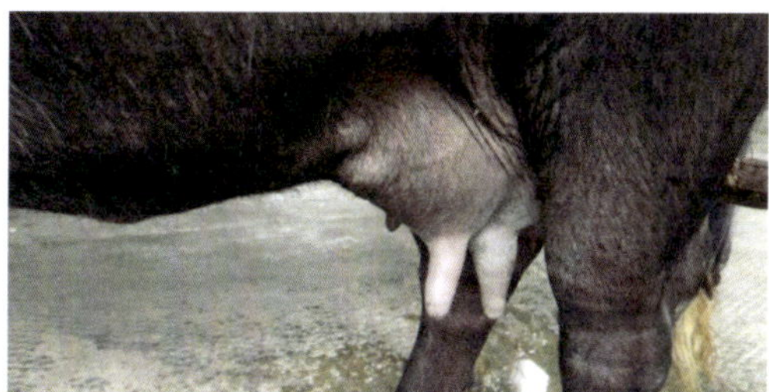

Fig. 3. A buffalo suffering from mastitis.

Clinical signs and symptoms: Clinical mastitis causes redness, swelling, heat, and discomfort in the udder of cows. The milk includes flakes, and there is a color change or the presence of fibrin clots. The cow may have fever as a systemic reaction to the illness. Other symptoms may include loss of appetite, oedema around the udder and decreased milk flow due to changes in milk consistency or udder tissue. Subclinical mastitis does not show noticeable signs of inflammation or changes in milk, but may cause temporary abnormalities.In subclinical mastitis, there will be increase in somatic cell count and it will be diagnosed with California Mastitis Test (CMT).

Diagnosis: At first mastitis will be diagnosed based upon sign and symptoms. The milk quality can be tested with strip cup test and CMT. Specific etiological agent can be identified by isolation and identification of bacteria.

Treatment: Treatment should be based upon sensitivity of the antibacterial agent. If animal is showing systemic symptoms, then parenteral antimicrobial drugs should be done.Udder should be cleaned with antiseptic agents. If only teats are affected with no systemic signs, then intramammary infusion can be done. To reduce the inflammation NSAIDs like flunixin meglumine can be given. Vitamins A, C and E can be given to reduce free radicals and boost epithelial health of udder.

Prevention and prophylaxis: To prevent occurrence of mastitis, proper hygiene should be maintained in herd. Milker's hand should be clean before milking and pre-as well as post germicidal dips should be used to remove existing pathogens on teat.

6. Black quarter

Black quarter (BQ) is a deadly disease caused by *Clostridium chauvoei* that affects cattle, sheep, goats, pigs, mink, and horses. It causes toxemia, inflammation, and necrosis of skeletal and cardiac muscles, with significant death rates. It is a soil-borne illness that enters through the alimentary mucosa after consuming contaminated feed or fodder. Spores are swallowed, multiply in the gut, pass the mucosa, and are carried away by macrophages. They are then deposited in numerous organs and tissues, including skeletal muscles.

Clinical signs and symptoms: It is acute fatal disease so many times animal will be find dead without showing any clinical signs. In subacute cases animal will show lameness due to necrosis of leg muscles. Muscles will show crepitating swelling. This disease is most common in well fed young animals. Along will muscle involvement animal will show anorexia, rumen stasis and fever.

Diagnosis: Diagnosis can be based on history of wound, symptoms. In laboratory microorganism can be seen under microscope with subterminal spores. Thiscondition should be differentiated from anthrax, lightning strike etc.

Treatment: Treatment should be done with Penicillin G @ 22000- 44000 IU/ Kg body weight Intramuscular for 5 days.

Prevention and prophylaxis: For prevention, annual vaccination can be done in enzootic areas during summer and spring season.The dead animal should be deep buried to avoid soil contamination and further spread.

7. Milk fever/parturient paresis, hypocalcemia, parturient apoplexy

It is a disease of adult dairy animals, severe deficiency of calcium (acute hypocalcaemia) causes acute to per acute, afebrile, flaccid paralysis most commonly at or just after parturition.

Clinical signs and symptoms: Clinical signs of milk fever are manifested in three stages. Stage 1 is a brief period of excitation and tetany, with hypersensitivity and muscular tremors in the brain and limbs; the animal does not feed and is hesitant to move. Rectal temperature is normally within physiological range or slightly elevated. Stage 2: Prolonged sternal recumbency with a lateral twist in the neck or the head bent towards the flank. The muzzle is dry, the skin and extremities chilly, and the rectal temperature is below normal. The absolute strength of the heart sounds decreases significantly, but the pace increases. Ruminal stasis, secondary bloating, and constipation are all rather prevalent. Stage 3: Lateral recumbency. The animal is almost comatose and the limbs may be stuck out with complete flaccidity on passive movement and cannot assume sternal recumbency on its own.

Diagnosis: Diagnosis is done based upon sign and symptoms. On serum examination the calcium concentration will be < 5 mg/dl.

Treatment: Treatment can be done with calcium borogluconate (400-800 mL, slow IV in large animals and 50-100 ml in small animals) or calcium magnesium borogluconate (450- 800 mL, slow IV in large animals and 50-100 ml in small animals), depending upon the severity of the condition. The same treatment can be repeated after 10-12 hours if needed.

Prevention and prophylaxis: Administer calcium gels orally at the time of parturition and vitamin D immediately before parturition to enhance the mobilization of calcium. Supplementation of the anionic salts such as calcium chloride, magnesium chloride, calcium sulphate, and magnesium sulphate before calving twice weekly during prepartum period can prevent milk fever.

8. Post Parturient Hemoglobinuria (PPH)

It is a disease of high yielding animals, characterized by acute intravascular hemolysis and ultimately hemoglobinuria and death. Dietary deficiency of phosphorous and feeding of cruciferous plant is known to be cause of this condition. High yielding animals are more prone to this condition.

Clinical signs and symptoms: There will be hemoglobinuria (fig. 3), inappetence, weakness, loss of milk production, pale mucus membrane, increased jugular pulse, tachycardia, dyspnea and jaundice in advanced cases.

Diagnosis: Diagnosis can be done on the basis of history and clinical signs such as hemoglobinuria, anemia. During blood examination there will be reduced concentration of phosphorous (0.5-1.5 mg/dl), red blood cells and hemoglobin.

Fig. 4. The characteristics coffee colour urine in PPH condition.

Treatment: Whole blood transfusion is essential in severe cases. Administration of phosphorus in acute cases (60 g sodium acid phosphate in 300 ml of distilled water IV initially, followed by SC injections at 12-hour intervals for three occasions) is needed. Additionally, haematinics (iron dextran) are recommended during the recovery period.

Prevention and prophylaxis: Animal should be fed with adequate amount of phosphorous and excess feeding of crucifer plants should be avoid to reduce chances of PPH.

9. Anaplasmosis/ Gall Sickness

It is a tick-borne disease caused by an obligate intracellular rickettsia parasite *Anaplasm amarginale.* Mostly high producing, pure bred, undernutrition and young cattle are affected. This disease is transmitted by hard ticks like *Rhipicephalus*, *Hyalomma*, *Dermacentor, Ixodes* etc. and blood sucking flies.

Clinical signs and symptoms: This is a chronic disease. Initially there will be fever but as disease progresses there will be severe anemia and body temperature will decline. Animal will be dull, anorectic, with increased heart and respiratory rate. There will be jugular pulsation, pale mucous membrane, jaundice, panting, reduced milk production.

Diagnosis: Initial diagnosis is done based on history of anemia and ticks. Blood and serum analysis will reveal decreased hemoglobin, red blood count and increased bilirubin levels subsequently. For definite diagnosis Haemoprotozoan examination is done with staining with Giemsa stain under microscopy. The presence of dot like intracellular parasite on the margin of RBC under microscope confirms anaplasmosis. Advanced testslike ELISA and PCR can also be performed for further diagnosis.

Treatment: Diseased animals can be treated with Oxytetracycline (OTC) @ 6-10 mg/kg Body weight intramuscularly for 3 to 5 days. Imidocarb @5 mg/

kg Body weight intramuscularly once can also be used instead of OTC. Along with this symptomatic treatment with haematinics and antipyretics may be given.

Prevention and prophylaxis: For prevention there should be control of tick population in the farm. Regular screening can be done to remove carrier animals so that further spread of disease can be stopped. Sterile needles and surgical instruments must be used to prevent mechanical transmission.

10. Babesiosis/ Red Water fever

Babesia, an intra-erythrocytic apicomplexan parasite, causes bovine babesiosis. In India, *Babesia bigemina* is mostly responsible for clinical illness in cattle. This disease is transmitted by *Boophilusmicroplus*tick.

Clinical signs and symptoms: Infected animals often exhibit high temperatures, anorexia, weakness, severe anemia, hemoglobinuria, icterus, depression, and gastrointestinal stasis. Severely diseased animals may also exhibit diarrhea or constipation, increased heart rate, and difficulty breathing. *B. bigemina* infection in pregnant animals can lead to abortion.

Diagnosis: Babesiosis is diagnosed based on clinical symptoms and the presence of piroplasms in peripheral blood smear. Giemsa-stained thin and thick smears aid in diagnosing clinical conditions. Furthermore, immunodiagnosis with ELISA or IFAT may be useful in herd level diagnosis.

Treatment: Treatment can be done with Diminazene aceturate @ 3.5-7 mg/kg BW intramuscularly as a single dose or Imidocarb dipropionate @ 2.2-4 mg/kg BW intramuscularly or subcutaneously every 72 hours for 2-4 times. Supportive therapy includes NSAIDs, fluid therapy. Blood transfusion should be done in severe cases.

Prevention and prophylaxis: To control *Babesia bigemina* infection, numerous measures can be used, such as segregating and treating diseased animals, protecting healthy ones, and controlling tick vectors.

11. Theileriosis/ Bovine Tropical Theileriosis

It is caused by an intracellular parasite named *Theileria annulate* which parasitizes RBCs and white blood cells. It is transmitted by hard tick of genus *Hyalomma.* It is more common in crossbred cattle.

Clinical signs and symptoms: Animal will be having enlarged superficial lymph nodes, high fever, depression, inappetence, anemia, jaundice, dyspnea, tachycardia and weakness. Hematological findings will have reduced RBC, leukopenia and lymphocytopenia. Severity of disease vary according to breed, age of the animal and strain of the parasite.

Diagnosis: Disease can be diagnosed by clinical signs and presence of the ticks. Hemoprotozoan examination can be done with Giemsa stain of blood or lymph node smear. Further tests like PCR, reverse PCR, ELISA can be done to detect disease in sub clinical cases.

Treatment: The drug of choice for this disease is buparvaquone @2.5 mg/kg BW by intramuscular route. If required, this drug can be repeated after 48 hours. Other drugs like imidocarb, Diminazene aceturate, 3% Trypan's blue, quinapyramine sulphate plus chloride and oxytetracycline can also be used. Supportive therapy like vitamin B complex and haematinics should be given.

Prevention and prophylaxis: For prevention, vaccination with tissue culture vaccine (Rakshavac-T), vector control and regular monitoring can be done. As ticks act as a vector so tick controlled should be done. Regular screening of herd and accordingly treatment of positive animals should be done.

Diseases of Poultry

1. New castle disease/ Ranikhet disease

Newcastle disease (ND) is a lethal respiratory, gastrointestinal, and neurological illness that affects both domestic and wild fowl, resulting in 100% morbidity and death. It is caused by avian paramyxovirus type-1 (APMV -1), genus *Orthoavulavirus*and family Paramyxoviridae. Infected birds can spread the virus by inhaled air, respiratory secretions, and feces, contaminatingthe surroundings. Transmission can occur by direct contact with feces and respiratory discharges, inhalation of minute infective particles formed by dried feces, or contamination of feed, water, footwear, clothes, utensils, equipment, and the surroundings.

***Clinical signs and symptoms*:** Infection with velogenic virus can lead to acute mortality. Other signs include severe sadness, prostration, nerve signs such as torticollis, head twitching, swelling of the head, inappetence, increased respiration, increasing weakening, and greenish diarrhea. Once shell-less or soft-shelled eggs develop, egg laying stops entirely. Symptoms of the mesogenic type include respiratory trouble, coughing, and nervousness. Mortality can potentially approach 50%. The lentogenic virus produces minor respiratory symptoms, an abrupt dip in egg production, and reduced feed intake. The death rate is low, and patients recover fully between 1-8 weeks.

Diagnosis: To diagnose ND, isolate the virus from embryonated chicken eggs at 9-11 days old. Molecular biological techniques, including RT-PCR and real-time PCR, are used to detect and identify NDV.

Treatment: There is no specific treatment for New Castle disease but symptomatic treatment can be done using antiviral disinfectants, immunomodulators like beta glucans, Vitamin E and Se can be given.

Prevention and prophylaxis: Regular vaccination should be done. Proper biosecurity measures should be followed. Movement of human can be restricted to the outbreak farm.

2. Avian Influenza/ Fowl Plague

Highly pathogenic avian influenza (HPAI) affects both domestic and wild birds, producing significant morbidity and death due to its high infectiousness and dynamic nature. It is caused by Influenza A virus of family Orthomyxoviridae.

Clinical signs and symptoms: HPAI symptoms includes sudden onset of disease, and significant morbidity and fatality rates of up to 100% within 3-10 days. In acute cases, birds will be dead with no clinical signs. Birds which survive 3-7 days show despondency, prostration, ruffled feathers, conjunctivitis, cyanosis of the head, combs, wattles, and shanks, facial puffiness, decreased feed and water consumption, extreme discomfort, rales, coughing, sneezing, and dyspnea. Nervous problems including head and neck tremors, torticollis, opisthotonus, nystagmus, paresis, and paralysis were seen.

Diagnosis: Diagnosis of this can be done by clinical sign and symptoms. Laboratory tests can be done with virus isolation and RT-PCR.

Treatment: There is no treatment for this disease. Affected birds should be culled and disposed off immediately to avoid further spread of disease.

Prevention and prophylaxis: For prevention, proper biosecurity measures and vaccination protocol should be followed. In case of disease outbreak higher authority should be informed immediately as it is a notifiable disease.

3. Marek's disease

It is highly contagious viral disease of chickens caused by Marek Disease virus, subfamily Alphaherpesvirinae. The illness is very infectious and often spreads by infected feather-follicle dander, fomites, and other sources. Infected birds are viraemic for life.

Clinical signs and symptoms: Affected birds may exhibit paralysis in legs, wings, and neck, weight loss, grey iris or uneven pupil, eyesight impairment, and follicular tumors. Affected birds have reduced immunity, making them more vulnerable to infectious diseases.

Diagnosis: Diagnosis can be done by clinical signs and post mortem lesions like enlargement of nerves.

Treatment: There is no particular therapy for this disease. Regular grading and culling of diseased chickens help prevent disease spread. Supplementing the diet with vitamin E, selenium, and MOS can boost immunity and reduce loss rates.

Prevention and prophylaxis: Vaccination and proper biosecurity measures should be followed for prevention of this disease.

4. Coccidiosis

In poultry, coccidiosis is caused by various species of genus *Eimeria* like E. *E. acervulina, E. maxima, E. tenella, E. necatrix, E. brunetti, E. preacox, E. mitis,* and E. *mivati.*Coccidiosis in poultry is normally host specific and different species have predilection for different portion of intestine.Disease is transmitted by ingestion of large quantity of sporulated oocysts. Clinically recovered and diseased birds shed oocysts in feces and it contaminates feed, dust, water, litter, and soil.

Clinical signs and symptoms: Different species of *Eimeria* exhibit different signs based upon location of intestine. But the usual signs are decreased growth rate and egg production, bloody diarrhea, decreased food and water intake and high mortality. Mild infections cause immunosuppression and predisposes the birds to other infections.

Diagnosis: Diagnosis is based on the location and appearance of lesions in the host, with oocyst size determining the species. It can be verified by identifying oocysts in feces or intestinal scrapings.

Treatment: Anticoccidial drugs in feed or water, immunization, or a combination of the two are used to prevent clinical symptoms. To prevent dehydration and subsequent bacterial infection, antibiotics and supportive care should be used as soon as clinical signs appear.

Prevention andprophylaxis: Poultry kept on wire flooring had fewer infections because of reduce contact with feces. Vaccination and anticoccidial medicines like inophores, can also be used for control measures. Proper biosecurity measures and vaccination protocol can be followed.

5. Salmonellosis

There are a number of serotypes prevailing for Salmonella genus but among these *S. pullorum* and *S. gallinarum* causes severe diseases in poultry. *S. pullorum* causes Pullorum disease which is an acute disease affecting all systems of body. *S. gallinarum*causes fowl typhoid which is an acute to chronic disease characterized by septicaemia. Pullorum disease mainly affects immature birds whereas fowl typhoid affects mature birds.

Salmonella is transmitted vertically from hens to chicks and rapidly spreads laterally in hatcheries and rearing units. The organism can survive outside the body for months and may be transmitted by red mites, rodents, and farm pets.

Clinical signs and symptoms: Signs of pullorum disease include high number of dead-in-shell chicks, depression, huddling, inappetence, respiratory distress, white pasty dropping. Mortality is upto 100 % in acute cases where subacute cases show signs like lameness, swollen joints and stunted growth. Clinical signs of fowl typhoid include anorexia, diarrhea, dehydration and ultimately death. It is more severe as compare to pullorum disease.

Diagnosis: To diagnose and confirm avian salmonellosis, isolate, identify, and serotype Salmonella strains. Serologic testing can identify infections in mature birds. Further confirmation can be obtained by necropsy findings, microbiological culture, and typing. The serological ELISA test is used to diagnose S. typhimurium and S. enteritidis.

Treatment: For treatment different antibiotics can be given orally like oxytetracycline, neomycin, tetracycline, polymyxin, and sulfadiazine plus trimethoprim etc.

Prevention and prophylaxis: Diseases can be controlled by proper vaccination and applying different biosecurity measures.

Table 1. Infectious diseases of livestock and their causative agents

S. No.	Name of the disease	Etiology of the disease
1	**Viral diseases**	
	i. Foot and mouth disease	*Aphthovirus*
	ii. Lumpy Skin Disease	LSD virus genus *Capripoxvirus*
	iii. Bovine Leukemia	*Deltaretrovirus*
	iv. Bovine Viral Diarrhoea	*Pestivirus*
	v. Infectious Bovine Rhinotracheitis	bovine herpesvirus 1
	vi. Malignant Catarrhal Fever	Gammaherpesvirinae subfamily
	vii. Bovine Ephemeral Fever	*Rhabdoviridae*Family
	viii. Rabies	*Lyssavirus* of family Rhabdoviridae
	ix. Peste des Petits Ruminants	*Morbillivirus*
	x. Contagious Ecthyma	parapox-virus
	xi. Bluetongue	*Orbivirus*
2	**Bacterial diseases**	
	i. Brucellosis	*B. abortus, B. melitensis*
	ii. Tuberculosis	*Mycobacterium bovis*
	iii. Colibacillosis	*Escherichia coli*
	iv. Anthrax	*Bacillus anthracis*
	v. Actinomycosis	*Actinomyces bovis*
	vi. Actinobacillosis	*Actinobacillus lignieresii*
	vii. Leptospirosis	*Leptospira interrogans*
	viii. Black leg	*Clostridium chauvoei*
	ix. Bacillary Hemoglobinuria	*Clostridium haemolyticum*
	x. Tetanus	*Cl. Tetani*
	xi. Botulism	*Cl. Botulinum*
	xii. Foot Rot	*Fusobacterium necrophorum*
	xiii. HaemorrhagicSepticaemia	*Pasteurella multocida*
	xiv. Paratuberculosis	*Mycobacterium avium* subsp.
	xv. Listeriosis	*Paratuberculosis*
	xvi. Caseous Lymphadenitis	*L. monocytogenes*
	xvii. Enterotoxaemia	*Corynebacterium pseudotuberculosis*
		Cl. Perfringens type D

3	**Fungal diseases** i.Aspergillosis ii.Dermatophytosis	 *Aspergillus. Fumigatus* *Trichophyton verrucosum*
4	**Parasitic diseases** i. Anaplasmosis ii. Babesiosis iii. Theileriosis iv. Hydatidosis v. Paramphistomosis vi. Fasciolosis vii. Trypanosomosis viii. Trichomonosis ix. Mange	 *Anaplasmamarginale* *Babesia bigemina* *Theileriaannulate* *Echinococcus granulosus* *Gastrothylaxcrumenifer* *Fasciola hepatica* *Trypanosoma evansi* *Tritrichomonas foetus* Mites like *Psoroptes*, *Chorioptes*, *Otodectes*etc

Table 2. Infectious diseases of poultry and their causative agents

S. No.	Name of disease	Etiology of the disease
1	**Viral diseases** i. Ranikhet Disease ii. Infectious Bronchitis iii. Avian Influenza iv. Infectious Laryngotracheitis v. Infectious Bursal Disease vi. Marek's Disease vii. Fowl-pox viii. Chicken infectious anaemia ix. Avian Encephalomyelitis x. Leukosis	 paramyxovirus type-1 Coronavirus Orthomyxovirus Gallid herpesvirus 1 *Avibirnavirus* Gallid herpesvirus 2 *Avipoxvirus* Gyrovirus Picornavirus Alpha Retrovirus
2	**Bacterial disease** i. Avian Mycoplasmosis ii. Colibacillosis *iii. Salmonellosis* iv. Infectious Coryza v. Fowl Cholera	 *Mycoplasma gallisepticum* *E. Coli.* *S. gallinarum, S. pullorum* *Avibacteriumparagallinarum* *Pasteurella multocida*
3	**Parasitic diseases** i. Coccidiosis ii. Red Mite infestation	 *Eimeria sps.* *Dermanyssus gallinae*mite
4	**Fungal diseases** i. Brooder pneumonia ii. Thrush	 *Aspergillus fumigatus* *Candida albicans*

Further reading

Sharma, R. D., Kumar, M. and Sharma, M. C. 2010. A textbook of Preventive Veterinary Medicine and Epidemiology. Indian Council of Agricultural Research, ne Delhi.

The Merck Veterinary Manual. 2016. 11th Edition.Merck & Co Inc.

USDA Animal and Plant Health Inspection Service. Detections of Highly Pathogenic Avian Influenza.

23

Prevention and Control of Important Diseases of Livestock and Poultry

Saurabh Zingare[1], Dayanidhi Jena[1] and Sakshi[2]

[1]*Faculty of Veterinary and Animal Sciences, I. Ag.Sc., RGSC-Banaras Hindu University (BHU), Barkachha, Mirzapur, Uttar Pradesh 231 001*

[2]*Division of Medicine, ICAR-Indian Veterinary Research Institute, Izzatnagar, Bareilly, Uttar Pradesh*

The prevention of diseases is a proactive and strategic approach focused on mitigating potential risks and addressing vulnerabilities to stop diseases from occurring. This method prioritizes early intervention and safeguards health by preventing outbreaks before they arise. Disease control measures are reactive approaches designed to contain and mitigate the spread of a diseases after its onset. These measures focus on limiting transmission within affected regions where an outbreak has already occurred.

General Disease Prevention and Control Measures

- Quarantine
  - It is the segregation of apparently healthy animals before introducing into the herd.
  - The objective is to identify any animal under incubation period of disease.
  - Quarantine period depends on the incubation period of particular disease but in practice quarantine period of about 30 days covers almost all diseases.
  - During quarantine period animals should be screened for ectoparasitic and endoparasitic infestations.
- Isolation of sick animals
  - Isolation is segregation of suspected or affected animals with contagious diseases from healthy animals (fig. 1).

Fig. 1. Isolation shed for diseases animals away from healthy animal shed.

- Isolated animals should be kept in isolation ward far away from healthy shed (fig. 2).

Fig. 2. Diseased animals are housed in sick animal shed away from healthy population.

- Feed and used equipments, if any, for isolated animals should be kept separate.

- **Elimination of carriers**
  - Carriers are animals which harbour pathogens in tissues after recovery from diseases and do not show any clinical signs.
  - Common diseases for which carriers have been observed in farm animals are Tuberculosis, Leptospirosis and Brucellosis.
  - Some of the commonly used screening tests are tuberculin test, Johnin test, agglutination test and test for detection of subclinical mastitis.
  - Carrier animals should be screened and diagnosed and should be culled from herd.

- **Vaccination**
  - Vaccination is a practice of artificially building up immunity in the animal body against specific infectious diseases by injecting biological agents called vaccines.
  - The term vaccine is used to denote an antigen (substance form organisms) consisting of a live, attenuated or dead bacterium, virus or fungus and used for the production of active immunity in animals.
  - The term also includes substances like toxins, toxoids or any other metabolites etc. produced by microbes and used for vaccination
  - The farm animals and young ones should be vaccinated at regular intervals at appropriate times.
- **Carcass Disposal**
  - Carcass should not be disposed off in or near stream of flowing water.
  - Carcass should not be kept for longer time in sheds.
  - All carcasses should be disposed properly by burial or burning method.

Control Measures Against Ectoparasites/Vectors

Ectoparasites are organisms that inhabit the surface of the host, either on or within the skin, for varying durations and can be harmful to the host. Arthropods such as flies, mites, lice, and ticks are the most economically significant ectoparasites. They cause a range of pathological effects, including anemia, hypersensitivity reactions, irritability, dermatitis, skin necrosis, secondary infections, focal hemorrhages, orifice blockages (e.g., ears), and, in rare cases, exsanguination. The following control measures can be applied for control of ectoparasites or vectors:

- Regular cleaning damp and dark corners and removal of manure, stagnant water is must to prevent breeding of parasites.
- Eggs of ticks and mites deposited in cracks and crevices in the walls, floors and wood work of the animal houses should be removed regularly with use of chemicals or flame gun.
- Regular dipping or spraying of animals with suitable insecticides to prevent lice, flies, fleas, mites and ticks on skin of animals.
- Fans can be installed in shed area to keep away flies.
- Heaped manure should be always kept covered to avoid flies and other arthropods.

- Integrated pest management (IPM) involves the selection and use of several methods to reduce pest population with expected ecological, economic and sociological costs and benefits.
- The fundamental principle of IPM is to treat only when necessary. Total elimination of pests is undesirable as they are part and parcel of ecosystem.

Chemical method to control ectoparasites

- Drugs and chemicals used to control ectoparasites are known as **ectoparasiticides**. They can be administered through various methods, including dips, sprays, pour-ons, spot-ons, dusting powders, and ear tags.
- Different classes of ectoparasiticides most commonly used are organochlorines like Methoxychlor, organophosphates like Malathion, carbamates like Carbaryl, formamidines like Amitraz, pyrethroids like Cypermethrin and macrocyclic lactones like Ivermectin.

Non-Chemical methods to control ectoparasites

- Semiochemicals are chemical signals of host/tick secreted in environment and are used to mediate tick behaviour. e.g. pheromones are used for mass trapping or mating disruptions of ectoparasites. These are highly species specific and environment friendly.
- Tick Vaccine – The recombinant vaccines (e.g. GAVAC and TickGARD) against R. (B.) microplus are available commercially.
- Zebra stripes – strips painted on animals may interfere with optic flow patterns needed by flying insect to land properly.
- Landscape intervention involves altering the landscape to increase sunlight and lower humidity so as to reduce ectoparasites.
- Phyto acaricides (Herbal) - Acaricidal property of plant extracts can provide a potential substitute to synthetic acaricides.

Prevention and Control Measures in important Livestock and Poultry Diseases

Foot and Mouth Disease

Foot and Mouth Disease (FMD) is viral vesicular disease of cloven- hoofed animals such as cattle, buffaloes, sheep, goats and pigs etc. and it is highly contagious in nature. Post FMD affected animals tend to have reduction in milk yield, infertility, decreased in growth rate, reduced working capacity in bullocks etc.

The following measures can be taken in enzootic areas or if outbreak is imminent

- Mass vaccinationof susceptible population of bovines, small ruminants (sheep and goats) and pigs at six-monthly intervals.
- Primary vaccination of calves and deworming one month prior to the vaccination.
- Disinfection of floors, premises should be done using sodium carbonate (4 %), sodium hydroxide (2 %) and citric acid (0.2 %).
- A foot bath or truck bathe at entry of farm or village. Lime powder or bleaching powder or equal proportion of both can be used in foot bath.
- Restriction to calves suckling to infected dams.
- Sero-surveillance/sero-monitoring of animal population and continuous disease surveillance programme in and around wildlife habitats.
- Restriction of animals at common grazing pastures and water sources during outbreak.
- Proper rinsing and cleaning of wounds with antiseptic solutions.
- Infected carcasses must be disposed via burial or burning methods along with manure. The burial site should be away from the water bodies.

Johne's Disease

Controlling Johne's disease in ruminants is challenging owing to the multifaceted nature of the disease like the widespread nature of the causative organism, its long incubation periodand the limited sensitivity of available laboratory tests for detecting subclinical infections. Most current control programs focus on keeping the disease at a low prevalence rather than completely eliminating it, as outlined below:

- *Identification and elimination of infected animals* –The seropositive or pathogen shedding animals along with their offspringare culled and sold for slaughter. Calves from infected animals are kept separate and grown and fed under feedlot conditions.
- *Prevention of introduction of infected animals into the herd* - Measures are taken to avoid the introduction of infected animals into the herd by maintaining a completely closed herd or by carefully screening purchased animals.
- *Prevention of exposure of susceptible animals to the infectious agent* – Involves Minimizing contact between young and older animals and from faecal contaminated feed and water, using separate equipment for handling feed and manure

Peste Des Petits

Peste des petits ruminants (PPR) is an acute, highly contagious and transboundary viral disease specifically occurs in sheep and goats. It is otherwise known as "Small ruminant plague" owing to the species it affects. Clinically, the disease is characterized by high fever (pyrexia), gastro enteritis, necrotizing and erosive stomatitis, oculonasal discharges, diarrhoea and bronchopneumonia, followed by either death of the animal or recovery from the disease.

The Prevention and Control Measures are as below

- Strict quarantine and control of animal movements.
- Quarantine of newly purchased or newly arriving animals for at least two to three weeks.
- PPR virus susceptible to most disinfectants, e.g. phenol, sodium hydroxide 2% for 24 h.
- Infected carcasses should be disposed with burial or burning method.
- Isolation of affected animals from the flock.
- Biosecurity measures like wearing personal protectives like gloves, apron etc., cleaning of premises.
- Effective way to control PPR is by mass vaccination of the animals.

The control measures during PPR outbreaks as below

- Strict quarantine of exposed animals & restriction of animal movements.
- Mass immunization of susceptible animals.
- Disinfection of the premises with common disinfectants.
- Proper disposal of carcasses and contact fomites like bedding must be burnt or buried deeply on site.
- Restriction on importation of sheep and goats from affected areas.

Classical Swine Fever

Classical swine fever is a contagious, often fatal, disease of pigs clinically characterized by high body temperature, lethargy, yellowish diarrhoea, vomiting, and a purple skin discoloration of the ears, lower abdomen, and legs.

The Prevention and Control Measures for Classical Swine Fever are as below:

- Quarantine of newly purchased pigs.
- Effective sterilization of waste food fed to pigs.
- Structured serological surveillance targeted to breeding sows and boars.

- Effective hygiene measures protecting domestic pigs from contact with wild boar.
- Use of foot baths with disinfectant solutions at the entrance of the farm.
- Proper disinfection of vehicles and farm equipment's before entering into the farm.
- Vaccination with modified live virus strains is effective in preventing losses in countries where classical swine fever is enzootic.
- CSFV is moderately fragile in the environment and does not spread far (≤1 km) when airborne. In faeces, CSFV may be detectable through day 42 or longer, depending on the strain.
- Prompt and correct disposal of infected carcasses and bedding has to be disposed by burial or incineration. Thorough disinfection of the pens and premises has to be done.

Rabies

Rabies is an acute viral infection of warm-blooded animals characterized by signs of abnormal behaviour, nervous disturbances, ascending paralysis and death.

The following measures can be undertaken to prevent and control rabies

- Control of stray dog population via animal birth control programmes.
- Mass vaccination of dogs in endemic areas is the method of choice to control rabies in both humans andanimals.
- Vaccinating 70% of dogs allows rabies to be eradicated from a given endemic area
- For farm animals, there are two useful techniques: the prevention of exposure and preexposure vaccination.
- Prevention of exposure to the virus achieved by controlling access of wildlife species that are likely to come into contact with the farm livestock in particular areas or through vaccination of the wildlife.
- Inactivated tissue culture cell vaccines given to cattle result in neutralizing antibodies in 1 month after the primary vaccination. A booster given 1 year later increases the titres, which are detectable 1 year after the booster.
- An effective postexposure protocol for unvaccinated domestic animals exposed to rabies includes immediate vaccination against rabies, a strict isolation period of 90 days, and administration of booster vaccinations during the third and eighth weeks of the isolation period.

Brucellosis

Brucellosis is a bacterial disease often associated with abortion storm in farms that leads to heavy economical losses for Indian farmers. It is also an important public health concern for people associated with the livestock industry as the infection can be transmitted from animals to humans (zoonosis).

The following control measures should be undertaken for brucellosis

- Preventing free grazing and movement along with frequent mixing of flocks of sheep and goats, use of semen from unscreened bulls for artificial insemination and poor farm hygiene which contribute to the high prevalence and wide distribution of brucellosis in animals
- The quarantine period should be followed ranging from 120 days to 1 year, or until all breeding animals have completed a gestation without test evidence of infection.
- In cattle, RBPT is used as screen test followed by testing positive sera with CFT for confirmation.
- The milk ring test (MRT) could be used for identifying infected dairy herd with good results followed by sero-testing individual animals.
- Direct contact of infected or suspected animals should be avoided, consumption of unpasteurized milk or milk products is not recommended. The pregnant lady should take special care as the infection may also cause abortion in humans.
- Different types of vaccines are available for prevention of brucellosis in animals such as Brucella abortus strain 19 (S19 most commonly used vaccine) and Brucella melitensis Rev 1, RB51, are live attenuated vaccines available for animals.
- Female calves aged between 4–8 months are recommended for vaccination by S19 vaccine.
- Vaccination of bulls is discouraged due to development of orchitis and the presence of *B. abortus* strain 19 in the semen.
- Deworming before vaccination should be done for better immune response.

Newcastle Disease

Newcastle disease (ND) is a virus borne disease of poultry birds which has capability to infect wide range of poultry birds including domestic, commercial layer, broiler and breeder flocks. It is one of the most virulent diseases in birds which is responsible for high morbidity and mortality among birds.

Prevention and control measures are as below

- Bird-proofing houses, feed and water supplies
- Pest control in flocks; insects and mice
- Avoidance of contact with birds of unknown health
- One age group per farm ('all in-all out') breeding
- Effective quarantines and movement controls during outbreak.
- Destruction of all infected and exposed birds;21 days before restocking
- Live virus-based vaccines administered via drinking water or by aerosol, eye or nostril droplets, or beak dipping, some mesogenic strains by wing-web intra-dermal inoculation
- Inactivated vaccines are formulated as oil emulsions and injected.

Haemorrhagic Septicaemia

Haemorrhagic Septicaemia is an acute bacterial disease of cattle and buffaloes which usually occurs during monsoon. Mortality rate may be as high as 80 %. The bacterial pathogens of this disease survive longer in humid and waterlogged conditions.

- Isolate the sick animal from healthy ones and avoid contamination of feed, fodder and water.
- Avoid crowding especially during wet seasons.
- Vaccinate all animals which are 6 months and above of age annually before the onset of monsoon in endemic areas.

Black Quarter

Black Quarter (BQ) is an acute disease of ruminants (cattle predominantly) characterized by emphysematous swelling usually in heavy muscles. Most often the animals which are well-fed are commonly become the victim. Buffaloes usually suffer from a milder form. Contaminated pasture appears to be major source of infection. Healthy animals in the age group 6 months to 2 years are generally affected.

- Struct vaccinate of all animals above 6 months of age annually before the onset of monsoon in endemic areas.
- In endemic areas, burning the upper layer of soil with straw to eliminate spores may be of help to control the spreading of the disease.
- Sprinkle lime over carcass at the time of burial.

Acute Mastitis

It is the severe inflammation of udder where physical changes are clearly visible in udder and milk. Nost often, the high yielding animals are affected and is mainly caused by group of organisms like bacteria, fungi, virus and rarely by algae.

- Before milking, clean the udder well with clean water and wipe dry with clean towel. Separate cloth towel should be used for each animal.
- Milking should be quick, complete and hygienic.
- Milk animals with chronic mastitis in the end.
- Carry out teat dipping or spray immediately after milking.
- Prevent the animal from sitting for at least 30-45 minutes after milking.
- Periodically check and treat for sub-clinical mastitis
- Keep the floor of the cattle shed without holes and as dry as possible.
- Continue teat dipping/spray 2 weeks after drying off and start the practice two weeks before calving.

Table 1. Vaccination Schedule for Cattle and Buffaloes

Sr. No.	Name of Disease	Age at first dose	Booster	Subsequent dose
1.	Foot and Mouth Disease (FMD)	4 month& above	1 month after primary	Six months interval
2.	Haemorrhagic Septicaemia (HS)	6 month& above	-	Annually in endemic area
3.	Black Quarter (BQ)	6 months and above	-	Annually in endemic areas.
4.	Brucellosis	4-8 months of age (Only female calves)	-	Once in a lifetime
5.	Theileriosis	3 months of age and above	-	Once in a lifetime. Only required for crossbred and exotic cattle.
6.	Anthrax	4 months and above	-	Annually in endemic areas.
7.	IBR	3 months and above	1 month after first dose	Six-month interval
8.	Rabies (Post bite)	Immediately after suspected bite.	-	3rd, 7th, 14th,28th , 90th (Optional) days

Ref.: NDDB Handbook of Good Dairy Husbandry Practices.

Table 2. Vaccination Schedule for Sheep and Goat

Sr. No.	Name of Disease	Primary Dose	Booster	Subsequent dose
1.	Foot and Mouth Disease	At 3 months age	3-4 weeks after primary	Six monthly
2.	PPR	At 3 months age	-	At 4th year
3.	Goat Pox	At 3 months age	3-4 weeks after primary	Annually
4.	Sheep Pox	At 3 months age	3-4 weeks after primary	Annually
5.	Enterotoxaemia (ET)	At 3 months age	3-4 weeks after primary	Annually (2 dose with 1-month interval
6.	Haemorrhagic Septicaemia	At 3 months age	3-4 weeks after primary	Annually

Ref.: Goat Management Practices, ICAR-CIRG.

Table 3. Vaccination Schedule for Swine

Sr. No.	Name of Disease	Primary Dose	Booster	Subsequent dose
1.	Swine Fever	25-30 days	1 month after primary	Six monthly
2.	Foot & Mouth Disease	42 days	1 month after primary	Six monthly
3.	Haemorrhagic Septicaemia	At 2 months age	3-4 weeks after primary	Annually

Table 4. Vaccination Schedule for Poultry (Broilers)

Sr. No.	Age	Vaccine	Route
1.	First day	Marek's Disease	SC
2.	5-7th day	RDV F1	I/O OR I/N
3.	14th day	IBD	I/O OR I/N
4.	21st day	RDV La Sota	Drinking Water
5.	28th day	IBD	Drinking Water

Ref.: Expert System on Poultry, ICAR-TANUVAS-TNAU.

Table 5. Vaccination Schedule for Poultry (Layers)

Sr. No.	Age	Vaccine	Route
1.	First day	Marek's Disease	SC
2.	5-7th day	RDVF	I/O OR I/N
3.	12-14 day	IBD	I/O OR I/N
4.	18-22 days	Infectious Bronchitis	Drinking Water
5.	24-27 day	IB booster	Drinking Water
6.	28-30 day	RD vaccine booster La Sota	Drinking Water
7.	6th week	Fowl pox/ Infectious coryza	SC

8.	8th week	RD Vaccine R2B	SC
9.	9th week	Fowl pox	Wing web
10.	12-13th week	IB booster	Drinking Water
11.	18th week	RD booster R2B	SC
12.	45-50th week	RD La Sota repeated once in 2 months	Drinking Water

Ref.: Expert System on Poultry, ICAR-TANUVAS-TNAU.

Deworming

- In hot and humid areas, livestock may be dewormed regularly with suitable anthelmintic.
- In places where heavy worm load is present, heifers are dewormed twice a year up to two years of age.
- Young animals should be dewormed every month. Older stock can be dewormed at 4-6 months intervals.
- Anthelmintic resistance in gastrointestinal (GI) helminths is a major concern nowadays in small ruminants and cattle. As eradication of GI parasites is unlikely, livestock producers must adopt sustainable management strategies.
- A key approach to mitigating resistance is slowing its development through methods such as **Targeted Selective Treatment (TST)**, which involves treating only selected individuals rather than entire groups.
- Three methods are commonly used to assess helminth infections in livestock. The first relies on parasitological indicators, with faecal egg count (FEC), the second method is based on pathophysiological indicators, including body weight, body condition score, anaemia, diarrhoea, and plasma pepsinogen levels and the third method focuses on performance-based indicators, such as milk yield and live weight gain, which reflect the impact of parasitic burden on animal productivity.

Table 6. Deworming Schedule for Cattle and Buffaloes

Roundworms/ Tapeworms	First dose at 3-4 days of age. Monthly interval up to 6-month age. Every 4-6 month or 2-3 times per year after 6-month age.
Liver flukes	Twice a year in endemic areas (before & after monsoon)

Table 7. Deworming Schedule for Sheep and Goat

Roundworms/ Tapeworms	Once a month between 1 to 6-month age Once in two months between 6 to 12-month age. Thrice a year thereafter.
Liver flukes	Twice a year in endemic areas (before & after monsoon)

Deworming Schedule for Poultry (Layers)

First deworming at 4 weeks age, second deworming at 8 weeks age and thereafter every 6-8 weeks depending on worm burden.

Further reading

Boral, R., Singh, M. & Singh, D. K. 2011. Status and strategies for control of brucellosis-A review. Indian J. Anim. Sci. 79 (12).

Edith, R., Harikrishnan, T. &Balagangatharathilagar, M. 2018. Targeted selective treatment (TST): A promising approach to combat anthelmintic resistance in farm animals. J. Entomol. Zool. Stud. 6 (1)844-847.

FMD Technical Bulletin 2021, ICAR- NRCM.

Ghosh, S. and Nagar, G. 2014. Problem of ticks and tick-borne diseases in India with special emphasis on progress in tick control research: A review. J. Vector Borne Dis. 51, 259.

PPR Technical Bulletin 2019, ICAR-NIVEDI.

SOP Manual 2016, USDA.

Index